LIGNIN - TRENDS AND APPLICATIONS

Edited by **Matheus Poletto**

Lignin - Trends and Applications
http://dx.doi.org/10.5772/intechopen.68464
Edited by Matheus Poletto

Contributors

Kenji Koido, Takahiro Iwasaki, Łukasz Klapiszewski, Tadeusz Jan Szalaty, Teofil Jesionowski, Jorge Aburto, Myriam Amezcua, Zahed Mahmood, Muhammad Jahangeer, Muhammad Yameen, Ana Lourenço, Helena Pereira, Jans Alzate-Morales, Ángel T. Martínez, Rodrigo Recabarren-Hurtado, Jefferson O Romero-Marulanda, Isabel Fuenzalida-Valdivia, Peter Papoh Ndibewu, Pierre Gerard Tchieta, Peter M Czermak, Stefan Schoenherr, Mehrdad Ebrahimi, Behzad (Benji) Ahvazi, Tri-Dung Ngo, Daniela Pufky-Heinrich, Bjoern Roessiger, Gerd Unkelbach, Aurélie Cayla, Mandlekar Neeraj, Francois Rault, Stéphane Giraud, Fabien Salaün, Giulio Malucelli, Guan Jinping

Published by InTech
Janeza Trdine 9, 51000 Rijeka, Croatia

Publishing Process Manager Martina Usljebrka
Technical Editor SPi Global
Cover InTech Design team

First published March, 2018
Printed in Croatia
Legal deposit, Croatia: National and University Library in Zagreb

Additional hard copies can be obtained from orders@intechopen.com

Lignin - Trends and Applications, Edited by Matheus Poletto
p. cm.
Print ISBN 978-953-51-3901-0
Online ISBN 978-953-51-3902-7

For EU product safety concerns:
IN TECH d.o.o., Prolaz Marije Krucifikse Kozulić 3, 51000 Rijeka, Croatia,
info@intechopen.com or visit our website at intechopen.com.

Meet the editor

Matheus Poletto is a professor and researcher at the Universidade de Caxias do Sul. He is an expert in composite science, working with thermoplastic composites and cellulosic materials. He has published over 100 scientific and conference papers and several book chapters. He obtained his degree in Chemical Engineering, master degree in Materials Science and Engineering from the Universidade de Caxias do Sul, Brazil, and PhD degree in Materials Engineering from the Universidade Federal do Rio Grande do Sul, Brazil, in 2014. Professor Matheus currently works with cellulosic and lignocellulosic materials studying the effect of fiber composition on the thermal and mechanical properties of polymer composite materials. He is also a member of the Polymer Laboratory research group from the Universidade de Caxias do Sul.

Contents

Preface

The growing concern about environmental issues and the depletion of fossil fuels promotes the search for using the natural materials. On the basis of this fact, lignin is the nature's dominant aromatic polymer and an emerging material in the bioeconomy. Lignin and its derivatives are alternative sources of chemicals, energies, and potential materials for several industries. The importance of lignin as a potential material is widely recognized, and the research in the lignin field is expanding around the world. Therefore, a solid knowledge is necessary to transform lignin in useful materials and chemicals. In this way, this book intends to contribute to a deep understanding about the main characteristics, current applications, and future trends of lignin and its derivatives.

Lignin—Trends and Applications consists of 11 chapters related to the lignin structure, modification, depolymerization, degradation process, computational modeling, and applications. This is a useful book for readers from diverse areas, such as physics, chemistry, biology, materials science, and engineering. It is expected that this book may expand the reader's knowledge about this complex natural polymer.

Matheus Poletto
Engineering of Processes and Technologies Postgraduate Program (PGEPROTEC)
Caxias do Sul University (UCS)
Brazil

Chapter 1

Depolymerization and Activation of Lignin: Current State of Knowledge and Perspectives

Łukasz Klapiszewski, Tadeusz J. Szalaty and Teofil Jesionowski

Additional information is available at the end of the chapter

http://dx.doi.org/10.5772/intechopen.70376

Abstract

A very important topic in present-day research is the depolymerization of lignin, meaning the multi-parametric decomposition of the biopolymer into low-molecular-weight products (monomers) by breaking of the intermolecular bonds. Depolymerization can occur under many different factors, such as high temperature or catalysts, which determine the mechanism of disintegration. In the case of lignin, this process is carried out in order to obtain many valuable low-molecular-weight compounds. It is becoming more and more popular as a result of the use of ionic liquids, but methods using alkaline, acidic, and metallic catalysts, as well as pyrolysis and supercritical fluids, are also known. All of these methods will be described in detail in this chapter.

Keywords: biopolymers, lignin, depolymerization, ionic liquids, modification

1. Introduction

Pro-ecological technologies for the production of activated, functional products with unique properties are being vigorously developed in the field known as green chemistry. Of particular significance is the production of novel materials based on polymers of natural origin, such as lignin and its derivatives.

Lignin is a byproduct in the paper and pulp industry and has been mostly used as a fuel to provide energy for technological processes. This biopolymer is a main source for the production of biofillers. It can also be used as an effective sorbent of organic compounds and hazardous metal ions, as a polymer filler, in the synthesis of polyethers, polyurethanes and epoxy resins, and as a component of abrasive tools. Lignin can be applied in its untreated state. However, to fully exploit its properties, its natural conformation has to be improved through certain structural modifications, which will be described in this chapter.

Activation of lignin is carried out to modify functional groups while leaving the aromatic base of the polymer intact. Originally, lignin was subjected to oxidation to better understand its structure and to identify its bonds. At present, it is known that the use of strong oxidants breaks up the aromatic ring of the polymer, whereas using milder ones changes only its functional groups. The lignin surface contains many hydroxyl groups, which can be oxidized to carbonyl groups. These functional groups are more reactive than the hydroxyl groups and hence the resulting compound can offer more potential applications, including in electrochemistry.

2. Lignin—an overview

There is an ever increasing importance to polymers of natural origin. They are widely used in many branches of science and industry, chiefly because of their distinctive or unique properties and their renewability. Of particular interest are materials obtained from biomass, broadly defined. These include lignin and its derivatives.

Lignin is a biopolymer that occurs in the cell walls of plants. It enables plants to retain rigidity, mechanical strength and integrity between the other natural polymers contained in the basic plant cell: cellulose and hemicellulose [1]. Depending on the species of tree, lignin may account for between 20 and 30% of the total mass of wood (lower values in tropical and subtropical hardwoods, and higher in softwood from conifers) [1]. It is estimated that lignin is the source of approximately 30% of the organic carbon occurring in nature [2].

Lignin has a complex structure, which is still not completely understood, and is thus currently the subject of intense research efforts among a large group of scientists. It is known that the basic monomers making up the structure of the biopolymer include *p*-coumaryl, coniferyl, and sinapyl alcohols, generally called monolignols (see **Figure 1**), which link together in an unsystematic manner [1–3]. The content of particular monolignols varies depending on the type of wood.

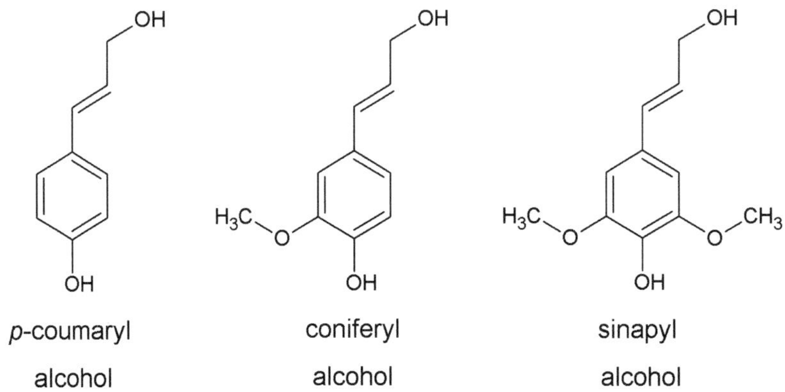

Figure 1. Basic monomers making up the structure of the biopolymer, based on [1–3].

http://dx.doi.org/10.5772/intechopen.70376

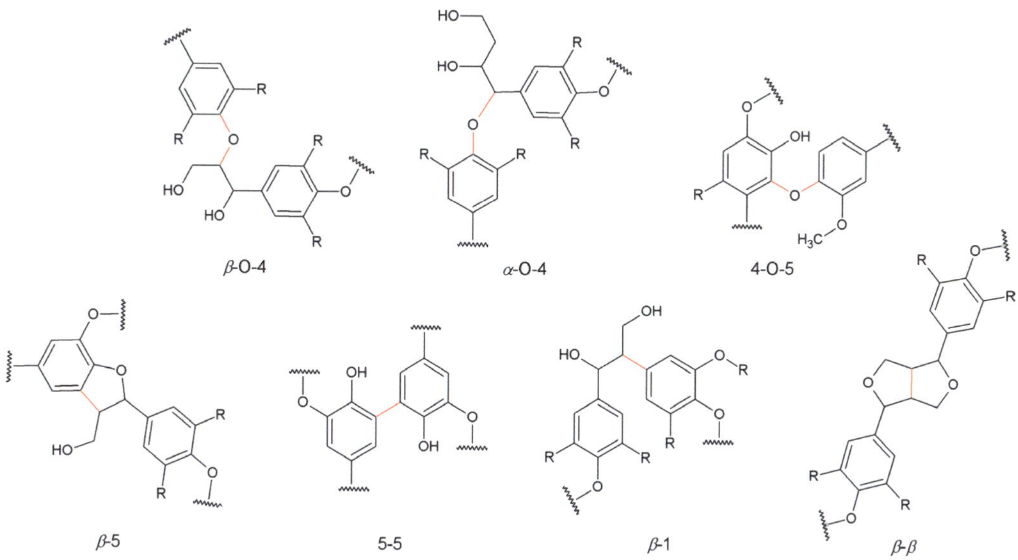

Figure 2. Characteristic bonds occurring in the structure of lignin, based on [2–4].

Characteristic bonds occurring in the lignin structure include carbon–oxygen bonds (β-O-4, α-O-4, 4-O-5), which account for almost two-thirds of all of the bonds present, as well as carbon–carbon bonds (β-5, 5-5, β-1, β-β)—see **Figure 2** [2–4]. Of the linkages occurring in the structure, the most easily decomposed are the β-O-4 bonds [5]. It should also be noted that the lignin macromolecules contain various functional groups that affect its reactivity. These are mostly methoxy and hydroxyl groups, as well as carbonyls and carboxyls. Only some of the hydroxyl groups are free to react, since most form bonds with neighboring structural units.

3. Application of lignin and its derivatives

Lignin, which is being produced in greater and greater quantities each year, not only as a byproduct of the paper industry but also as a valuable raw material obtained from biofuel production, is becoming the object of increasing interest in many branches of science and industry.

Approximately, 98% of the lignin produced by the paper industry is burnt for energy recovery, with only 2% being used for commercial purposes. At present, this market consists mainly of low-value products, such as binders and emulsifiers, low-quality fuel, carbon fibers, and phenolic resins. The complex structure of lignin and the fact that methods for its modification and depolymerization are relatively expensive and hard to carry out on an industrial scale, limit possibilities of its wider use [6]. Nonetheless, lignin is constantly finding new potential fields of application.

In recent years, lignin has come to be commonly used in the preparation of multifunctional hybrid systems. There are reports in the literature concerning silica–lignin materials, which may be used as effective and relatively cheap sorbents of environmentally hazardous metal ions [7], polymer fillers [8–11], innovative systems with antibacterial properties [12], components in abrasive products [13, 14], and substances used in the catalytic reduction of synthetic dyes, in sensors, and in surface-enhanced Raman spectroscopy [15, 16]. There are also reports of systems in which lignin is combined with titanium dioxide [17], magnetite [18], and the oxide systems $MgO{\cdot}SiO_2$ [19, 20] or $TiO_2{\cdot}SiO_2$ [17], as well as chitin, another natural polymer occurring as a waste product [21, 22]. All of these systems have been used as effective sorbents of environmentally hazardous metal ions, and the chitin-lignin system can also be used as an enzyme carrier.

4. Depolymerization of lignin

Lignin, which has a complex chemical structure, is often subjected to a process of depolymerization. This has the aim of decomposing the biopolymer into low-molecular-weight products (monomers) by breaking the intermolecular bonds. Depolymerization of lignin may be carried out under a number of factors, including catalysts or high temperatures, which determine the mechanism of decomposition. The depolymerization process may lead to numerous useful low-molecular-weight compounds—see **Figure 3** [23, 24].

Due to depolymerization of lignin, the following fractions are obtained: waste carbon, residual lignin, an aqueous phase, and oil, which consists of a mixture of phenolic monomers (such as 2-hydroxy-1,3-dimethoxybenzene and guaiacol) and oligomers. This last is the most desired fraction, as it contains the greatest quantity of valuable chemical compounds which can be used in industry. The residual lignin consists of particles which have repolymerized or which have not undergone decomposition. The waste carbon fraction contains unwanted products of the depolymerization reaction. Any gaseous products will chiefly be a mixture of oxides of carbon and hydrocarbons. There may also be an aqueous phase, which may contain alcohols, as well as hydrophilic aromatic compounds which cannot be isolated. The quantity and composition of each fraction depends on the process conditions [24–26].

Numerous research projects have helped determine the most effective methods of depolymerization, which lead to large quantities of phenolic monomers. The most important of these methods includes (i) depolymerization in supercritical fluids, (ii) pyrolysis, (iii) the use of metallic catalysts, (iv) acidic catalysts, (v) alkaline catalysts, and (vi) the use of ionic liquids.

4.1. Depolymerization in supercritical liquids

It is becoming increasingly common for the depolymerization of lignin to be carried out with the use of supercritical liquids [27–30]. The thermochemical depolymerization of lignin in supercritical methanol with various catalysts was studied by Singh et al. [27]. Zeolites, sodium hydroxide, and iron filings were used in the process. Analysis of the results led to suggested mechanisms of the decomposition reaction. Depolymerization and demethoxylation of lignin result from the donation of a proton to ether bonds. The proton donor is the solvent, which as a result forms a formaldehyde capable of further reaction, particularly with ring compounds. The presence of NaOH in the reaction system not only accelerates the reaction but also enables

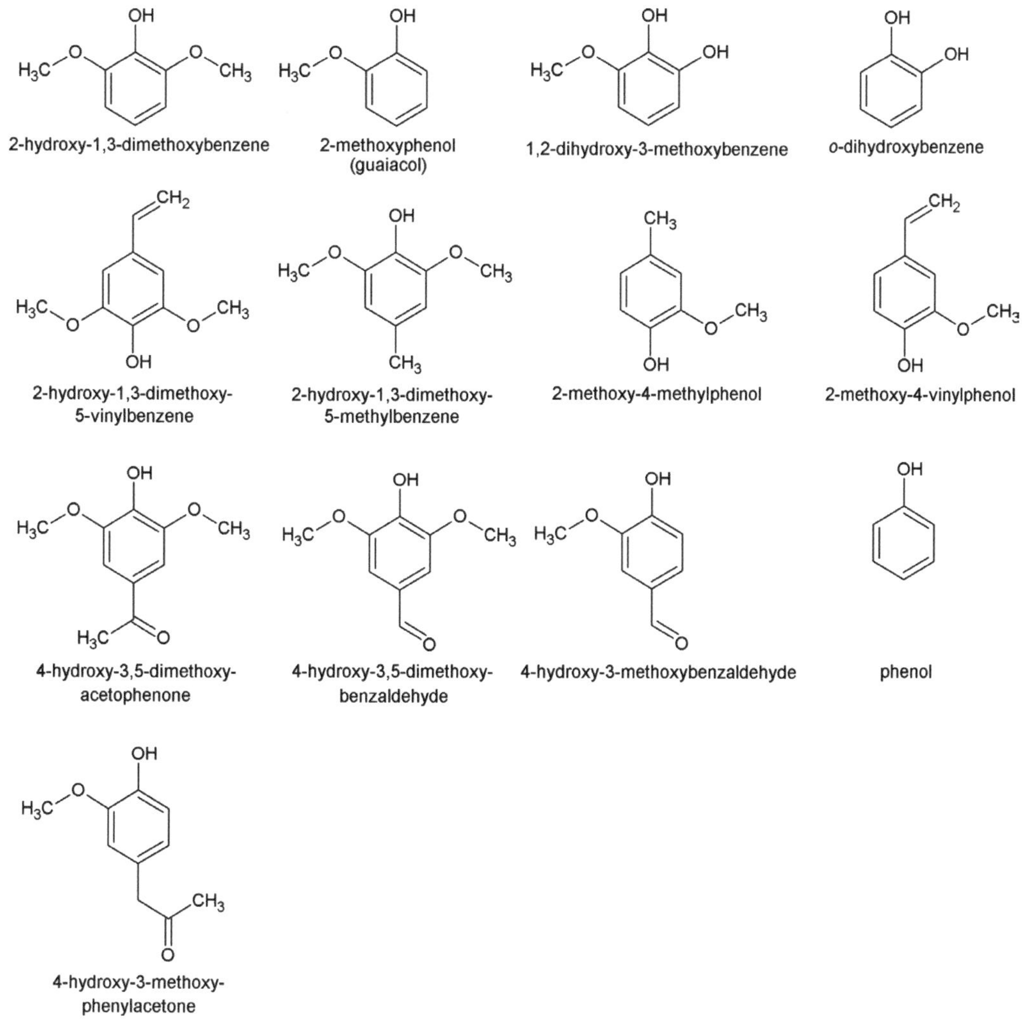

Figure 3. The most important low-molecular-weight compounds obtained from the depolymerization of lignin, based on [24–26].

the initiation of a demethylation reaction. The intermediate products undergo a condensation reaction, leading to a final product which is rich in solid carbon residues. This phenomenon is not observed in the case of zeolites, due to the presence of acid centers capable of catalyzing the direct methylation of the aromatic ring, which retard or prevent any potential condensation reactions [27].

In another study, Erdocia et al. used as the supercritical phase a triple system of methanol, ethanol, and acetone, determining their impact on the decomposition of lignin [28]. The largest content of monomeric phenol derivatives was found in the product obtained in an acetone environment. Detailed analysis confirmed the high degree of degradation of lignin, leading to a high content of solid carbon residues and catechol and cresol in the sample, which resulted from a reaction involving the detachment of alkyl and methoxy groups.

Depolymerization of the biopolymer in supercritical fluids was also studied by Kim et al. [29]. The process was conducted at 350°C in the presence of gaseous hydrogen and at a pressure of 13–19 MPa, using systems consisting of various alcohols in supercritical state—methanol, ethanol, and propan-2-ol—and metallic catalysts deposited on active carbon. The best-performing system proved to be Pd/C/ethanol, with which a very large quantity of oil and the lowest quantity of byproducts were obtained. This can be explained by the catalyst's high surface area and the fact that ethanol is the most effective hydrogen donor among the alcohols used.

Gosselink et al. used a system of carbon dioxide, acetone, and water as the supercritical phase, carrying out depolymerization of lignin in an organic solvent using formic acid as catalyst [30]. The process was carried out at the temperatures of 300 and 370°C under a pressure of 10 MPa. The quantity of products obtained was found to be affected by the depolymerization reactions taking place, which led to the formation of byproducts.

4.2. High-temperature depolymerization of lignin

Depolymerization of lignin also takes place at very high temperatures without the presence of oxygen, in a process known as pyrolysis. The high temperature causes cracking of the ether and carbon–carbon bonds, and the absence of oxygen prevents oxidation of the products. The process leads to, among others, liquid products containing monomeric phenolic compounds, gaseous products consisting mainly of hydrocarbons and oxides of carbon, and solid carbon residues. The proportions of individual compounds depend on the process conditions, including the temperature. Above 500°C, bonds are broken in the aromatic ring and hydrogen is formed. The process carried out at such high temperatures is called gasification, and gases account for more than 80% of the products in that case. There are two ways in which pyrolysis may be carried out—slow (conventional) and fast. In conventional pyrolysis, the lignin is heated more slowly, and the products consist of approximately 40% gases, 30% carbon, and only 10% oil containing monomeric compounds. Fast pyrolysis leads to as much as 60% oil, with only about 10% carbon and 15% gases, and the time of the process is shortened from 5–30 min to 2 s. A disadvantage is the presence of a large number of oxygen compounds in the resulting oil, which means that it cannot be used as a fuel. Research is being done, however, with the aim of improving the pyrolysis process by reducing the number of organic oxide compounds (pyrolysis with the addition of polyolefin) and increasing the content of phenolic compounds in the oil (pyrolysis with the addition of substances whose chain contains compounds of phenol, such as polystyrene) [25, 31].

4.3. Depolymerization of lignin using metallic catalysts

The depolymerization of lignin is often carried out at a high temperature in the presence of hydrogen (hydrogenolysis) with the use of metallic catalysts. Catalysts used include palladium, platinum, rhodium, nickel, ruthenium, and copper with the addition of carbon or aluminum oxide. At temperatures as high as 450°C, in the presence of hydrogen gas and under pressures of 2–35 MPa, various substances are obtained, including cyclohexanols, phenolic decomposition products, and cyclic alkanes.

Kloekhorst and Heeres carried out depolymerization in the presence of hydrogen on various different metal catalyst systems, without the use of a solvent. Analysis of the resulting products showed a Ru/TiO_2 system to be the most promising, in view of the variety of monomeric phenolic products obtained. Gaseous products, including methane, were also formed during depolymerization [32].

Depolymerization of lignin without the addition of a solvent was also investigated by Kumar et al. [26]. As catalysts, they used NiMo and CoMo deposited on various basic and acidic media. Apart from lignin and the catalyst, the reaction mixture also contained dimethyldisulfide, which served to create the system S-NiMo/medium or S-CoMo/medium. The introduction of sulfur into the catalyst causes the formation of new active sites with catalytic ability. It was found that the depolymerization was influenced by the type of medium, as well as by other factors. Acid media led to the largest quantities of repolymerization products. The best medium for the catalyst proved to be MgO-La_2O_3, which makes it possible to obtain a large number of monomers (particularly alkylphenols), while reducing the content of residual lignin in the post-reaction mixture [26]. Based on the results of the study, a mechanism for the process was proposed (see **Figure 4**).

Wang and Rinaldi investigated the ability of nickel catalysts to decompose a model lignin molecule—diphenyl ether, having a strong ether bond [33]. Various media were used for the metallic catalyst, and their impact on the hydrodeoxygenation of the compound was studied. A high degree of transformation of the ether was found to be linked to the acidity of the medium and its surface area.

Das et al. carried out depolymerization using niobium(V) oxide as catalyst and hydrogen peroxide and potassium permanganate as oxidizing agents [34]. The main decomposition products

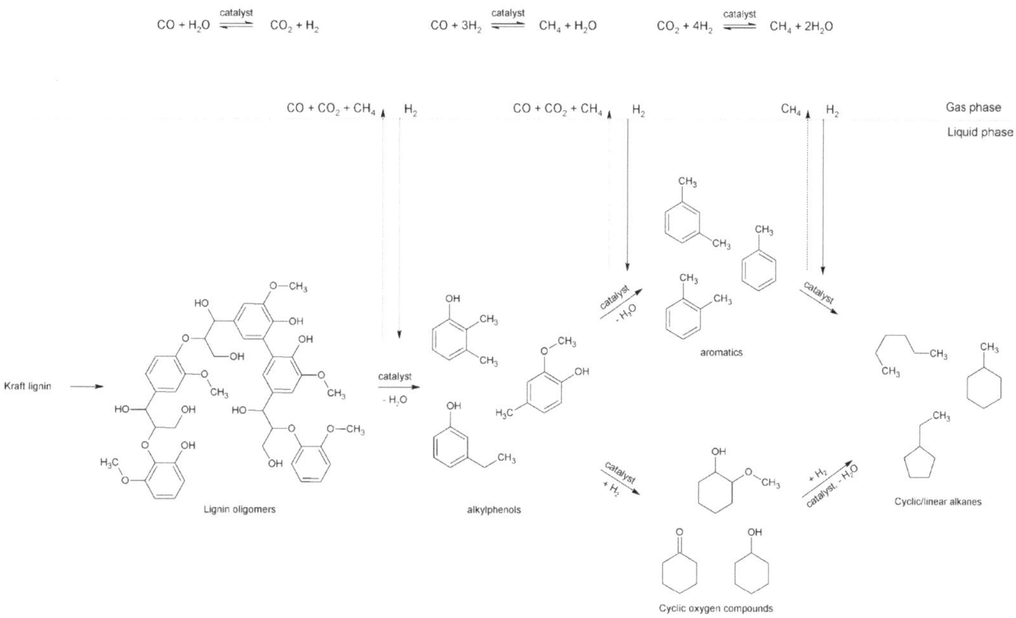

Figure 4. Mechanism of the process of depolymerization of lignin, adapted from Kumar et al. [26].

were vanillin and its derivatives. Due to the presence of trace quantities of acids in the post-reaction mixture, the possibility was suggested that the lignin is adsorbed by the catalyst and aldehydes are formed as a result of reactions taking place on the catalyst's surface.

4.4. Depolymerization of lignin using acidic catalysts

Valuable low-molecular-weight products can also be obtained from lignin with the use of catalysts of acid origin.

Sturgeon et al. investigated the mechanism of decomposition of the β-O-4 bond, adding the following model molecules to a 0.2 M solution of H_2SO_4 at 150°C: 2-phenoxy-1-phenylethanol (PE), 2-phenoxy-1-phenylpropan-1,3-diol (PD), 1-(4-hydroxyphenyl)-2-phenoxypropan-1,3-diol (HH), and 1-(4-hydroxy-3-methoxyphenyl)-2-(2-methoxyphenoxy) propan-1,3-diol (GG), which represent the phenolic (HH and GG) and non-phenolic (PE and PD) groups [35]. Depolymerization was shown to take place by way of an ionic mechanism, via protonation of the hydroxide group on the α-carbon, followed by the separation of a water molecule and the formation of a carbocation as an intermediate stage (see **Figure 5**). It was also observed that the presence of a phenolic hydroxyl group accelerates the decomposition of the β-O-4 bond.

Deepa and Dhepe carried out depolymerization of lignin using a solid acid catalyst [36]. Several types of lignin were selected to undergo depolymerization in a mixture of water and methanol in a nitrogen atmosphere, at a temperature below 250°C, using various types of zeolites, aluminosilicates, and metal oxides as catalysts. It was found that a very large quantity of monomers could be obtained, particularly when zeolites were used (the yield in that case reached 60%). However, the susceptibility of zeolites to poisoning with metal ions (such as sodium) led to their deactivation and also caused instability during the reaction. It was therefore decided to focus further attention on the possible use of the amorphous SiO_2-Al_2O_3 catalyst. Following optimization of the process, a method was proposed for obtaining a large quantity of monomers and for recovering the catalyst for reuse [36].

There also exists the possibility of using organic acids as catalysts in the decomposition of lignin. Rahimiet al. used formic acid and sodium formate for this purpose, carrying out depolymerization on previously oxidized lignin extracted from aspen [37]. Desirable properties of this system include the relatively low process temperature (110°C) and the very large quantity of low-molecular-weight aromatic compounds produced (approximately 61% of the initial mass of lignin).

Güvenatam et al. investigated the possibility of carrying out depolymerization of lignin and its model compounds using trifluoromethane sulfonates of various metals [38]. These

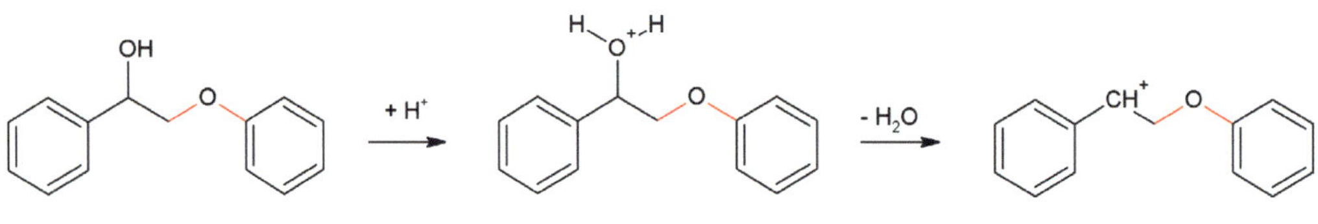

Figure 5. Mechanism of the depolymerization of lignin as established by Sturgeon et al. [35].

compounds are known for their high acidity. The process took place in a mixture of water and ethanol in a supercritical state. It was found that a significant quantity of ether bonds in the model lignin molecules was broken. It was also discovered that ethanol not only served as a solvent but also played an active part in the reaction, attaching itself to the aromatic ring and to the alkyl chain of the molecules. Further investigation confirmed the ability of liquids with a trifluoromethane sulfonate anion to catalyze the decomposition of lignin, proving the significant role played by that anion in the depolymerization process. In the proposed mechanism, the catalyst is capable of interacting with various oxygen functional groups in lignin, causing them to regroup and forming a carbocation, which can react further with nucleophiles, and also with the solvent. There exists a possibility of further dehydration and hydrogenation of the resulting compound (see **Figure 6**). The chief products of the reaction were hydrocarbons, both aromatic and non-aromatic, accompanied by a small quantity of monomeric phenolic derivatives. The absence of residual lignin in the post-reaction mixture was explained by the interaction of ethanol with the carbocation, which prevented repolymerization reactions [38].

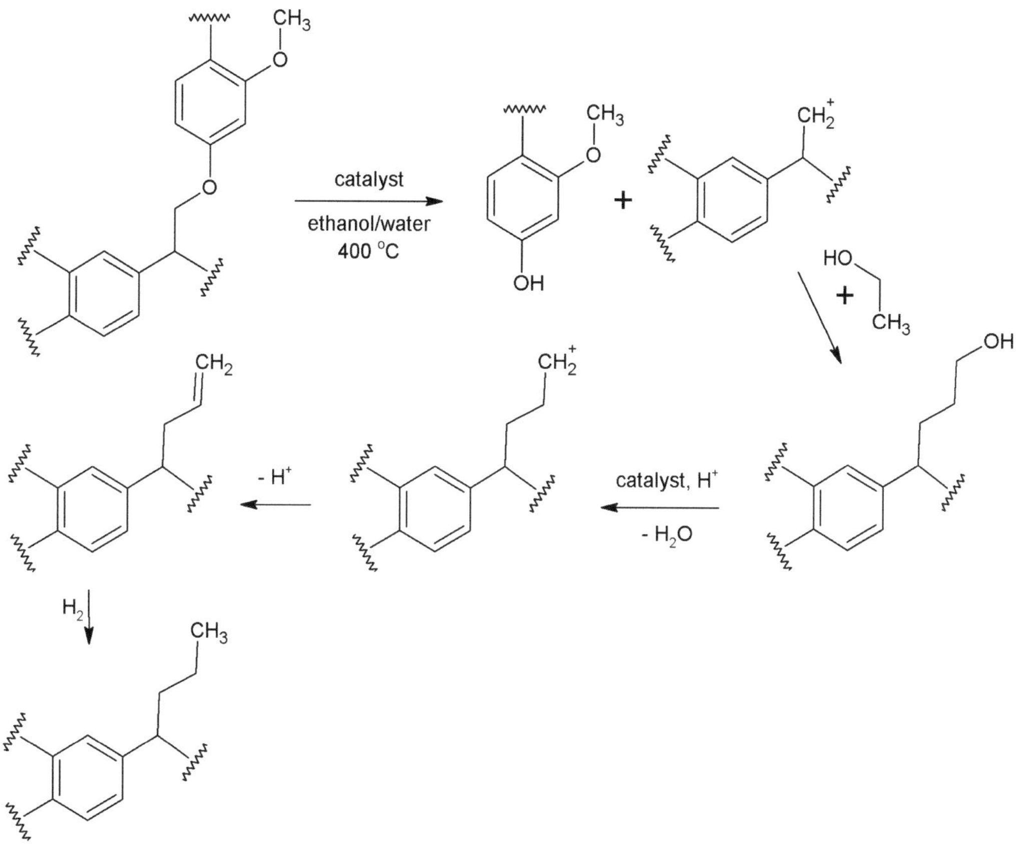

Figure 6. Mechanism of the process of depolymerization of lignin proposed on the basis of studies by Güvenatam et al. [38].

4.5. Depolymerization of lignin using basic catalysts

The reaction of the decomposition of lignin into low-molecular-weight products is also carried out using inorganic basic catalysts such as NaOH [24, 39]. During the process, alkyl-aryl ether bonds are broken under the action of the sodium ion, which polarizes the bond, making it easier to break (see **Figure 7**).

Optimization of the depolymerization process using a basic catalyst was performed by Roberts et al., who investigated the effect of temperature, the quantities of NaOH and lignin, and the pressure and reaction time on the quantities of monomer particles formed [24]. The largest quantity of monomers as a percentage of the original quantity of lignin was obtained using 4% NaOH. The process was most successful when carried out at a temperature of 300°C for 4 min. With increasing reaction pressure, the concentration of monomers decreased, possibly as a result of the stronger intermolecular interactions. The quantity of depolymerized lignin behaved analogously. It was also shown that the process temperature has a strong impact on the quantity of particular monomers occurring in the mixture [24].

A study by Erdocia et al. also investigated how the decomposition products depend on the type of lignin used [28]. Three types of lignin dissolved in organic solvents (acetone, formaldehyde, and a mixture of the two) were subjected to the action of NaOH at a temperature of 300°C and a pressure of 9 MPa. A large quantity of residual lignin (approximately one quarter of the original quantity) was present following the depolymerization process, due to repolymerization of the compounds formed during that process. These recombine with each other or with the original lignin, forming new C–C and C–O bonds, and this process is promoted by the presence of formic acid. The quantity of monomeric phenolic particles formed was found to depend on the average molecular weight of the lignin used—the smaller the molecular weight, the greater the quantity of particles formed during depolymerization [28].

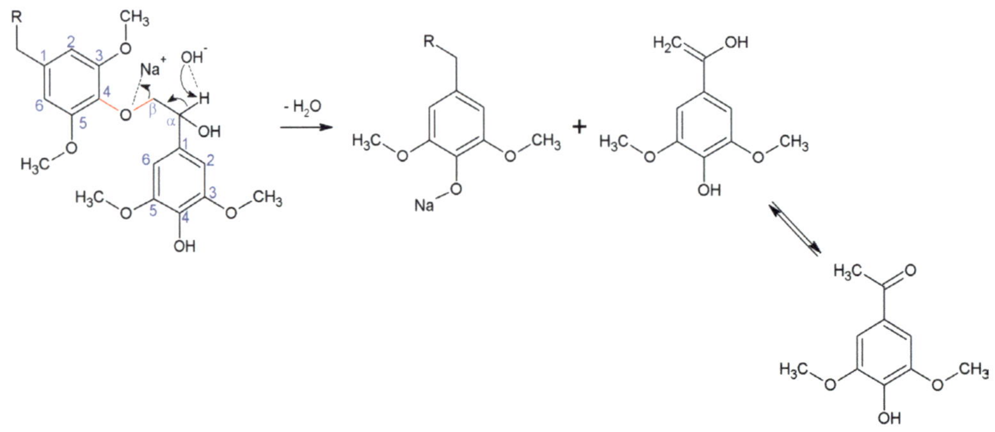

Figure 7. Low-molecular-weight products resulting from the depolymerization of lignin with the use of NaOH, based on [24].

Santos et al. also investigated the depolymerization of the biopolymer using NaOH [40]. Based on the analysis of the oil produced, it was shown that the principal decomposition products were catechol and phenol. Small amounts of cresol, guaiacol, and 2-hydroxy-1,3-dimethoxy-benzene were also formed. The authors not only analyzed the product but also suggested a potential application: the antiseptic and insecticidal properties of catechol, in combination with the preserving and disinfecting action of phenol, meaning that the mixture might potentially be used as an antifungal treatment for wood. Preliminary studies showed that the oil indeed increased the resistance of wood panels to fungal action, thus confirming its potential for use as a wood preservative [40].

An investigation of the effect of the catalyst on the depolymerization of lignin was carried out by Toledano et al. [39]. Lignin was dissolved in an organic solvent with the addition of potassium, calcium, lithium and sodium hydroxides and sodium carbonate, and the biopolymer underwent depolymerization at 300°C under a pressure of 90 MPa. The large differences in the quantities of phenolic particles (from around 10% for $Ca(OH)_2$ to 20% for NaOH) and residual lignin (above 37% for almost all of the catalysts apart from calcium carbonate) show that the catalyst used affects the mechanism of the reactions taking place during depolymerization.

Although metal hydroxides would appear to be the most popular catalysts, they are not the only compounds used in the depolymerization process. Long et al. used magnesium oxide, together with the solvent tetrahydrofuran, which increases the power of the catalyst [41]. The quantity of phenolic monomers produced (more than 13%) and their increased ease of separation from the mixture mean that this is a promising catalyst for the decomposition of lignin.

In turn, Widyay et al. studied the impact of different catalysts (magnesium and calcium oxides, potassium acetate, and potassium hydroxide) on the depolymerization of lignin in ethanol [42]. In spite of the very good results obtained for the potassium catalysts, in view of their high solubility in the reaction medium, further tests were carried out using magnesium oxide. At the next stage, an analysis was made of the effect of the polarity of the solvent on the products obtained. Among the various solvents used, the best results were obtained for ethanol. The post-reaction mixture contained large quantities of hydroxyl (aromatic and non-aromatic) and carboxyl groups, probably due to the oxidation reactions taking place. Also noted was the degradation of some of the carbon–carbon bonds and the structural units of lignin. The authors further drew attention to the possibility of reusing the catalyst without detriment to its activity or structure [42].

4.6. Depolymerization of lignin using ionic liquids

The unique properties of ionic liquids are exploited in a number of new multifunctional applications, including their possible use in the depolymerization of lignin. A number of important studies of such depolymerization processes have been carried out by Jia et al. [43]. Since the most frequently occurring bond in lignin is the β-O-4 bond, the model compounds used were 1-(4-hydroxy-3-methoxyphenyl)-2-(2-methoxyphenoxy)propan-1,3-diol and 1-(3,4-dimethoxyphenyl)-2-(2-methoxyphenoxy)propan-1,3-diol (VG). Although both of these contain the β-ether bond, in view of the different substitutions, they represent, respectively, the phenolic and non-phenolic character of the subunits of the biopolymer. One of the experiments was carried out in the presence of the ionic liquid [HMIM][Cl] (acting as both catalyst and solvent) with varying quantities of water. A significant percentage (more than 70%) of the ether bonds was broken, and the product formed most often and in the largest quantities was guaiacol. It was also found that a larger quantity of water favors the decomposition of the β-O-4 bond. Importantly, it was confirmed that the ionic liquid can be reused without loss of activity [43]. Further studies were made to analyze the ability of the N-base-[BDMIM][Cl] system to break β-O-4 bonds. Following heating of the systems to a temperature of 150°C, the principal decomposition product was found to be enol-ether (EE) [44].

The same researchers have also evaluated the effect of the anion on the depolymerization process. Using the aforementioned model substances (VG and GG), they carried out a depolymerization reaction at 150°C for the first substance and at 110, 130, and 150°C for the second, selecting appropriate ionic liquids with the 1-methylimidazole cation and various anions. A determination was also made of the Hammett acidity of particular ionic liquids (with the use of 3-nitroaniline), showing them to be strongly acidic. Although no dependence was found between the acidity of the liquid and the effectiveness of depolymerization, the results indicated a link between the possible formation of anion–molecule hydrogen bonds and hydrolysis of the β-O-4 bond. This was found to be a factor that favored the decomposition of the bond, and thus the depolymerization of the molecule [45].

Yan et al. also studied the depolymerization of lignin using a liquid with the imidazole cation, confirming the dependence between the anion of the ionic liquid used and the resulting depolymerization products [46]. Moreover, factors affecting the process of decomposition of lignin were found to include the pH of the reaction mixture, the quantity of ionic liquid used, and the presence in the system of a cosolvent. In this case, however, the addition of water to the reaction system was found to limit depolymerization. It was concluded that there is a clear need to seek new model lignin molecules which better imitate the structure of the biopolymer.

Liu et al. developed an innovative catalyst containing choline methane sulfonate and palladium deposited on carbon [47]. This produced a system which combined the acidic properties of the ionic liquid with the oxidizing properties of the metal and was at the same time a very good solvent of lignin. Studies carried out using kraft lignin confirmed the selectivity of the hydrogenolysis process, as well as the high content of phenol and its derivatives, particularly 2-hydroxyphenol. In addition, in an experiment using a model lignin molecule, the reaction mechanism was investigated. It was found that the hydrolysis of the substrate leads to breaking of the ether bond and the formation of phenol derivatives. The intermediate products may subsequently undergo hydrogenolysis or a retro-aldol reaction (see **Figure 8**) [47].

Another method used to carry out the depolymerization of lignin used a model lignin molecule in an environment of 1-benzyl-3-methylimidazolium bis(trifluoromethylsulfonyl) imide, with phosphoric(V) acid as catalyst [48]. Following thorough analysis of the product, which was rich in phenol and benzoic acid, a possible mechanism of the decomposition was proposed, according to which the ionic liquid not only interacts with the molecule undergoing decomposition but also has the ability to take part in a reaction with atmospheric oxygen leading to the formation of radicals, these being capable of further interaction with the

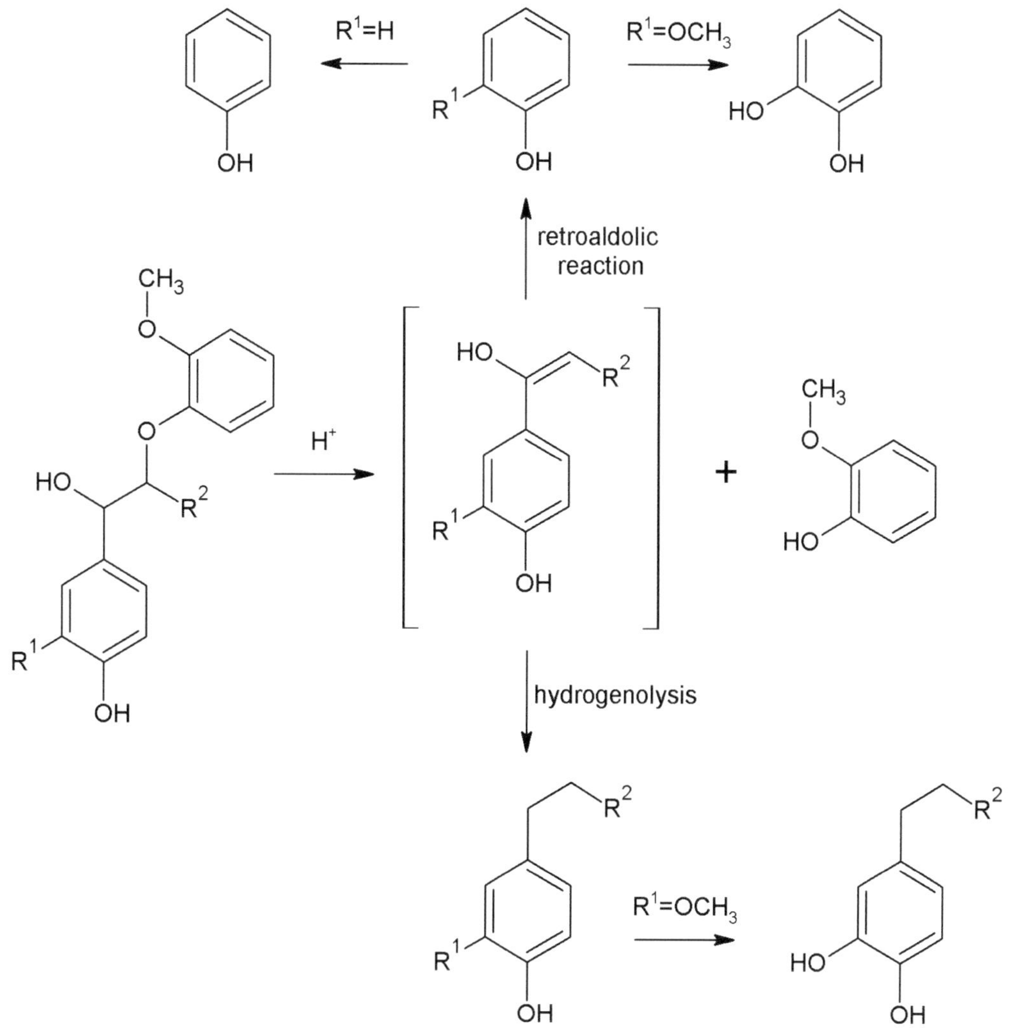

Figure 8. Mechanism of the depolymerization reaction of a model lignin molecule with the use of choline methane sulfonate, adapted from [47].

depolymerized compound. As a result, peroxy acid is formed, which via acidic hydrolysis is transformed into the final decomposition products—phenol and organic acids [48].

In the search for effective solvents, Rashid et al. investigated the ability of protonic ionic liquids to dissolve lignin [49]. Various pyridine ionic liquids with methanoate, ethanoate, and propanoate anions were synthesized. Among the systems obtained, the largest quantity of lignin was dissolved in pyridine methanoate, demonstrating that an increase in the length of the alkyl chain in the anion has an unfavorable effect on the solubility of lignin. In the next stage, the regeneration of lignin was carried out and its properties were analyzed. The regenerated lignin was found to have a significantly smaller average molecular weight, showing that the material underwent defragmentation during dissolution. There was also a decrease in the quantity of hydroxyl groups linked to the aromatic ring and an increase in the quantity of quinone groups. This meant that the properties of the material were more favorable for further chemical treatment. An effective process of regeneration of the solvent was also carried out, obtaining a liquid suitable for reuse [49].

An innovative depolymerization solution was proposed by Caiet al., who carried out the process in an emulsion reactor [50]. They used a solution of water and an ionic liquid—1-butyl-3-(butyl-4-sulfone)imidazolium hydrogen sulfate(VI)—as a catalyst and an n-butanol/n-hexane system as the oil phase. Lignin was used as a surfactant, in view of the presence in that material of both hydrophilic and hydrophobic groups. The lipophobic nature of the ether groups in lignin facilitates contact between the compound and the catalyst and thus favors the decomposition process. Detailed analysis of the results confirmed the high effectiveness of the process and the significant increase in the quantity of monomeric phenolic products obtained compared with traditional techniques, thereby confirming the good potential of the method for practical application [50].

The ability of ionic liquids to dissolve lignin is exploited not only in depolymerization processes but also in extraction from lignocellulose materials. Prado et al. combined both processes, carrying out the depolymerization of lignin present in the black liquor formed during one of the stages of delignification. For this purpose, butylimidazolium hydrogen sulfate(VI) and triethylammonium hydrogen sulfate(VI) were used, with hydrogen peroxide as an oxidizing agent. The oils obtained were rich in phenol derivatives, particularly acids, including vanillic and benzoic acids [51]. In a subsequent study, titanium(IV) oxide was used as an oxidizing agent alongside hydrogen peroxide. Analysis of the products showed the inorganic oxide to be an effective catalyst, enabling a large quantity of monomeric phenolic compounds to be obtained. The product also contained no significant quantity of molecules originating from the process of oxidation of carbohydrates, which indicates that titanium(IV) oxide exhibits greater selectivity than hydrogen peroxide in the oxidation of lignin [52].

Stärk et al. investigated the effect of systems of various ionic liquids and salts of iron(III), copper(II), and manganese(II), used as catalysts, on the degree of conversion of lignin in organic solvent [53]. Aerobic depolymerization was carried out for over 10 h at a temperature of 100°C under increased pressure. Among 40 systems investigated, the best parameters were obtained for [EMIM][OTf] combined with manganese(II) nitrate(V) as catalyst; this produced a conversion yield of more than 50%.

Nanayakkara et al. carried out an experiment to determine the effect of adding 4-tert-butyl-2,6-dimethylphenol (TBDMP, blocked in the ortho and para positions to prevent a polymerization reaction) on the products obtained from depolymerization via a redistribution reaction [54]. Using Klason lignin, an organic solvent and a Cu/EDTA complex as catalyst, the depolymerization reaction was carried out at 180°C with the addition of TBDMP in two ionic liquids: [EMIM][ABS] and [BMIM][$MeSO_4$]. The process produced numerous oligomers which could not be obtained without the addition of the catalyst and TBDMP, which demonstrates that these two components are essential for the depolymerization of the biopolymer in the proposed ionic liquids [54].

Binder et al. investigated the possible depolymerization of lignin in an organic solvent and of its model molecules—eugenol (2-methoxy-4-(2-propenyl)phenol), 1-phenoxy-2-phenylethane, and 4-ethyl-2-methoxyphenol—at temperatures below 200°C [55]. A Brønsted acid catalyst was used, with the ionic liquids [EMIM][OTf] and [EMIM][Cl]. From the first two model compounds, guaiacol and phenol, respectively, were obtained, but in the case of 4-ethyl-2-methoxyphenol and lignin itself, no depolymerization reaction could be achieved. This indicates the need to seek better model substances and to gain a more precise understanding of the mechanism of depolymerization.

The depolymerization of lignin (extracted from oak) was also investigated in a study by Cox and Ekerdt, who used acidic [HMIM][Cl] [56]. It was shown that this ionic liquid may serve both as a solvent and as a catalyst for the depolymerization reaction, and confirmation was obtained for a proposed mechanism of decomposition via hydrolysis of the alkyl-aryl ether bond.

It can therefore be concluded, based on an exhaustive survey of the latest literature, that the catalytic properties of ionic liquids, in combination with their ability to dissolve lignin, give them great potential for practical applications in the conversion of lignin materials. They are also made more attractive in comparison with existing methods by their low toxicity and the ability to create liquids with specified parameters. It is expected that research efforts in this area will lead to further progress in the near future.

5. Modification of the lignin structure

In the preceding chapters, attention has been drawn to the potential for practical applications of lignin and of the products obtained as a result of its depolymerization. Nonetheless, this biopolymer may alternatively be used following preliminary processing or modification of its structure. Such an operation can be used to introduce new functional groups into lignin by way of chemical reactions of many kinds and also to achieve mild oxidation of the surface hydroxyl groups [57]. These groups, which are substituted not only on the aromatic ring but also in aliphatic chains, are of decisive importance for the modification of lignin. Modification of the structure of functional groups serves to increase the chemical reactivity of the biopolymer, improve its solubility in polar or non-polar solvents, facilitate the creation of a network of interactions between lignin and polymers in composites or inorganic compounds in functional materials, and simplify the processing of the final product. These goals may be achieved by creating new active centers. By this means, it is possible to increase the biopolymer's reactivity using, for example, hydroxyl groups (inductive effect) or substituents in the ortho position of the aromatic ring (mesomeric effect) [57]. It is significant that hydroxyl groups occur in lignin both in aliphatic chains in the $C\alpha$ and $C\gamma$ positions and as substituents on the aromatic ring. In view of their weak acidity, it is phenyl groups that determine the reactivity of the whole lignin molecule [57]. The types of reaction used to obtain new active centers include sulfonation [1, 58], hydroxyalkylation [59], nitration [57], amination [60, 61], halogenation [57], and alkylation/dealkylation [57].

The sulfonation of lignin and of its previously sulfonated equivalents enables improvement of its solubility in aqueous solutions, irrespective of their pH. In the reaction, the biopolymer is acted on by a 95% solution of sulfuric(VI) acid, with intense mixing, at a temperature of 40°C [62]. Lignosulfonates can also be included in the group of modified lignins produced as a result of sulfite pulping [1]. An alternative method of sulfonation was described by Ouyang et al. [58]. The first stage involved hydroxymethylation of alkaline lignin with the use of formaldehyde in a basic environment (see **Figure 9**). Following this substitution, the lignin underwent sulfonation with the use of Na_2SO_3, where the hydroxymethyl substituent undergoes a reaction. The mechanism of this modification of the lignin structure is shown in **Figure 9** [58]. Compared with the lignosulfonates commonly used in concrete mixtures, lignin prepared in this way exhibits a higher degree of sulfonation, a lower surface tension, and a stronger interaction with the surface of cement particles. It also produces desirable electrostatic repulsion forces among the cement particles, providing further confirmation of its potential for use in concrete mixtures [58].

The hydroxymethylation of lignin was also investigated in a study by Sen et al. [63], where a sample of the biopolymer was dissolved in sodium hydroxide and then subjected to the action

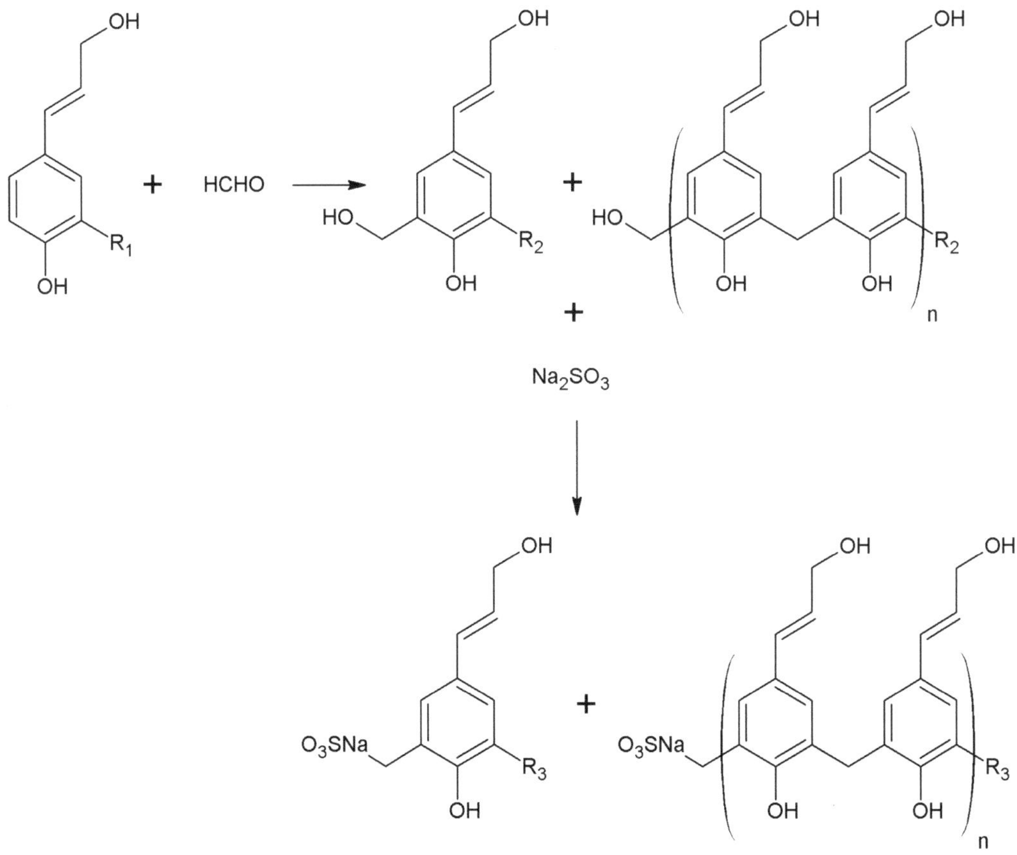

Figure 9. Mechanism of hydroxymethylation and sulfonation of lignin, adapted from [58].

of formaldehyde. In this way, the content of aliphatic hydroxyl groups and carbonyl groups was increased, as was confirmed by various analytical techniques. Based on the results, it was proposed that the derivative obtained might be used in glues or for wood preservation [63]. A similar modification of the lignin structure was studied by Malutan et al. [64]. They used lignin extracted from annual plants, subjecting it to hydroxymethylation with formaldehyde in an alkaline environment at room temperature. It was found that the reaction took place chiefly in the ortho position of the aromatic ring. At higher temperatures, methylene bonds are formed between the substituted lignin fragments. The desired reaction is the Lederer–Manasse reaction, in which a hydroxymethyl group is substituted on the aromatic ring, increasing the reactivity of the molecule. Based on literature data, it can be stated that the hydroxymethylation reaction is dependent on the type of lignin used (the raw material and the method of isolation) and also on the process conditions.

In a study by Du et al., amination of lignin took place via the Mannich reaction [60]. The authors also investigated the impact of preliminary phenolation of the biopolymer structure on the effectiveness of the amination reaction. For a successful Mannich reaction, it is very important to select a lignin with a free C-5 position in the aromatic ring of the guaiacyl units. This carbon has very high electron density, which promotes the introduction of amino-alkyl groups in the ortho position relative to the hydroxyl group. Irrespective of the pH of the reaction, although it may proceed according to different mechanisms, these lead to the same product, with substitution at C-5. The mechanism of the reaction of amination of lignin is shown in **Figure 10** [60]. The study also showed that preliminary phenolation of the lignin sample increases the content of unsubstituted C-5 carbons, which subsequently improves the yield of the amination reaction. However, in view of the toxicity of this reactant, it must be considered whether such pretreatment is reasonable [60].

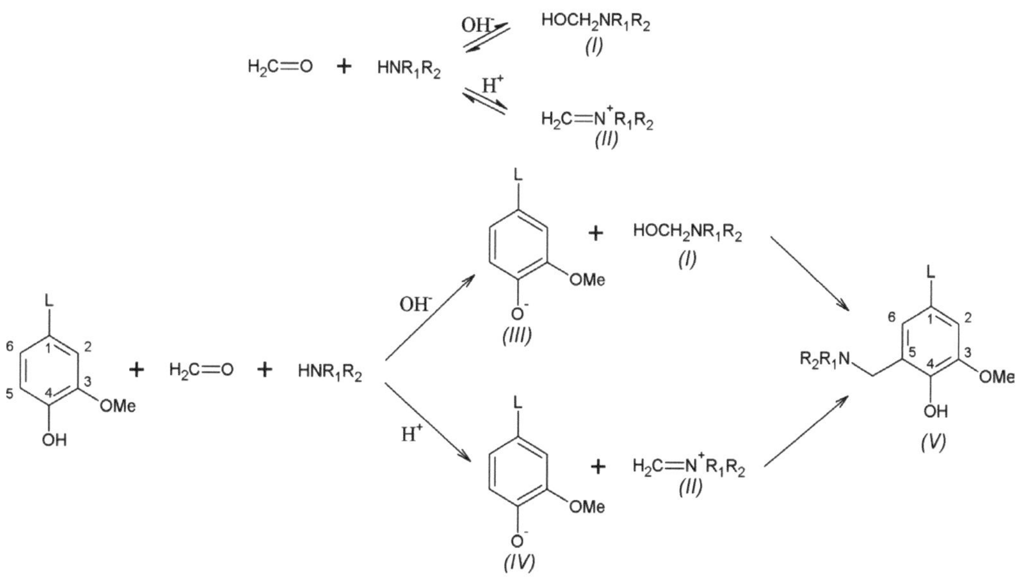

Figure 10. Mechanism of the reaction of amination of lignin in a basic or an acidic environment, adapted from [60].

Corresponding results on the formation of aminated lignin were published by Ge et al. [61]. They used alkaline lignin, which they modified at pH 8–13 with methylamine, adding appropriate doses of formaldehyde to the reactor. The product was evaluated as a possible sorbent of lead ions from aqueous solutions. Aminated lignin may be used successfully as a surfactant, in polycationic materials, in slow-release fertilizers [60], and in the sorption of lead ions from waste water [61].

The aforementioned use of the phenolation of lignin to unblock a greater number of C-5 carbons in the aromatic ring was also investigated by Podschun et al. [65]. The purpose of the study was to determine optimum conditions for the modification of lignin with phenol, to give a product with better solubility in various thermosetting resins. The reaction took place in an acidic environment at an elevated temperature, via the mechanism shown in **Figure 11**. In addition, some of the ether bonds are broken during the process, which leads to a reduction in molecular weight [66]. In the process used by Hu et al., the phenolation of lignin took place in ethanol at 70°C, over a time of several hours [66].

Lignin may also be successfully subjected to alkylation or dealkylation, by which means alkyl groups can be introduced into the biopolymer molecule. Here, the alkylating agent reacts with the nucleophilic centers of the biopolymer (oxygen atoms). In the case of lignin, atoms with an excess of electrons are attacked not only in the aromatic ring but also in the aliphatic chains. Sen et al. carried out methylation of kraft lignin with the use of dimethyl carbonate (DMC) [63]. They confirmed the possibility of controlling the degree of methylation by varying the quantity of DMC used. There was no drop in the thermal stability of the methylated lignins, although the glass transition temperature was reduced. This was achieved by a reduction in the number of sites capable of forming intermolecular hydrogen bonds [63]. It was also shown that the use of DMSO enables the reaction to be carried out at higher temperatures, thus increasing the degree of methylation of lignin. Moreover, the use of a polar aprotic solvent further catalyzed the substitution of the nucleophilic center [63].

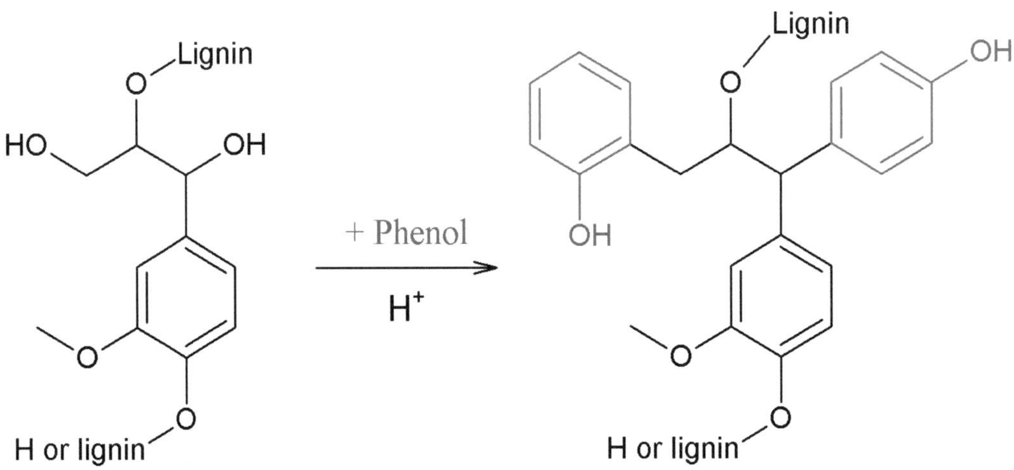

Figure 11. Mechanism of the reaction of phenolation of lignin, adapted from [65].

Lignin has also been successfully subjected to cationization, which leads to a material that may function as a flocculant of dyes from aqueous solutions. Cationic modification of the lignin structure was performed with the use of glycidyl-trimethylammonium chloride, which in a basic environment (pH=12.6) is joined to the phenol group of the aromatic ring in the biopolymer. The resulting material can be subjected to hydrolysis, with separation of the biopolymer from the quaternary ammonium salt. The mechanism of the modification of the biopolymer and hydrolysis is shown in **Figure 12** [67]. The resulting material may interact with the dye being removed, and in addition, the aromatic part of the lignin may develop hydrophilic/hydrophobic interactions with the pollutant [67].

Another method for modifying the structure of lignin is epoxidation of the phenol groups of the aromatic ring, carried out in an alkaline environment [59, 68–70]. Pan et al. performed such an action using epichlorohydrin, which was mixed with alkaline lignin in an appropriate ratio at 50°C [68]. The solid product was further subjected to amination using propandiamine. This Lewis base reacts with the ether oxygen, lengthening the chain of the alkyl substituent. The use of diamine makes possible a reaction with a further epoxy substituent, which finally increases the degree of crosslinking of the biopolymer. The epoxy lignin prepared in this way, or aminated crosslinked derivative, can be used as an active bioorganic additive to epoxy resins. This solution is friendly to the environment and can reduce the costs of resin production [68]. In a similar study by Feng and Chen [69], lignin was obtained by pulping of raw material in acetic acid, followed by phenolation of the biopolymer. This pretreatment enables an increase in the content of phenolic hydroxyl groups, which in the next stage are attacked by the epoxidation agent. Moreover, phenolation of lignin enables reduction of the molecular weight and thermal stability, compared with the unmodified biopolymer [66, 69]. It was shown that the addition of epoxidated lignin to an epoxy resin mixture increases its adhesive shear strength and water-absorbing power. Based on the results of the study, it was proposed that lignin might be used in the manufacture of cheap epoxy resins [69]. An oxypropylation reaction can also be used in the case of other biopolymers, such as chitin and chitosan [70]. A general mechanism for this reaction is given schematically in [70], see **Figure 13**.

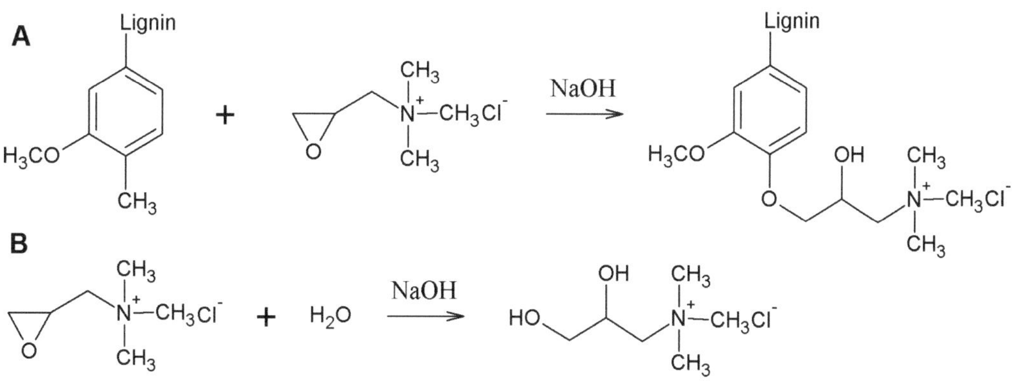

Figure 12. Mechanism of the cationization of lignin (A) and its hydrolysis (B), adapted from [67].

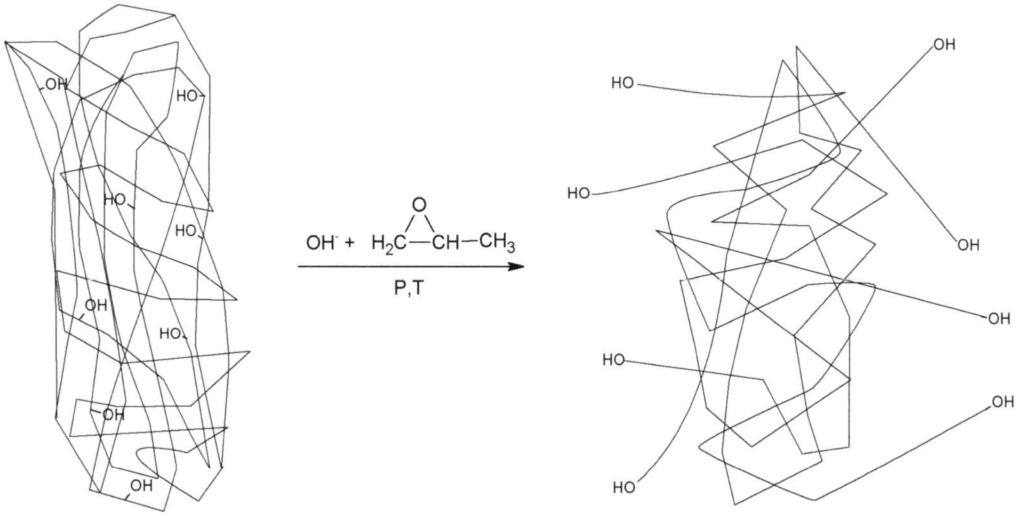

Figure 13. Schema for the oxypropylation (epoxidation) of hydroxyl groups in biopolymers, adapted from [70].

Esterification of lignin was investigated by Thielemans and Wool [71]. This modification was catalyzed using 1-methylimidazole, where acetic anhydride acted on the hydroxyl groups of lignin. At this stage, a cation is formed from N-alkyl-N`-methyl imidazolium, which acts on the hydroxyl group (aliphatic or aromatic). This leads to a protonated catalyst and acetylated lignin. The study showed that phenyl hydroxyl groups undergo acetylation significantly more easily. It was also found that the modification improves the solubility of lignin in styrene and butanoic acid, confirming its potential for use in the processing of thermosets [71].

An alternative focus of research is the optimization of the structure and degree of polymerization of the lignin molecule. For example, Duong et al. carried out polycondensation from sebacoyl chloride catalyzed with triethylamine [72]. The resulting material offered better thermal stability than kraft lignin, giving it potential for use in environmentally friendly composites [72].

Interest in the use of lignin in many areas of chemistry and everyday life is motivated by the high potential for its reproduction by plant organisms. The annual production of the biopolymer is certainly large enough not only to meet some of the demands for low-molecular-weight organic compounds obtained in refineries but also to serve as a functional material. This would also enable a reduction in atmospheric CO_2, which is taken up by plants and transformed into more complex compounds via photosynthesis. In addition, the use of lignin would appear to be favorable not only in terms of the trend toward policies focused on protecting the natural environment but also for economic reasons. Fluctuations in the prices of petroleum-based products therefore make lignin and other renewable biopolymers into attractive alternatives to the products currently in common use [70].

6. Summary and a look at the future

In recent years, many techniques and methods have been developed for the depolymerization and modification of lignin, making use of a variety of substances offering catalytic properties or the ability to dissolve the biopolymer. Nonetheless, their use in technological processes on a wider scale remains very limited. This is primarily a result of the lack of sufficient fundamental information on the structure of lignin itself, which makes it harder to gain a precise understanding of the mechanism of depolymerization and thus to propose appropriate means to enable its degradation, in addition to further optimization of the process. It is therefore important that wide-ranging research into lignin be continued, enabling its detailed physicochemical properties to be determined and consequently the most effective and efficient techniques to be developed for converting it into valuable products.

Acknowledgements

This work was supported by the Polish Ministry of Science and Higher Education research project number IP2015 032574 (Iuventus Plus) in the years 2016–2019.

Author details

Łukasz Klapiszewski*, Tadeusz J. Szalaty and Teofil Jesionowski

*Address all correspondence to: lukasz.klapiszewski@put.poznan.pl

Faculty of Chemical Technology, Institute of Chemical Technology and Engineering, Poznan University of Technology, Poznan, Poland

References

[1] Calvo-Flores FG, Dobado JA. Lignin as renewable raw material. ChemSusChem. 2010; **3**(11):1227-1235. DOI: 10.1002/cssc.201000157

[2] Boerjan W, Ralph J, Baucher M. Lignin biosynthesis. Annual Review of Plant Biology. 2003;**54**:519-546. DOI: 10.1146/annurev.arplant.54.031902.134938

[3] Vanholme R, Demedts B, Morreel K, Ralph J, Boerjan W. Lignin biosynthesis and structure. Plant Physiology. 2010;**153**(3):895-905. DOI: 10.1104/pp.110.155119

[4] Neutelings G. Lignin variability in plant cell walls: Contributions of new models. Plant Science. 2011;**181**(4):379-386. DOI: 10.1016/j.plantsci.2011.06.012

[5] Wong DWS. Structure and action mechanism of ligninolytic enzymes. Applied Biochemistry and Biotechnology. 2009;**157**(2):174-209. DOI: 10.1007/s12010-008-8279-z

[6] Chatel G, Rogers RD. Review: Oxidation of lignin using ionic liquids—An innovative strategy to produce renewable chemicals. ACS Sustainable Chemistry & Engineering. 2014;**2**(3):322-339. DOI: 10.1021/sc4004086

[7] Klapiszewski Ł, Bartczak P, Wysokowski M, Jankowska M, Kabat K, Jesionowski T.Silica conjugated with kraft lignin and its use as a novel 'green' sorbent for hazardous metal ions removal. Chemical Engineering Journal. 2015;**260**:684-693. DOI: 10.1016/j.cej.2014.09.054

[8] Bula K, Klapiszewski Ł, Jesionowski T. A novel functional silica/lignin hybrid material as a potential bio-based polypropylene filler. Polymer Composites. 2015;**36**(5):913-922. DOI: 10.1002/pc.23011

[9] Klapiszewski Ł, Pawlak F, Tomaszewska J, Jesionowski T. Preparation and characterization of novel PVC/silica-lignin composites. Polymers. 2015;**7**(9):1767-1788. DOI: 10.3390/polym7091482

[10] Grząbka-Zasadzińska A, Klapiszewski Ł, Bula K, Jesionowski T, Borysiak S. Supermolecular structure and nucleation ability of polylactide-based composites with silica/lignin hybrid fillers. Journal of Thermal Analysis and Calorimetry. 2016;**126**(1):263-275. DOI: 10.1007/s10973-016-5311-3

[11] Borysiak S, Klapiszewski Ł, Bula K, Jesionowski T. Nucleation ability of advanced functional silica/lignin hybrid fillers in polypropylene composites. Journal of Thermal Analysis and Calorimetry. 2016;**126**(1):251-262. DOI: 10.1007/s10973-016-5390-1

[12] Klapiszewski Ł, Rzemieniecki T, Krawczyk M, Malina D, Norman M, Zdarta J, et al. Kraft lignin/silica-AgNPs as a functional material with antibacterial activity. Colloids and Surfaces B. 2015;**134**:220-228. DOI: 10.1016/j.colsurfb.2015.06.056

[13] Strzemiecka B, Klapiszewski Ł, Voelkel A, Jesionowski T. Functional lignin-SiO_2 hybrids as potential fillers for phenolic binders. Journal of Adhesive Science and Technology. 2016;**30**(10):1031-1048. DOI: 10.1080/01694243.2015.1115602

[14] Strzemiecka B, Klapiszewski Ł, Jamrozik A, Szalaty TJ, Matykiewicz D, Sterzyński T, et al. Physicochemical characterization of functional lignin–silica hybrid fillers for potential application in abrasive tools. Materials. 2016;**9**(7):517-530. DOI: 10.3390/ma9070517

[15] Konował E, Modrzejewska-Sikorska A, Motylenko M, Klapiszewski Ł, Wysokowski M, Bazhenov VV, et al. Functionalization of organically modified silica with gold nanoparticles in the presence of lignosulfonate. International Journal of Biological Macromolecules. 2016;**85**:74-81. DOI: 10.1016/j.ijbiomac.2015.12.071

[16] Milczarek G, Motylenko M, Modrzejewska-Sikorska A, Klapiszewski Ł, Wysokowski M, Bazhenov VV, et al. Deposition of silver nanoparticles on organically-modified silica in the presence of lignosulfonate. RSC Advances. 2014;**4**(94):52476-52484. DOI: 10.1039/C4RA08418G

[17] Klapiszewski Ł, Siwińska-Stefańska K, Kołodyńska D. Preparation and characterization of novel TiO_2/lignin and TiO_2-SiO_2/lignin hybrids and their use as functional biosorbents for Pb(II). Chemical Engineering Journal. 2017;**314**:169-181. DOI: 10.1016/j.cej.2016.12.114

[18] Klapiszewski Ł, Zdarta J, Antecka K, Synoradzki K, Siwińska-Stefańska K, Moszyński D, et al. Magnetite nanoparticles conjugated with lignin: A physicochemical and magnetic study. Applied Surface Science. 2017;**422**:94-103. DOI: 10.1016/j.apsusc.2017.05.255

[19] Ciesielczyk F, Bartczak P, Klapiszewski Ł, Jesionowski T. Treatment of model and galvanic waste solutions of copper(II) ions using a lignin/inorganic oxide hybrid as an effective sorbent. Journal of Hazardous Materials. 2017;**328**:150-159. DOI: 10.1016/j.jhazmat.2017.01.009

[20] Ciesielczyk F, Klapiszewski Ł, Szwarc-Rzepka K, Jesionowski T. A novel method of combination of Kraft lignin with synthetic mineral support. Advanced Powder Technology. 2014;**25**(2):695-703. DOI: 10.1016/j.apt.2013.10.016

[21] Wysokowski M, Klapiszewski Ł, Moszyński D, Bartczak P, Szatkowski T, Majchrzak I, et al. Modification of chitin with kraft lignin and development of new biosorbents for removal of cadmium(II) and nickel(II) ions. Marine Drugs. 2014;**12**(4):2245-2268. DOI: 10.3390/md12042245

[22] Zdarta J, Klapiszewski Ł, Wysokowski M, Norman M, Kołodziejczak-Radzimska A, Moszyński D, et al. Chitin-lignin material as a novel matrix for enzyme immobilization. Marine Drugs. 2015;**13**(4):2424-2446. DOI: 10.3390/md13042424

[23] Pandey MP, Kim CS. Lignin depolymerization and conversion: A review of thermochemical methods. Chemical Engineering & Technology. 2011;**34**(1):29-41. DOI: 10.1002/ceat.201000270

[24] Roberts VM, Stein V, Reiner T, Lemonidou A, Li X, Lercher JA. Towards quantitative catalytic lignin depolymerization. Chemistry—A European Journal. 2011;**17**(21):5939-5948. DOI: 10.1002/chem.201002438

[25] Erdocia X, Prado R, Corcuera MA, Labidi J. Base catalyzed depolymerization of lignin: Influence of organosolv lignin nature. Biomass & Bioenergy. 2014;**66**:379-386. DOI: 10.1016/j.biombioe.2014.03.021

[26] Kumar CR, Anand N, Kloekhorst A, Cannilla C, Bonura G, Frusteri F, et al. Solvent free depolymerization of Kraft lignin to alkyl-phenolics using supported NiMo and CoMo catalysts. Green Chemistry. 2015;**17**(11):4921-4930. DOI: 10.1039/C5GC01641J

[27] Singh SK, Nandeshwar K, Ekhe JD. Thermochemical lignin depolymerization and conversion to aromatics in subcritical methanol: Effects of catalytic conditions. New Journal of Chemistry. 2016;**40**(4):3677-3685. DOI: 10.1039/C5NJ02916C

[28] Erdocia X, Prado R, Fernández-Rodríguez J, Labidi J. Depolymerization of different organosolv lignins in supercritical methanol, ethanol and acetone to produce phenolic monomers. ACS Sustainable Chemistry & Engineering. 2016;**4**(3):1373-1380. DOI: 10.1021/acssuschemeng.5b01377

[29] Kim JY, Park J, Kim UJ, Choi JW. Conversion of lignin to phenol-rich oil fraction under supercritical alcohols in the presence of metal catalysts. Energy & Fuels. 2015;**29**(8):5154-5163. DOI: 10.1021/acs.energyfuels.5b01055

[30] Gosselink RJA, Teunissen W, van Dam JEG, de Jong E, Gellerstedt G, Scott EL, et al. Lignin depolymerisation in supercritical carbon dioxide/acetone/water fluid for the production of aromatic chemicals. Bioresource Technology. 2012;**106**:173-177. DOI: 10.1016/j.biortech.2011.11.121

[31] Zhao C, Chen A, Jiang E, Qin L. Pyrolysis of industrial waste lignin: Analysis of product yields and character. Energy Sources, Part A. 2017;**39**(5):458-464. DOI: 10.1080/15567036.2016.1217293

[32] Kloekhorst A, Heeres HJ. Catalytic hydro treatment of Alcell lignin using supported Ru, Pd and Cu catalysts. ACS Sustainable Chemistry & Engineering. 2015;**3**(9):1905-1914. DOI: 10.1021/acssuschemeng.5b00041

[33] Wang X, Rinaldi R. Bifunctional Ni catalysts for the one-pot conversion of Organsolv lignin into cycloalkanes. Catalysis Today. 2016;**269**:48-55. DOI: 10.1016/j.cattod.2015.11.047

[34] Das L, Kolar P, Sharma-Shivappa R, Classen JJ, Osborne JA. Catalytic valorization of lignin using niobium oxide. Waste and Biomass Valorization. 2016;**1**:1-8. DOI: 10.1007/s12649-016-9717-8

[35] Sturgeon MR, Kim S, Lawrence K, Paton RS, Chmely SC, Nimlos M, et al. A mechanistic investigation of acid-catalyzed cleavage of aryl-ether linkages: Implications for lignin depolymerization in acidic environments. ACS Sustainable Chemistry & Engineering. 2014;**2**(3):472-485. DOI: 10.1021/sc400384w

[36] Deepa AK, Dhepe PL. Lignin depolymerization into aromatic monomers over solid acid catalysts. ACS Catalysis. 2015;**5**(1):365-379. DOI: 10.1021/cs501371q

[37] Rahimi A, Ulbrich A, Coon JJ, Stahl SS. Formic-acid-induced depolymerization of oxidized lignin to aromatics. Nature. 2014;**515**(7526):249-252. DOI: 10.1038/nature13867

[38] Güvenatam B, Heeres EHJ, Pidko EA, Hensen EJM. Lewis acid-catalyzed depolymerization of soda lignin in supercritical ethanol/water mixtures. Catalysis Today. 2016;**269**:9-20. DOI: 10.1016/j.cattod.2015.08.039

[39] Toledano A, Serrano L, Labidi J. Organosolv lignin depolymerization with different base catalysts. Journal of Chemical Technology and Biotechnology. 2012;**87**(11):1593-1599. DOI: 10.1002/jctb.3799

[40] Dos Santos PSB, Erdocia X, Gatto DA, Labidi J. Bio-oil base-catalyzed depolymerization of organosolv lignin as an antifungal agent for wood. Wood Science and Technology. 2016;**50**(3):599-615. DOI: 10.1007/s00226-015-0795-8

[41] Long J, Zhang Q, Wang T, Zhang X, Xu Y, Ma L. An efficient and economical process for lignin depolymerization in biomass-derived solvent tetrahydrofuran. Bioresource Technology. 2014;**154**:10-17. DOI: 10.1016/j.biortech.2013.12.020

[42] Widyay VT, Vo HT, Dahnum D, Lee H. Magnesium oxide-catalyzed oxidative depolymerization of EFB lignin. Bulletin of the Korean Chemical Society. 2016;**37**(4):515-521. DOI: 10.1002/bkcs.10715

[43] Jia S, Cox BJ, Guo X, Zhang ZC, Ekerdt JG. Cleaving the β-O-4 bonds of lignin model compounds in an acidic ionic liquid, 1-H-3-methylimidazolium chloride: An optional strategy for the degradation of lignin. ChemSusChem. 2010;**3**(9):1078-1084. DOI: 10.1002/cssc.201000112.

[44] Jia S, Cox BJ, Guo X, Zhang ZC, Ekerdt JG. Decomposition of a phenolic lignin model compound over organic N-bases in an ionic liquid. Holzforschung. 2010;**64**(5):577-580. DOI: 10.1515/hf.2010.075

[45] Cox BJ, Jia S, Zhang ZC, Ekerdt JG. Catalytic degradation of lignin model compounds in acidic imidazolium based ionic liquids: Hammett acidity and anion effects. Polymer Degradation and Stability. 2011;**96**(4):426-431. DOI: 10.1016/j.polymdegradstab.2011.01.011

[46] Yan B, Li K, Wei L, Ma Y, Shao G, Zhao D, et al. Understanding lignin treatment in dialkylimidazolium-based ionic liquid-water mixtures. Bioresource Technology. 2015;**196**:509-517. DOI: 10.1016/j.biortech.2015.08.005

[47] Liu F, Liu Q, Wang A, Zhang T. Direct catalytic hydrogenolysis of kraft lignin to phenols in choline-derived ionic liquids. ACS Sustainable Chemistry & Engineering. 2016;**4**(7):3850-3856. DOI: 10.1021/acssuschemeng.6b00620

[48] Yang Y, Fan H, Song J, Meng Q, Zhou H, Wu L, et al. Free radical reaction promoted by ionic liquid: A route for metal-free oxidation depolymerization of lignin model compound and lignin. Chemical Communications. 2015;**51**(19):4028-4031. DOI: 10.1039/C4CC10394G

[49] Rashid T, Kait CF, Regupathi I, Murugesan T. Dissolution of kraft lignin using protic ionic liquids and characterization. Industrial Crops and Products. 2016;**84**:284-293. DOI: 10.1016/j.indcrop.2016.02.017

[50] Cai Z, Li Y, He H, Zeng Q, Long J, Wang L, et al. Catalytic depolymerization of organosolv lignin in a novel water/oil emulsion reactor: Lignin as the self-surfactant. Industrial & Engineering Chemistry Research. 2016;**54**(46):11501-11510. DOI: 10.1021/acs.iecr.5b03247

[51] Younesi-Kordkheili H, Pizzi A, Honarbakhsh-Raouf A, Nemati F. The effect of soda bagasse lignin modified by ionic liquids on properties of the urea-formaldehyde resin as a wood adhesive. Journal of Adhesives. 2016. DOI: 10.1080/00218464.2016.1188284

[52] Prado R, Brandt A, Erdocia X, Hallet J, Welton T, Labidi J. Lignin oxidation and depolymerisation in ionic liquids. Green Chemistry. 2016;**18**(3):834-841. DOI: 10.1039/C5GC01950H

[53] Stärk K, Taccardi N, Bösmann A, Wasserscheid P. Oxidative depolymerization of lignin in ionic liquids. ChemSusChem. 2010;**3**(6):719-723. DOI: 10.1002/cssc.200900242

[54] Nanayakkara S, Patti AF, Saito K. Lignin depolymerization with phenol via redistribution mechanism in ionic liquids. ACS Sustainable Chemistry & Engineering. 2014;**2**(9):2159-2164. DOI: 10.1021/sc5003424

[55] Binder JB, Gray MJ, White JF, Zhang ZC, Holladay JE. Reactions of lignin model compounds in ionic liquids. Biomass & Bioenergy. 2009;**33**(9):1122-1130. DOI: 10.1016/j.biombioe.2009.03.006

[56] Cox BJ, Ekerdt JG. Depolymerization of oak wood lignin under mild conditions using the acidic ionic liquid 1-H-3-methylimidazolium chloride as both solvent and catalyst. Bioresource Technology. 2012;**118**:584-588. DOI: 10.1016/j.biortech.2012.05.012

[57] Laurichesse L, Averous L. Chemical modification of lignins: Towards biobased polymers. Progress in Polymer Science. 2014;**39**(7):1266-1290. DOI: 10.1016/j.progpolymsci.2013.11.004

[58] Ouyang X, Ke L, Qiu X, Guo Y, Pang Y. Sulfonation of alkali lignin and its potential use in dispersant for cement. Journal of Dispersion Science and Technology. 2009;**30**:1-6. DOI: 10.1080/01932690802473560

[59] Malutan T, Nicu R, Popa VI. Lignin modification by epoxidation. BioResources. 2008;**3**(4):1371-1376. DOI: 10.15376/biores.3.4.1371-1376

[60] Du X, Li J, Lindstrom ME. Modification of industrial softwood kraft lignin using Mannich reaction with and without phenolation pretreatment. Industrial Crops and Products. 2014;**52**:729-735. DOI: 10.1016/j.indcrop.2013.11.035

[61] Ge Y, Song Q, Li Z. A Mannich base biosorbent derived from alkaline lignin for lead removal from aqueous solution. Journal of Industrial and Engineering Chemistry. 2015;**23**:228-234. DOI: 10.1016/j.jiec.2014.08.021

[62] Peter Dilling, A sulfonation of lignins. Patent US 5049661. 1989

[63] Sen S, Patil S, Argyropoulos DS. Methylation of softwood kraft lignin with dimethyl carbonate. Green Chemistry. 2015;**17**(2):1077-1087. DOI: 10.1039/c4gc01759e

[64] Malutan T, Nicu R, Popa VI. Contribution to the study of hydroxymethylation reaction of alkali lignin. BioResources. 2008;**3**(1):13-20. DOI: 10.15376/biores.3.1.13-20

[65] Podschun J, Saake B, Lehnen R. Reactivity enhancement of organosolv lignin by phenolation for improved bio-based thermosets. European Polymer Journal. 2015;**67**:1-11. DOI: 10.1016/j.eurpolymj.2015.03.029

[66] Hu L, Pan H, Zhou Y, Zhang M. Methods to improve lignin`s reactivity as a phenol substitute and as replacement for other phenolic compounds: A brief review. BioResources. 2011;**6**(3):1-11. DOI: 10.15376/biores.6.3.3515-3525

[67] Kong F, Parhiala K, Wang S, Fatehi P. Preparation of cationic softwood kraft lignin and its application in dye removal. European Polymer Journal. 2015;**67**:335-345. DOI: 10.1016/j.eurpolymj.2015.04.004

[68] Pan H, Sun G, Zhao T. Synthesis and characterization of aminated lignin. Industrial Journal of Biological Macromolecules. 2013;**59**:221-226. DOI: 10.1016/j.ijbiomac.2013.04.049

[69] Feng P, Chen F. Preparation and characterization of acetic acid lignin-based epoxy blends. BioResources. 2012;**7**(3):2860-2870. DOI: 10.15376/biores.7.3.2860-2870

[70] Anicento JPS, Portugal I, Silva CM. Biomass-based polyols through oxypropylation reaction. ChemSusChem. 2012;**5**(8):1358-1368. DOI: 10.1002/cssc.201200032

[71] Thielemans W, Wool RP. Lignin esters for use in unsaturated thermosets: Lignin modification and solubility modeling. Biomacromolecules. 2005;**6**(4):1895-1905. DOI: 10.1021/bm0500345

[72] Duong LD, Luong ND, Thanh Binh NT, Park I, Lee SH, Kim DS, et al. Chemical and rheological characteristics of thermally stable kraft lignin polycondensates analyzed by dielectric properties. BioResources. 2013;**8**(3):4518-4532. DOI: 10.15376/biores.8.3.4518-4532

Chapter 2

Lignin Degradation Processes and the Purification of Valuable Products

Stefan Schoenherr, Mehrdad Ebrahimi and
Peter Czermak

Additional information is available at the end of the chapter

http://dx.doi.org/10.5772/intechopen.71210

Abstract

Lignin is a heterogeneous, phenolic and polydisperse biopolymer which resists degradation due to its aromatic and highly branched structure. Lignin is the most abundant renewable source of aromatic molecules on earth. The valorization of lignin could therefore provide a sustainable alternative to petroleum refineries for the production of valuable aromatic compounds. Even so, paper mills and lignocellulose feedstock biorefineries treat lignin largely as a waste product. In paper mills, 98% of technical lignin is incinerated for internal energy recovery while only 2% is used commercially (e.g. for the production of aromatics such as vanillin). The reasons for the underutilization of lignin include its recalcitrance to degradation and the challenge of separating mixtures of numerous degradation products. The successful valorization of lignin in the future thus depends on a broad understanding of biological and technical degradation processes, and the implementation of efficient product purification strategies. This article describes enzymatic, photocatalytic and thermochemical lignin degradation processes and considers purification methods for valuable lignin-derived degradation products. We focus on the potential of membrane-based separation technology, including data from our own recent research.

Keywords: lignin valorization, lignin degradation, enzymatic lignin degradation, thermochemical lignin degradation, photocatalytic lignin degradation, enzyme membrane reactor system (EMRS), membrane filtration, ceramic membranes

1. Introduction

Lignin, cellulose and hemicellulose are the main constituents of lignocellulosic biomass, with lignin accounting for ~10–40% by weight [1]. The biological functions of lignin in vascular plants are to provide rigidity and to protect the structural polysaccharides cellulose and hemicellulose from enzymatic hydrolysis [2–4]. Lignin is a heterogeneous, crosslinked and highly polydisperse phenolic copolymer of three different monolignols: *p*-coumaryl alcohol, coniferyl alcohol and sinapyl alcohol [3, 5]. When integrated into the lignin macromolecule, the building blocks based on these monolignols are the *p*-hydroxyphenyl unit (H), guaiacyl unit (G) and syringyl unit (S) (**Figure 1**) [6]. The distribution of these monomers varies in different plant species and tissues. Generally, lignin from grasses is a roughly equimolar mixture of G, S and H units, whereas lignin from hardwood contains approximately equal quantities of G and S units but relatively small amounts of the H unit, and lignin from softwood is mainly composed of G units (up to 90%) [7]. However, variations within these groups of species have been reported, for example, lignin from silver birch (*Betula pendula*) and several *Eucalyptus* species contains 65–80% S units, rather than the roughly equimolar mixture of G and S units typical for hardwoods [8, 9].

The natural polymerization of monolignols (lignification) starts with the oxidative formation of phenoxy radicals catalyzed by peroxidases and/or laccases, and is followed by combinatorial

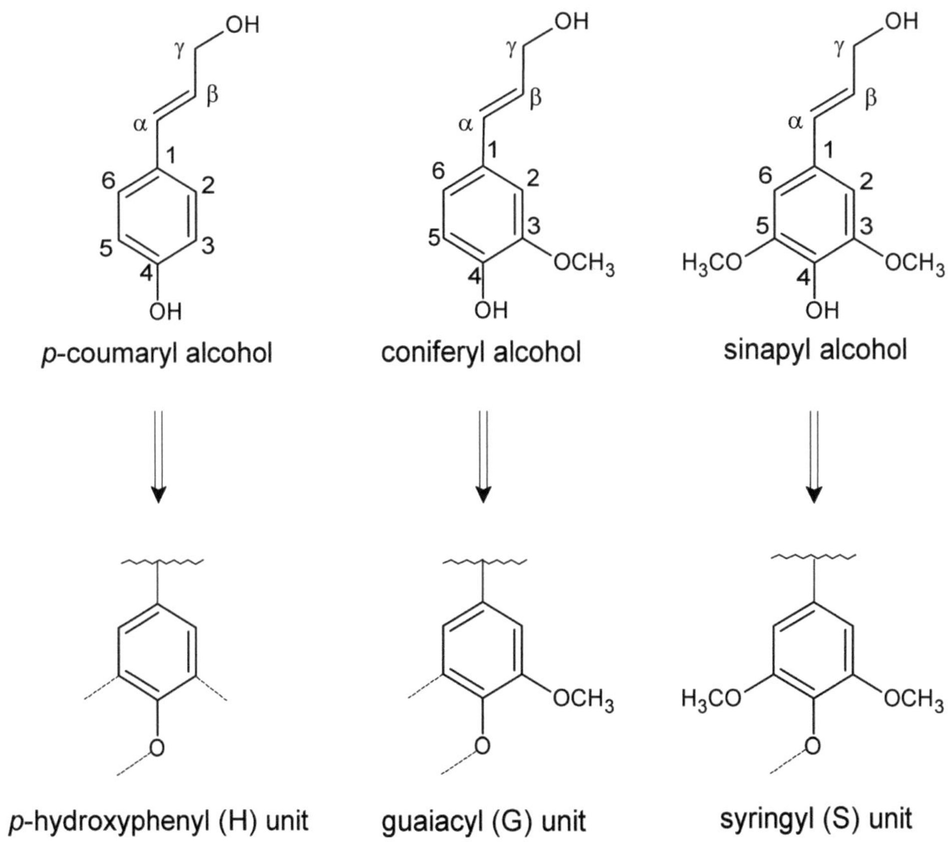

Figure 1. Chemical structure of monolignols and the corresponding building blocks in lignin.

radical coupling, generating carbon–carbon and carbon–oxygen (ether) bonds [6, 10]. The emerging crosslinked network and aromatic nature of lignin explain its recalcitrance towards degradation [11]. The most frequent linkages formed during lignification are *β*-O-4 bonds, which represent ~50% of all linkages between monolignol building blocks in hardwood and softwood lignin (**Table 1**) [3, 12]. The main linkage types in hardwood and softwood lignin are summarized in **Table 1** and are represented schematically in **Figure 2**.

	[%]	
Linkage type	**Hardwood lignin**	**Softwood lignin**
β-O-4	50–65	43–50
β-5 + cyclic α-O-4	4–6	9–12
noncyclic α-O-4	4–8	6–8
4-O-5	6–7	4
β-β	3–7	2–4
5–5	4–10	10–25
β-1	5–7	3–7

Table 1. Distribution of principal lignin building block linkages in hardwood and softwood (%), adapted from [13].

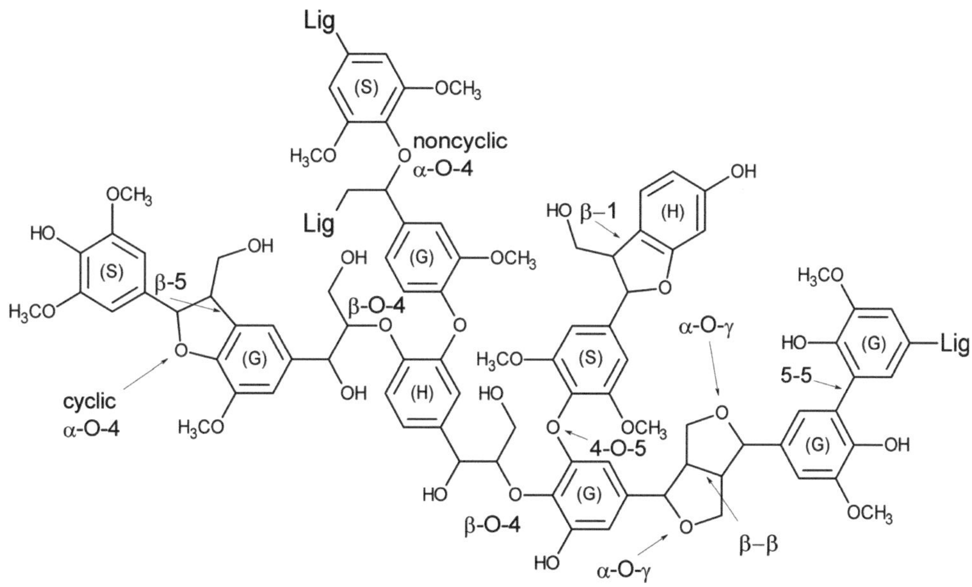

Figure 2. Schematic representation of a hypothetical lignin structure, including principal linkage types and building blocks (H, G and S).

Lignin is the most abundant renewable source of aromatic molecules and the second largest renewable source of carbon on earth, after cellulose [3, 14]. The estimated total amount of lignin on earth is ~300 × 10^9 tons, and the annual amount produced by natural lignification is ~20 × 10^9 tons [15].

There is great scientific and industrial interest in lignin due to the growing threat of climate change and the urgent need to reduce relevant emissions [16–18]. Lignin offers a sustainable alternative to the finite fossil resources currently used to produce aromatic compounds, and it is widely regarded as a waste product and is therefore inexpensive [19, 20]. One possible area of application for lignin-derived aromatic products is conventional jet fuel, which contains 20% fossil-derived aromatic compounds that could be replaced with lignin-derived compounds [21]. As well as the presence of valuable aromatic constituents, one of the main reasons for the interest in lignin is its availability, either directly from plants or as a byproduct of industrial conversion processes [22]. The pulp and paper industries, as well as lignocellulose feedstock biorefineries producing liquid fuels such as bioethanol, provide major examples of these processes [23, 24]. Lignin derived from industrial conversion processes is described hereafter as industrial or technical lignin.

Approximately 50 million tons of technical lignin is generated annually by the pulp and paper industries [15]. Depending on the pulping process, different forms of lignin are produced with distinct chemical properties, including kraft lignin, lignosulfonate, organosolv lignin, soda lignin and residual lignin from the bleaching stage (**Figure 3**) [19]. Even more industrial lignin is expected to be produced in lignocellulose feedstock biorefineries. The United States

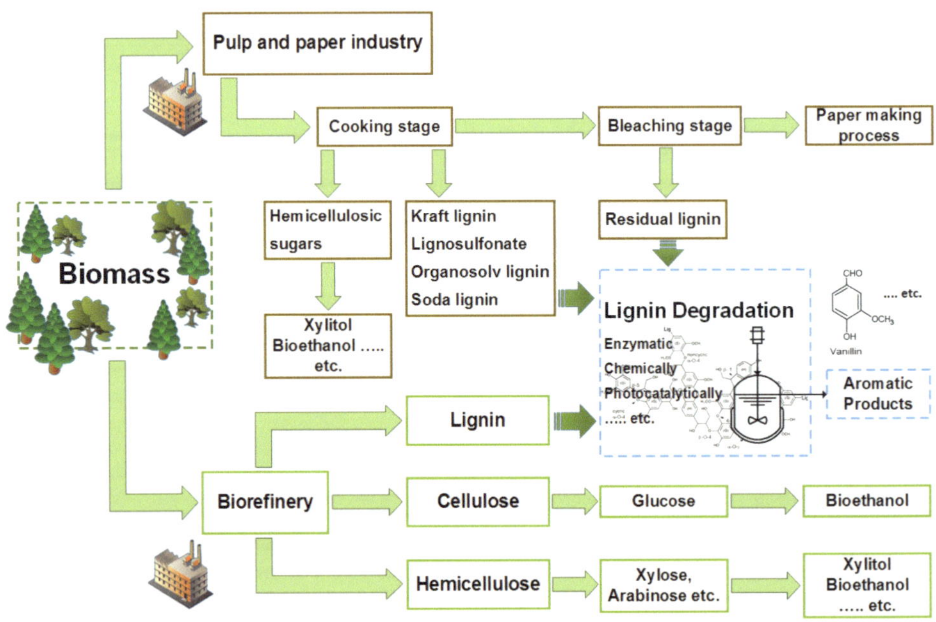

Figure 3. Lignin sources and pathways to generate valuable products from lignocellulosic biomass (highly simplified representation of process steps in the pulp and paper industry).

Department of Energy has set a goal to replace 30% of the transportation fuel supply with biofuels by 2030. This would require ~750 million tons of biomass and would release ~225 million tons of lignin in the United States alone [25]. However, process streams containing technical lignin originating from the pulp and paper industries and biorefineries tend to be treated as waste, and are therefore significantly underutilized. More than 98% of the industrial lignin produced by the pulp and paper industry is incinerated for internal energy recovery while only 2% is used commercially, mostly for niche products [15, 26, 27]. The contrast between lignin's huge economic potential as a source of value-added aromatic products and its classification as a waste product mainly reflects its recalcitrance to degradation, its heterogeneity and the challenge of isolating valuable molecules from a complex mixture of degradation products.

Lignin biodegradation clearly occurs in nature, otherwise all carbon would eventually become trapped in lignocellulosic biomass [11]. However, it has not yet been possible to replicate these enzymatic processes on an industrial scale. Instead, industrial lignin degradation is currently based on chemical processes. One example is the production of vanillin from lignosulfonates, which involves lignin degradation via alkaline oxidation [27]. The existence of such industrial processes for lignin valorization indicates that large-scale valorization and the full exploitation of technical lignin as a source of aromatic fine chemicals should be possible in the future, thus replacing the current use of unsustainable fossil resources. In addition to improving the efficiency of chemical processes, it should also be possible to scale up enzymatic lignin-degradation processes, including those based on an enzyme membrane reactor system (EMRS). This article discusses current research on the topic of industrial lignin degradation and valorization, including enzymatic, thermochemical and photocatalytic lignin-degradation methods. We also consider separation techniques for lignin-containing process streams, focusing on membrane technology.

2. Lignin degradation processes

Lignin degradation is often investigated using model lignin compounds rather than real lignin-containing substrates. The basic principles of lignin degradation, such as the reaction kinetics and conversion pathways, are easier to analyze using simple model compounds rather than complex lignin solutions because real substrates are heterogeneous, resulting in multiple overlapping degradation reactions.

2.1. Enzymatic lignin degradation

As stated above, one function of lignin in lignocellulosic biomass is to protect cellulose and hemicellulose from enzymatic hydrolysis. Non-lignified cellulose and hemicellulose are accessible for enzymatic hydrolysis, which enables their conversion into monosaccharides. Hydrolases (cellulases and hemicellulases) can produce monosaccharides in large quantities, and the industrial valorization of cellulose and hemicellulose by hydrolytic enzymatic processes is already well established (e.g., cellulosic bioethanol production and xylitol from xylose-rich hemicelluloses) [1, 28, 29]. When these polysaccharides are found in complex connection with lignin, as in lignocellulosic biomass, they are strongly resistant to hydrolysis [30]. Given that lignin inhibits the degradation of cellulose and hemicellulose, microbes that utilize these molecules have evolved pathways for lignin decomposition or modification in order to gain access to the polysaccharides, which they use as sources of carbon and energy [4].

2.1.1. Fungal lignin degradation

The fungal decomposition of lignocellulose relies on two types of extracellular enzyme systems. First, fungi can deploy hydrolases (cellulases and hemicellulases) for the degradation of structural lignocellulosic polysaccharides (cellulose and hemicellulose). Second, they possess a unique extracellular ligninolytic system that degrades or modifies lignin [31]. The prevalence of these systems differs with the type of fungus. Three types of fungal wood decay can be distinguished: white-rot, brown-rot and soft-rot [32]. Wood decayed by white-rot fungi is pale and often takes on a fibrous texture [33, 34]. Common white-rot fungi include *Phanerochaete chrysosporium, Pleurotus ostreatus* and *Ceriporiopsis subvermispora* [35]. Kirk and Farrell [3] claimed—based on literature analysis—that white-rot fungi were the best lignin degraders among all studied microorganisms, achieving degradation more rapidly and more comprehensively, even to the extent of complete mineralization. The fungi invade the lumens of wood cells and secrete enzymes that degrade lignin and the associated polysaccharides [3]. White-rot fungi predominantly degrade deciduous wood (hardwood). In contrast, brown-rot fungi grow primarily on coniferous wood (softwood) and are therefore mainly softwood degraders. Both white-rot and brown-rot fungi show host preferences [35]. Brown-rot fungi are an exception to the typical observation that lignocellulose must be lignified before access to lignocellulosic polysaccharides is granted, and degradation made possible [4]. They do not degrade lignin directly, but rather modify it by partial oxidization and then preferentially degrade the polysaccharides [36, 37]. Wood decomposed by brown-rot fungi is therefore characterized by a brown color arising from the residual lignin [33]. In addition to color changes, wood decayed by brown-rot fungi shrinks, breaks into brick-shaped pieces and crumbles to a brown powder [38]. Typical examples of brown-rot fungi include *Gloeophyllum trabeum, Postia placenta* and *Fomitopsis palustris* [36, 39]. The third group of wood-decaying fungi are the soft-rot fungi [4]. Finally, Findlay and Savory [40] introduced this term for wood decay based on the deterioration of timbers in water-cooling towers. Soft-rot fungi are responsible for wood decay in water-saturated systems or areas with fluctuating moisture levels. Typical examples include *Chaetomium cochliodes, Chaetomium globosum* and *Ophiostoma coerulescens* [41]. They are slower and less aggressive than white-rot and brown-rot fungi, and preferentially utilize polysaccharides, but they can also decompose lignin to a certain degree. Research has focused more on brown-rot and especially white-rot fungi than soft-rot fungi [4].

2.1.2. Enzymes involved in fungal lignin degradation

Successful lignin decomposition can only be achieved by the application of multiple enzymes due to the complexity and heterogeneity of lignin and the diverse chemical linkages it contains [19].

Therefore, the biodegradation of wood constituents (ligninolysis) is widely understood as a multi-enzymatic process that produces many intermediates [30]. However, fungal ligninolysis does not involve one specific set of dedicated enzymes [42]. Instead, the composition of the multi-enzymatic mixture depends on the type of fungus [11].

Enzymes responsible for lignin degradation are known as ligninases [43]. The most common fungal ligninases are copper-containing laccases and heme peroxidases, the latter defined further as lignin peroxidases (LiP), manganese peroxidases (MnP), versatile peroxidases (VP) and dye-decolorizing (DyP-type) peroxidases [44, 45]. These major lignin-degrading enzymes are assisted by fungal accessory enzymes. One group of accessory enzymes are oxidases, e.g. aryl-alcohol oxidase and glyoxal oxidase [46, 47]. They produce peroxide, which is essential for peroxidase activity. Other accessory enzymes include dehydrogenases that reduce lignin-derived compounds (e.g., aryl-alcohol dehydrogenase and quinone reductase) and those that modify lignin by generating hydroxyl radicals (e.g., cellobiose dehydrogenase) [48–50]. The fungal ligninases are described briefly below and more detailed information can be found in reviews focusing on this topic [34, 35, 45, 51, 52]. A generalized scheme for lignin biodegradation that summarizes the enzymatic reaction cycles is provided by Martínez et al. [53].

Laccases catalyze the oxidation of polyphenols and methoxy-substituted phenols by generating free radicals [54]. The oxidation is typically coupled with the reduction of molecular oxygen to water [51]. In the presence of redox mediators, laccases can even catalyze the breakdown of non-phenolic lignin structures, including the cleavage of β-O-4 linkages [55, 56]. Typical mediators in laccase-mediator systems include 2,2′-azino-bis(3-ethylbenzothiazoline-6-sulfonic acid) (ABTS), 1-hydroxybenzotriazole (1-HBT) and the natural mediator acetosyringone (4-hydroxy-3,5-dimethoxyacetophenone) [55–58].

Lignin peroxidases catalyze the oxidative depolymerization of lignin with H_2O_2 acting as the oxidizing agent. These enzymes are relatively nonspecific, and can therefore oxidize phenolic aromatic substrates as well as various non-phenolic lignin model compounds [59–61]. The importance of lignin peroxidases for lignin depolymerization reflects their ability to catalyze the cleavage of α,β and β-ether bonds (including β-O-4 linkages) leading to the efficient degradation of lignin into mono-aromatic structures, which has been demonstrated using lignin model compounds [61]. For example, synthetic hardwood lignin (molecular weight > 1800 Da) was degraded into fragments with an average molecular weight as low as ~170 Da [62].

Manganese peroxidases use H_2O_2 to oxidize Mn^{2+}, which is present in wood and soils, thus generating reactive Mn^{3+} ions. The latter are stabilized by chelators and act as diffusive charge-transfer mediators which are able to oxidize a large number of phenolic substrates [45]. The catalytic mechanism consists of two consecutive one-electron oxidation steps with intermediate cation radical formation [35]. In contrast to lignin peroxidases, manganese peroxidases under normal conditions cannot catalyze the oxidation of more recalcitrant non-phenolic structures [52]. However, a purified manganese peroxidase from *P. chrysosporium* can oxidize even non-phenolic lignin model compounds in the presence of the detergent Tween-80, which acts as radical mediator [63].

Versatile peroxidases combine the properties of lignin peroxidases and manganese peroxidases, conferring the catalytic versatility inferred by their name. They can oxidize Mn^{2+} to Mn^{3+} like manganese peroxidases, but can also oxidize non-phenolic compounds in the same manner as lignin peroxidases [64, 65].

In contrast to the other three classes of heme peroxidases described above, DyP-type peroxidases are not members of the classical plant/microbial peroxidase superfamily, due to differences in sequence, structure and function [66, 67]. Therefore, they represent an additional superfamily of heme-containing peroxidases. They were first identified in fungi and later in bacteria [68, 69]. DyP-type peroxidases show considerable potential for the degradation of lignin because they oxidize dyes, non-phenolic lignin model compounds (including veratryl alcohol and adlerol) and non-phenolic lignin model compounds containing β-O-4 linkages [66, 68].

2.1.3. Bacterial lignin degradation

Whereas fungal lignin degradation has been investigated extensively, bacterial ligninolysis and the corresponding enzymes have received comparatively little attention [70, 71]. Bacterial wood decay was initially considered more as a challenge for forestry than a biotechnological opportunity [72]. The classes of ligninolytic enzymes found in fungi play no role in bacterial ligninolysis, apparently reflecting the complexity of the fungal proteins and the need for post-translational modifications, which are largely absent in bacteria [11]. Genomic and proteomic analysis in ligninolytic bacteria has confirmed the absence of LiP, MnP and VP enzymes [73, 74], whereas bacterial laccases and DyP-type peroxidases have been identified [75, 76]. DyP-type peroxidases are less complex than the other heme peroxidases and are common among bacteria [11, 44], including the extracellular enzyme systems of *Thermobifida fusca* and *Rhodococcus jostii* [69, 77]. Several bacterial laccases have also been identified [78–80] including multiple laccase-like multi-copper oxidases in *Agromyces salentinus* and *Sinorhizobium morelense* [78]. Even if the knowledge concerning bacterial peroxidases and laccases has grown recently, additional—as yet undiscovered—enzymes may be required for bacterial lignin degradation, for example, oxidases for the production of H_2O_2.

Whereas large-scale lignin degradation by fungal enzymes has been hampered by the complexity of the enzymes and the challenges of fungal genetic modification and protein expression, the simpler bacterial DyP-type peroxidases and laccases are promising targets for enzyme development [11, 19, 75]. In contrast to fungal peroxidases, several bacterial DyP-type peroxidases and laccases have been expressed at high levels in *Escherichia coli* [81, 82] and efficient lignin degradation has been demonstrated for DyP-type peroxidases [83]. For example, a bacterial DyP-type peroxidase expressed in a gene deletion strain of *Rhodococcus jostii* (its native producer) efficiently degraded wheat straw lignocellulose and produced vanillin and small amounts of ferulic acid and 4-hydroxybenzaldehyde [83]. Such processes have proven feasible in the laboratory, and the next challenge is to scale up production to industrial levels.

2.1.4. Applications of enzymatic lignin degradation

As stated above, lignin is widely treated as a waste product, and industrial ligninolytic processes have until recently focused on removing lignin and valorizing the remaining cellulose and hemicellulose rather than exploiting the lignin itself. The applications of industrial scale enzymatic lignin degradation (ELD) also tend to focus on lignin removal. Biopulping in pulp and paper industries or for ethanol production in biorefineries is an example for such an ELD process [84, 85].

Biopulping can be defined as a pre-treatment method of wood chips with lignin-degrading fungi prior to the main pulping process [85, 86]. The main aim of biopulping is to soften the wood before the main pulping process through lignin modification or its removal [87]. The establishment of biopulping as a pre-treatment prior to mechanical and chemical pulping offers the potential to make the entire process more economical and sustainable, because biopulping can reduce the consumption of chemicals and energy (shorter cooking time) in the subsequent chemical pulping stage, and can also improve pulp quality and paper strength [87, 88]. White-rot fungi such as *C. subvermispora* are typically used in this context [89].

Another industrial ELD process discussed for industrial application is biobleaching. This process applies enzymes or ligninolytic fungi for the bleaching of pulps. Bleaching removes color from the wood pulp (much of which is derived from lignin and its degradation products) and biobleaching involves the application of ligninolytic microbes or their enzymes to achieve this outcome. Biobleaching is less environmentally harmful than conventional chlorine-based bleaching because it does not involve the use of bleaching chemicals and the formation of hazardous byproducts. The success of bleaching is measured by the decrease in lignin content (kappa number) and the increase in pulp brightness. Promising results have been achieved by using laccase-mediator systems on eucalyptus kraft pulp, e.g. bacterial laccase produced by *Streptomyces cyaneus* CECT 3335 in the presence of the mediator ABTS [58, 90].

The simultaneous removal and valorization of lignin can be achieved by enzymatic lignin degradation and product recovery in an EMRS (**Figure 4**). Compared to conventional reaction systems, an EMRS offers advantages such homogeneous catalysis, the use of free enzymes and complete biocatalyst retention and recycling [91]. An EMRS also allows the continuous separation of low-molecular weight products from high-molecular weight substrate and catalyst molecules, thus preventing the re-polymerization of target products. The biocatalyst is simple to replace (according to enzyme inactivation rates), which achieves stable substrate conversion rates [92]. EMRS processes have also been tested for the continuous removal of endocrine disrupting compounds from wastewater, using versatile peroxidases and polyethersulfone (PES) ultrafiltration membranes [93, 94].

An ERMS comprises two main parts: a reactor unit for enzymatic conversion and a filtration unit for enzyme/substrate retention and product removal. Therefore, the development of an EMRS process for enzymatic lignin degradation and the purification of aromatic monomers must focus on both the enzyme kinetics of the reactor unit and the filtration performance of the separation unit.

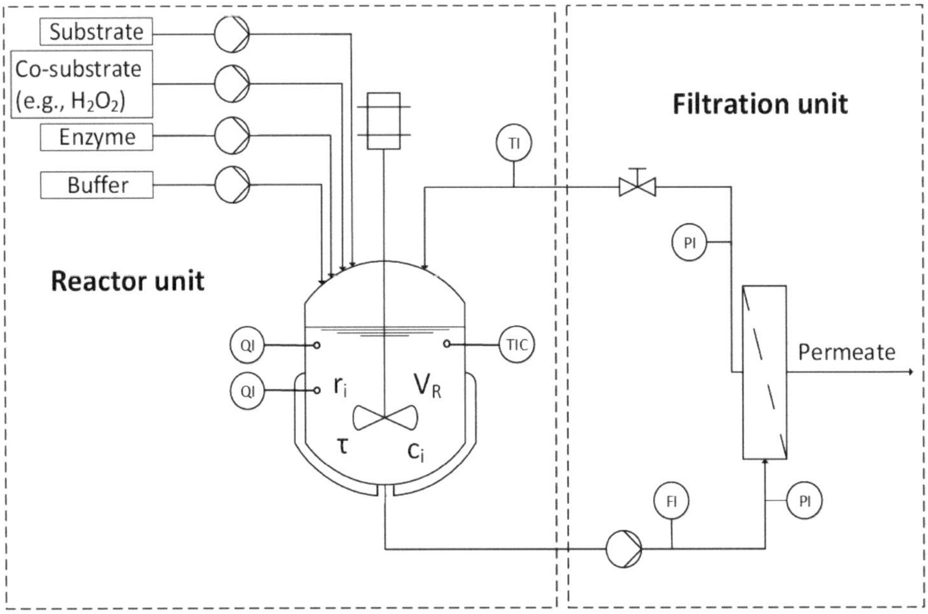

Figure 4. Schematic representation of an enzyme membrane reactor system (EMRS) including a continuous stirred-tank reactor and a filtration unit in crossflow operational mode (adapted from [91]).

Enzyme kinetics: successful EMRS bioprocess design requires the enzyme reaction kinetics to be understood in detail. The reaction kinetics of a crude versatile peroxidase from *Bjerkandera adusta* as a ligninolytic model peroxidase was recently investigated during the degradation of adlerol (a β-O-4 lignin model dimer) [95]. The crude enzyme showed saturation kinetics for adlerol degradation following the Michaelis-Menten equation, and was sensitive to H_2O_2. However, inactivation by H_2O_2 could be reduced by increasing the amount of adlerol substrate. The authors recommended that H_2O_2 concentrations should be kept below 50 μM at a substrate/H_2O_2 ratio of 15:1. Continuous lignin degradation in such a reactor would therefore require carefully controlled slow H_2O_2 feeding rates [95].

Filtration performance: an appropriate membrane is crucial for the design and productivity of EMRS processes because the filtration performance influences process parameters such as the hydraulic retention time (τ), the input of substrate and oxidizing agent (H_2O_2 for peroxidases), and therefore the enzyme reaction kinetics and substrate conversion rates [92, 96]. Importantly, the performance and productivity of membrane filtration systems are reduced by fouling, which increases membrane resistance and reduces flux due to the formation of unwanted deposits on the membrane surface and/or in the membrane pores [97, 98]. Extensive fouling may be caused, among others, by proteins, polysaccharides and polydisperse technical lignin, all of which are present during the enzymatic degradation of lignocellulosic biomass [92, 97]. Accordingly, the filtration performance of membranes used in delignification reactions must be robustly tested, including their fouling and separation characteristics. Such tests were recently reported for a ceramic ultrafiltration membrane exposed to a model reaction mixture containing a commercial polydisperse lignosulfonate substrate, guaiacol as a model degradation product, and ovalbumin as a model biocatalyst in place of the genuine versatile peroxidase [92]. As anticipated, the flux declined significantly during filtration but reached steady-state levels within 1 h, and the flux could be improved by 50% by increasing the crossflow velocity from 0.2 to 1 m/s. The latter indicated that fouling was reversible rather than irreversible, as confirmed by the rapid and efficient cleaning. The retention and separation characteristics of the filtration system were controlled primarily by the substrate concentration rather than the protein load and hydrodynamics. The retention of lignosulfonates using a 5 kDa membrane remained stable at 80–90% under all conditions, whereas the retention of the model product guaiacol increased with the lignosulfonate load. The authors noted that the retention of degradation products should be avoided because this can lead to undesirable enzyme inactivation or polymerization reactions if real enzymes are used instead of the model ovalbumin [92]. Based on the results summarized above [92, 95] *in silico* investigations were carried out to predict the flux needed to maximize the product yield during the enzymatic oxidation of lignin model compounds in a continuous EMRS process. The authors evaluated an optimal permeate flux of 4 L m^{-2} h^{-1} for the previously used 5 kDa membrane [96].

More recently, the H_2O_2-dependent reaction/inactivation kinetics of the crude versatile peroxidase from *B. adusta* [95] and the filtration performance data described above (retention coefficients and optimum permeate flux) [92, 96] were combined to model the design and operational mode of a continuous membrane reactor for enzymatic lignin modification [91]. Fed-batch experiments were used to determine starting concentrations for adlerol and H_2O_2, allowing the dynamic analysis of the kinetics with and without the continuous addition of the enzyme. The resulting model EMRS for the modification of lignin-related substrates by versatile peroxidases simulated the behavior of fed-batch and continuous processes, enabling process development, process design and process optimization. The authors concluded that the modification of technical lignin using versatile peroxidase in an EMRS is a promising approach, with membrane fouling and H_2O_2 sensitivity in an acidic milieu as major challenges that remain to be addressed [91]. Ceramic membranes were preferred in these studies [91, 92, 96] due to their advantages over polymeric membranes, particularly their greater chemical and thermal stability which allows the filtration of acids, bases, solvents and hot feeds, and allows them to survive the harsh cleaning conditions that are often necessary to restore membrane permeability. Ceramic membranes also maintain physical stability during the filtration of abrasive media and generally last longer than polymeric membranes [99, 100].

2.1.5. Conclusion and outlook

A comprehensive understanding of natural lignin degradation pathways is necessary for the development of effective bioprocesses based on ELD. Much is already known about fungal lignin degradation and the corresponding enzymes, but more work is required to understand bacterial lignin degradation. Biopulping and biobleaching are useful applications of ELD, but these are used to replace or support conventional pulping and bleaching processes rather than to valorize lignin by isolating its degradation products. ELD and product recovery can be achieved using an EMRS, which allows continuous processing and product removal. Busse and colleagues have studied EMRS-based lignin degradation in detail, including the use of a crude versatile peroxidase [91, 92, 95, 96], but the industrial application of peroxidase-based bioprocesses has not yet been realized [101]. Progress is hampered by the low availability of peroxidases in sufficient amounts, H_2O_2-related enzyme instabilities and high enzyme cost [102, 103]. Future research should focus on the development of cost-effective enzyme production systems (homologous and/or heterologous), and protein engineering to improve enzyme performance and reduce H_2O_2 sensitivity [101, 102]. Potential expression systems include bacteria (*E. coli*), yeast (*Saccharomyces cerevisiae* and *Pichia pastoris*) and fungi (particularly white-rot fungi and *Aspergillus* spp.) [104, 105] but none of these systems yet fulfills the requirements for industrial applications [105]. Alternatively, bacterial DyP-type peroxidases could be used for lignin valorization. The promising lignin degradation abilities of bacterial DyP-type peroxidases and the possibility to produce them—potentially more efficient than fungal peroxidaes—by recombinant expressions systems makes them interesting candidates for lignin utilization.

2.2. Photocatalytic degradation

2.2.1. Basic principles of photocatalysis

Photocatalysis is one of a special class of oxidation technologies known as advanced oxidation processes, which are characterized by the production of extraordinarily reactive hydroxyl radicals (OH radicals). These attack most organic molecules with rate constants usually in the order of 10^6–10^9 M^{-1} s^{-1}, which is several times higher than other reactive oxidizing agents such as ozone (**Table 2**). In addition to their reactivity, hydroxyl radicals are relatively nonselective, which makes advanced oxidation processes ideal for applications involving heterogeneous systems such as wastewater and industrial lignin [106].

Heterogeneous photocatalysis requires a semiconductor material that is irradiated with light. The absorption of light on the semiconductor surface generates electron-hole pairs by exciting electrons from the valence band to the conduction band of the semiconductor (**Figure 5**) [109]. Once the electron-hole pairs form, there is a period of time in the range of nanoseconds during which charge can be transferred to adsorbed species on the semiconductor surface [110]. The energy of the applied light source must be equal to or greater than the band-gap energy of the

	Rate constant (M^{-1} s^{-1})	
Organic compound	**OH radical**	**Ozone (O_3)**
Benzene	7.8×10^9	2
Chlorobenzene	4.5×10^9	0.75
Toluene	6.8×10^9	14
Styrene	6×10^9	3×10^5
n-Butanol	4.6×10^9	0.6
n-Propanol	2.2×10^9	0.4

Table 2. Rate constants for the reaction of hydroxyl radicals and ozone with different organic compounds (reproduced from Refs. [107, 108]).

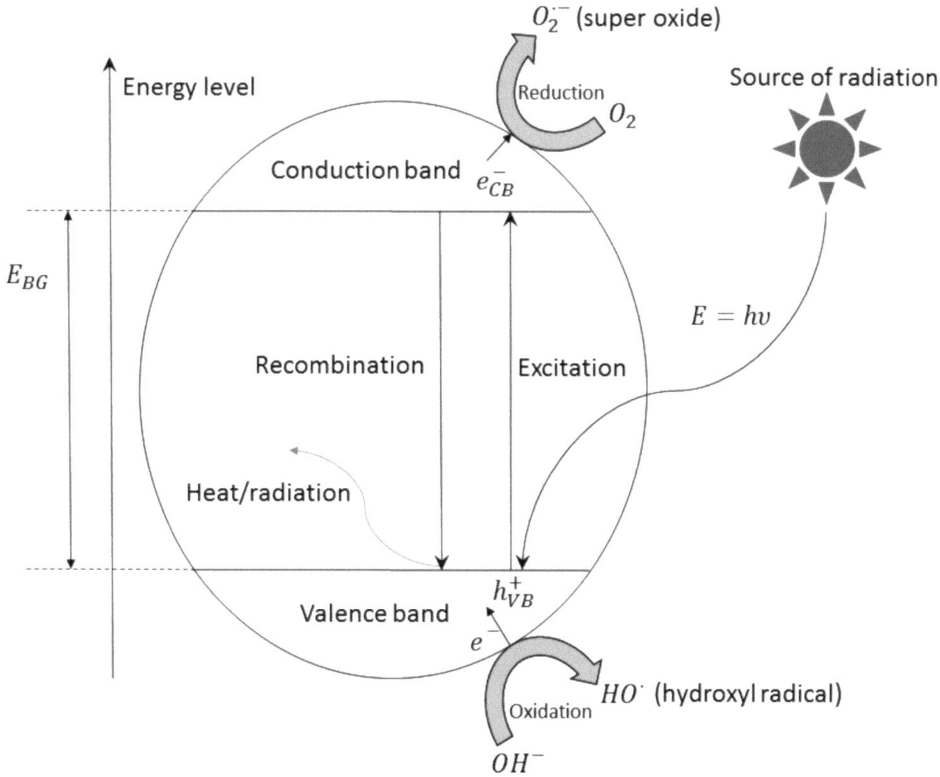

Figure 5. Schematic representation of primary steps during photocatalysis, adapted and modified from [111].

semiconductor (E_{BG}, energy difference between the valence and conduction bands) in order to form the electron-hole pairs [109].

Photocatalysis is typically carried out using semiconductor metal oxides, usually TiO_2, which is the most efficient and stable photocatalyst, and also the least expensive among the available materials. TiO_2 is nontoxic, and chemically and biologically inert, making it the only material suitable for current and probably also future industrial applications [112]. The three main types of TiO_2 are rutile, anatase and brookite, with most semiconductors based on anatase or brookite [113]. Band-gap energies are approximately 3.0 eV for rutile and 3.2 eV for anatase [114]. ZnO photocatalysts have also been used in several studies [115–117].

The fundamental reaction principle of photocatalysis is based on the generation of highly reactive species by the reaction of photogenerated electrons or holes with water, hydroxyl ions or oxygen, resulting in the oxidation and degradation of organic substrates [118]. Photogenerated electrons and holes can follow different reaction pathways, and multiple substrate-degrading mechanisms may co-occur depending on the nature of the substrates, the semiconductor material and several other parameters. A useful overview of the pathways involved in photocatalytic reactions is provided by Hoffmann et al. [119]. Photogenerated holes (Eq. (1)) partake in oxidation reactions with water or adsorbed hydroxyl ions to generate hydroxyl radicals (Eqs. (2) and (3)) that oxidize organic substrates (Eq. (5)). Another possibility is the direct oxidation of adsorbed substrate molecules by photogenerated holes (Eq. (4)) yielding oxidized substrate radicals, which may in turn partake in multiple consecutive reactions (e.g., Eq. (6)) [111].

Photogenerated electrons can chemically reduce dissolved molecular oxygen in aerated aqueous solutions (Eq. (7)) and in doing so generate superoxide radicals or singlet oxygen (Eqs. (7) and (8)). Hydrogen peroxide and hydroxyl radicals can also be formed via oxygen reduction pathways. In consecutive reactions, the reactive species created by photogenerated electrons can oxidize organic substrates as well as photogenerated holes [113, 119]. In addition to desirable degradation reactions, electrons and holes may recombine, which is associated with the dissipation of heat and/or radiation (Eq. (9)). Electron-hole recombination reduces the number of electron-hole pairs, and therefore the photocatalytic efficiency. This undesirable effect has been demonstrated for TiO_2-based catalysts [120].

$$TiO_2 \xrightarrow{h\upsilon \geq E_{BG}} TiO_2\left(e_{CB}^- + h_{VB}^+\right) \tag{1}$$

$$TiO_2\left(h_{VB}^+\right) + H_2O \rightarrow TiO_2 + HO^{\cdot} + H^+ \tag{2}$$

$$TiO_2\left(h_{VB}^+\right) + OH^- \rightarrow TiO_2 + HO^{\cdot} \tag{3}$$

$$TiO_2\left(h_{VB}^+\right) + S \rightarrow TiO_2 + S^{\cdot+} \tag{4}$$

$$S + HO^{\cdot} \rightarrow S^{\cdot+} + OH^- \tag{5}$$

$$S^{\cdot+} + O_2 \rightarrow SO_2^{\cdot+} \tag{6}$$

$$TiO_2\left(e_{CB}^-\right) + O_2 \rightarrow TiO_2 + O_2^{\cdot-} \tag{7}$$

$$HO^{\cdot} + O_2^{\cdot-} \rightarrow OH^- + {}^1O_2 \tag{8}$$

$$TiO_2\left(e_{CB}^-\right) + TiO_2\left(h_{VB}^+\right) \rightarrow heat/radiation \tag{9}$$

2.2.2. *Photocatalytic lignin degradation*

The precise comparison of studies involving photocatalytic lignin degradation (PLD) is difficult due to differences in the lignin source, photocatalyst type, sample pH, reactor design, illumination source, radiation intensity and analytical method used in each study. However, it is possible to make general statements about the fundamental scientific findings.

The first interest in PLD resulted from the knowledge that lignin and its derivatives are the main source of chemical oxygen demand (COD) in the wastewater from pulp and paper mills and are also responsible for the undesirable brown color of these effluents. Because both the COD and color intensity of discharged effluents are subject to strict regulations, the degradation of lignin in wastewater is an important requirement. Furthermore, applied activated sludge processes cannot completely decompose lignin or its degradation products. The undesirable brown color remaining after the activated sludge processes can only be reduced by repeating the process, which is expensive. For these reasons, PLD has been investigated as a possible alternative for color and COD removal in wastewater [117, 121].

2.2.3. *Lignin derived from the cooking stage*

The first studies dealing with heterogeneous PLD applied to technical lignin solutions from pulp and paper mills [117, 121] provided fundamental insights into lignin degradation by photocatalysis. The authors observed a significant decrease or complete removal of COD and color, suggesting the complete degradation of lignin into inorganic compounds such as carbon dioxide, carbon monoxide and water. Therefore, they proved that semiconductor photocatalysis efficiently decomposes kraft and alkaline lignins from the cooking stage. The basic principles of PLD were also confirmed in these pioneering studies. Ohnishi et al. [117] found that semiconductor photocatalysis applied in the dark without a source of radiation did not remove color at all, and combined with that lignin degradation and color loss did not occur in the absence of photocatalyst, even with prolonged illumination. These results have been confirmed in more recent studies [115, 116, 122, 123]. The important role of oxygen in PLD was also confirmed [117, 121]. Reducing the oxygen supply resulted in slower lignin degradation, and in the absence of oxygen there was no degradation at all. Oxygen therefore appears to promote the degradation process by accepting electrons from the conduction band of the semiconductor material (**Figure 5** and Eq. (7), Eq. (8)) [117, 121].

The photocatalytic degradation of alkaline lignin using different photocatalytic materials (e.g., TiO_2, ZnO and CdS) has also been investigated, applying a 500 W high-pressure mercury lamp as the radiation source [117]. The authors focused on complete lignin degradation rather than the formation of valuable mono-aromatic compounds during photocatalysis. They found that the most suitable semiconductor material depends on the pH of the treated lignin solution, and that catalytic efficiency improved when noble metals such as Pt, Ag or Au were combined with TiO_2 photocatalysts [117].

2.2.4. *Lignin from the bleaching stage*

Following the cooking stage, the bleaching process also creates process streams containing lignin (**Figure 3**). Bleaching is carried out to remove residual lignin from the pulp because paper containing lignin has a light brown color and the residual lignin causes undesirable photo-yellowing as the paper ages [124]. Bleaching is often achieved using oxidative chlorine compounds, but more environmentally sustainable options include total chlorine-free bleaching, which involves oxygen, hydrogen peroxide or ozone [125] and the biobleaching process discussed above [58, 90]. Bleaching with chlorine compounds is environmentally hazardous particularly due to the formation of absorbable organic halogens [126], including chlorinated lignin-related compounds [127]. Bleached effluents must therefore be treated before discharge into the environment. Several different setups have been used to achieve the photocatalytic oxidation of organic compounds in bleaching effluents from pulp and paper mills [122, 123, 128]. For example, a combination of suspended TiO_2 particles and H_2O_2 has been used for the photocatalytic oxidation of softwood kraft pulp bleach effluents from different stages (raw, biologically pretreated and coagulated) [128]. Photocatalytic oxidation achieved more efficient COD reduction in pretreated wastewater samples than raw kraft bleaching effluents. In contrast to the early studies discussed above which used lignin from the cooking stage [117, 121], the complete removal of COD from bleached effluents was not possible, and complete color removal was only achieved for coagulated wastewater samples [128]. Furthermore, free TiO_2 and ZnO or silica gel/porcelain-supported ZnO achieved considerable color clearance as well as the removal of phenols [123]. The efficiency of color clearance differed between the free and supported catalysts, mainly reflecting the better physical adsorption of organic compounds to suspended particles compared to those supported on a substrate [123]. Tanaka et al. [129] revealed that photocatalysis is accompanied by the adsorption of organic molecules onto the photocatalyst surface. Lignin was rapidly adsorbed onto the TiO_2 catalyst surface and subsequent illumination caused simultaneous desorption and degradation. Therefore adsorption, desorption and degradation must all be taken into account when designing a photocatalytic degradation process [129]. Yeber et al. [122] tested TiO_2 and ZnO supported on glass Raschig rings, and observed the complete removal of color, a substantial reduction in phenol levels and significantly lower toxicity following treatment. All these data indicate that the applicability of heterogeneous photocatalysis is not limited to lignin solutions from the cooking stage but can also be applied to effluents from the bleaching stage. Interestingly, the first applications of immobilized photocatalyst systems were introduced in this context [122, 123] as discussed in more detail below.

2.2.5. Immobilized photocatalytic systems

Immobilized systems combine a photocatalytic layer with a support material. The main advantages of immobilized systems compared to suspended photocatalyst particles include the reuse of the catalyst without downstream recovery steps (which increase upfront and operating costs) and reduced catalyst leakage. Potential disadvantages include the lower photocatalytic efficiency due to mass transfer limitations [130]. However, well-designed hydrodynamic operating conditions can address mass transfer limitations and restore high photocatalytic efficiency [115].

A particularly successful application of immobilized photocatalysts for PLD involved a ZnO photocatalyst fixed on pumice stone for the degradation of lignin solutions prepared by the precipitation of lignin from black liquor (kraft lignin). The experiments were conducted under solar illumination and achieved 98% lignin degradation after 2.5 h [116]. These operating conditions are cost efficient for two reasons: the immobilized catalyst removes the need for a downstream recovery step and the solar illumination is free, making such systems highly suitable for the treatment of wastewater produced by the pulp and paper industry. A comprehensive comparative analysis of TiO_2 coatings on sintered glass particles (grain size = 200–400 μm) packed in a borosilicate tube (26 cm long, 1 cm outer diameter, 6 mm internal diameter) was used to study the degradation of lignosulfonates supplied by a paper company [115]. The study compared several catalysts (TiO_2-P25-SiO_2 + Pt, TiO_2-P25-SiO_2, ZnO + TiO_2-P25-SiO_2 and $TiOSO_4$_30.6 wt%) using established synthesis and coating procedures [131]. The study set out not only to address the lignin-related problems of COD and color removal in pulp and paper mill wastewater, but also to evaluate the formation of valuable products during photodegradation. The operation mode included an integrated solid-phase extraction and dialysis membrane in order to directly extract smaller molecules, preferably aromatic and phenol-like compounds, from the aqueous solution (**Figure 6**).

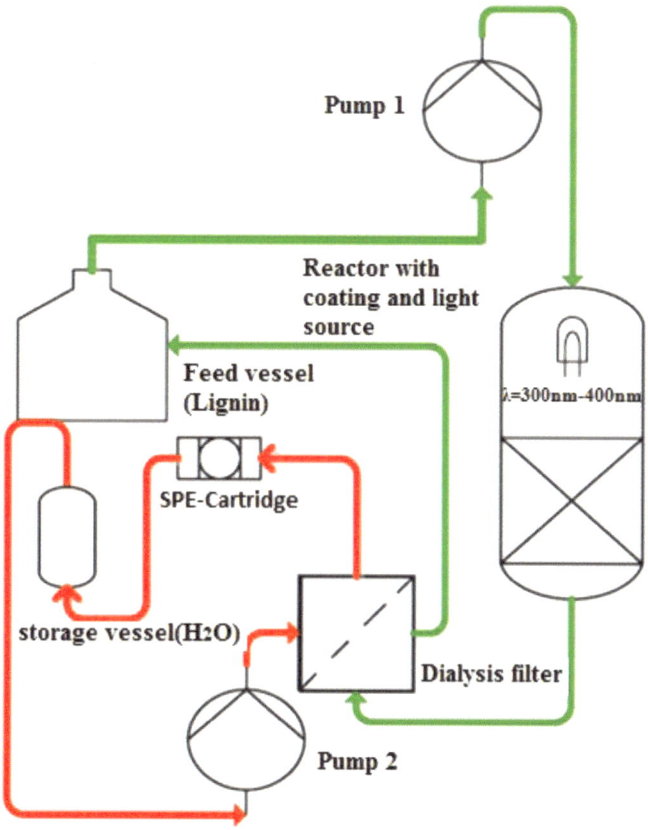

Figure 6. Photocatalytic lignin degradation with integrated solid-phase extraction and dialysis membrane (reprinted with permission [115]).

This study confirmed the effective degradation of relatively high concentrations of lignosulfonates (500 mg/L) and indicated that potential high-value degradation products were formed, although these products were not identified or quantified [115]. The strategy of continuously isolating target products from the reaction mixture during photocatalytic degradation using a combination of solid-phase extraction and dialysis is an interesting approach for the production of aromatic products by PLD. Further research is needed to reveal the nature of the isolated molecules, and quantification would help to determine the economic viability of such processes.

2.2.6. Mechanistic aspects and potential products

The formation of mono-aromatic products during PLD has been described in several studies. Miyata et al. [132], using defatted *Picea glehnii* wood flour (softwood) as a substrate, confirmed the formation of vanillin during PLD process using H-NMR analytics. The amount of vanillin was not determined, but the authors suggested a PLD reaction mechanism including the cleavage of Cα-Cβ bonds [132]. In another study focusing not only on COD removal but also on the potential formation of intermediates such as vanillin, a TiO_2 (rutile) photocatalyst was applied to softwood-derived kraft lignin [121]. Aromatic compounds originating from the kraft lignin solution (e.g., vanillin and catechol) could not be recovered because most aromatic monomers were dimerized at the beginning of photocatalysis, and were completely decomposed as the reaction proceeded [121]. Similar results were reported for a TiO_2 (anatase) photocatalyst applied to lignosulfonates from coniferous wood (softwood): lignin fractions were first polymerized and then converted into smaller molecules during the photocatalytic process, ultimately leading to complete mineralization [129]. In the same study, Fourier-transform infrared spectroscopy showed that the aromatic moiety was transformed more rapidly than the aliphatic side chains [129]. The opposite was found for photocatalytic degradation of lignosulfonate using UV–Vis spectroscopy: the absorption peaks at 203 and 280 nm were associated with unsaturated chains and unconjugated phenolic hydroxyl groups/aromatic rings, respectively [115].

2.2.7. Conclusion and outlook

The large-scale photocatalytic production of aromatic compounds from industrial lignin has not yet been reported. Most PLD research has focused on COD and color removal during wastewater treatment, whereas few reports consider the recovery of valuable aromatic products. Lekelefac and Czermak [115] used a combination of solid-phase extraction and dialysis for the continuous isolation of desired products from the reaction mixture during PLD, but this is an exceptional approach. The development of heterogeneous PLD processes that also produce mono-aromatic compounds will require intense scientific research because PLD ultimately results in the complete mineralization of lignin, and the mechanistic principles discovered thus far imply a polymerization-depolymerization mechanism in which the reaction conditions and especially the residence times have a strong impact on the yields of desirable mono-aromatic products. Given the heterogeneity of lignin and the multiple PLD reaction pathways, a broad mixture of products can be expected. Efficient and cost-effective separation technologies for product recovery would therefore play a key role in the successful development of PLD processes. In the foreseeable future, the main application of heterogeneous PLD is likely to remain the complete mineralization of lignin to reduce COD and remove color from wastewater.

2.3. Thermochemical lignin degradation

2.3.1. Lignin pyrolysis

Pyrolysis is the thermal treatment of a feedstock in absence of oxygen with or without a catalyst, similar to the cracking process applied in petroleum refineries. The aim of pyrolysis is to break down the molecular structure of the feedstock to smaller components while the absence of oxygen prevents combustion to carbon dioxide and water [5, 133]. The pyrolysis of biomass results in the formation of gases, liquid oils and a thermally stable solid fraction known as char [133]. The pyrolysis of lignin generates gaseous hydrocarbons including carbon monoxide and carbon dioxide, as well as monolignols (e.g., coniferyl alcohol), monophenols (e.g., phenol and guaiacol), polysubstituted phenols and char [5]. The ratio of products depends on parameters such as the heating rate, temperature, pressure and type of lignin feedstock (e.g., lignocellulosic biomass, kraft lignin or lignosulfonate) [134–136].

2.3.2. Hydrogenolysis

Pyrolysis in the presence of hydrogen is described as hydrogenation or hydrogenolysis. Gaseous hydrogen and hydrogen-donating solvents such as tetralin or common alcohols have been used for the hydrogenolysis of lignin and lignocellulosic biomass [137–141]. During this process, the cleavage of bonds is assisted by the addition of hydrogen, which results in the reductive conversion of lignin. These reductive reaction characteristics of hydrogenolysis tend to reduce or remove functionality in lignin ultimately leading to simpler phenols [2]. Therefore, hydrogenolysis is a promising method for the production of bulk phenolic chemicals from lignin [2, 5, 142]. A large number of heterogeneous metal-based catalysts (e.g., Ni, Rh, Ni-Mo, Co-Mo and Pd) and a significant smaller number of homogeneous catalysts have been combined with the hydrogenolysis of different lignin models, technical lignins or lignocellulosic biomass [2, 141, 143]. An exhaustive review of applied catalyst was published by Zakzeski et al. [2].

2.3.3. Lignin oxidation

Unlike photocatalytic lignin oxidation as described above, thermal oxidation involves the application of heat as well as oxidizing agents such as oxygen or hydrogen peroxide, and the reaction conditions and catalysts are distinct. Compared to hydrogenolysis, lignin oxidation leads to more complex aromatic compounds with greater functionality [2]. Lignin oxidation plays an important role in the conversion of lignin to aldehydes such as vanillin and syringaldehyde [13, 144–146]. However, the spectrum of potential products also includes carboxylic acids, aliphatic alcohols and hydrocarbon gases, with the product ratios depending on the lignin source, type of catalyst and reaction conditions [2, 5]. Catalytic lignin oxidation has also been reviewed by Zakzeski et al. [2]. This process offers immense scope for the production of bulk platform chemicals as well as value-added fine chemicals, as shown by the alkaline oxidation of lignosulfonates in air for the production of vanillin [27, 147].

2.3.4. Separation technology based on ceramic membranes

Like ELD and PLD, thermochemical lignin conversion produces complex mixtures of degradation products including residual non-degraded lignin substrate. These must be separated to achieve the full valorization of technical lignin derived from industrial processes. Membrane technology is ideal for the recovery of technical lignin from various industrial processes given its unique combination of scalability, economic efficiency and versatility [148]. Therefore, the membrane-based fractionation of reaction mixtures from thermochemical lignin degradation processes could be used to separate non-degraded lignin from valuable mono-aromatic compounds [149]. The key advantages of ceramic membranes include their chemical and thermal stability, which allow the filtration of acids, bases, solvents and hot feeds, and make the membranes resistant to the harsh cleaning conditions that may be necessary to restore permeability [99, 100]. These properties match the requirements of filtration processes involving reaction mixtures from thermochemical degradation, including extreme pH and the presence of harsh solvents.

We have investigated the fractionation of lignin derivatives (produced using oxygen as the oxidizing agent and methanol as the solvent) using ceramic membranes in order to separate residual lignosulfonate from mono-aromatic compounds. The ceramic membrane we used was a 30-nm hollow fiber membrane (MANN + HUMMEL) with three fibers, a molecular weight cut-off (MWCO) of ~800 kDa (based on our own measurements using polyethylene glycol) and a surface area of 0.0035 m^2. The filtration test bench is shown schematically in **Figure 7** and comprised two operational modes: dead-end filtration (valve 1 closed) and crossflow filtration (valve 1 open). Tests were carried out in crossflow filtration mode. The permeate flux was measured using an electronic balance and the transmembrane pressure (TMP) was calculated according to Eq. (10) using the pressure gauges P_1I (p_1) and P_2I (p_2). The pressure on the permeate side (p_3) was negligible. The tangential crossflow velocity (v_{CF}) was determined by flowmeter data (FI) and membrane geometry.

$$TMP = \frac{p_1 + p_2}{2} - p_3 \tag{10}$$

The filtration experiments were carried out in batch mode, and a filtration run was deemed complete when 50% of the original feed was present on the permeate side. The resulting flux

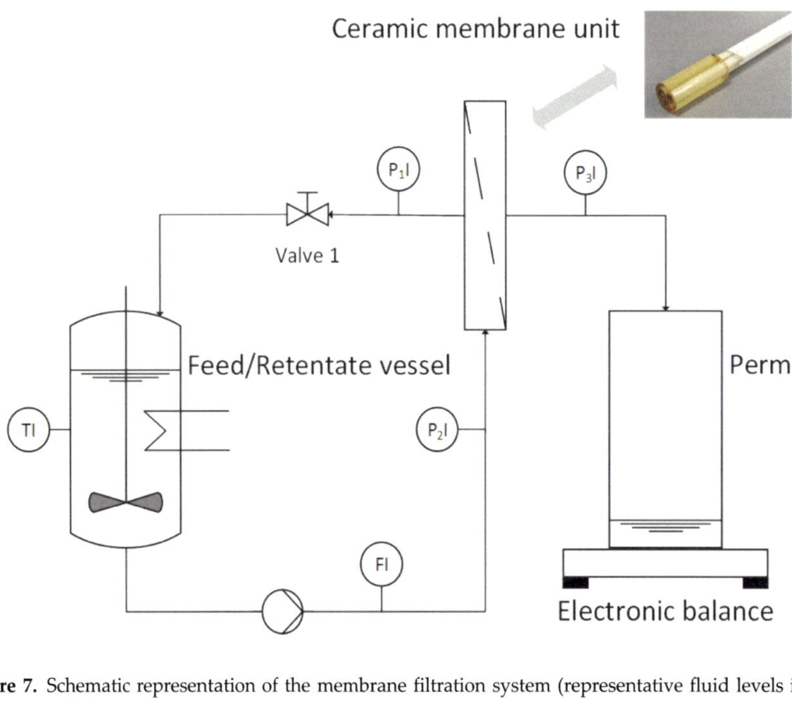

Figure 7. Schematic representation of the membrane filtration system (representative fluid levels in feed/retentate and permeate tanks).

http://dx.doi.org/10.5772/intechopen.71210

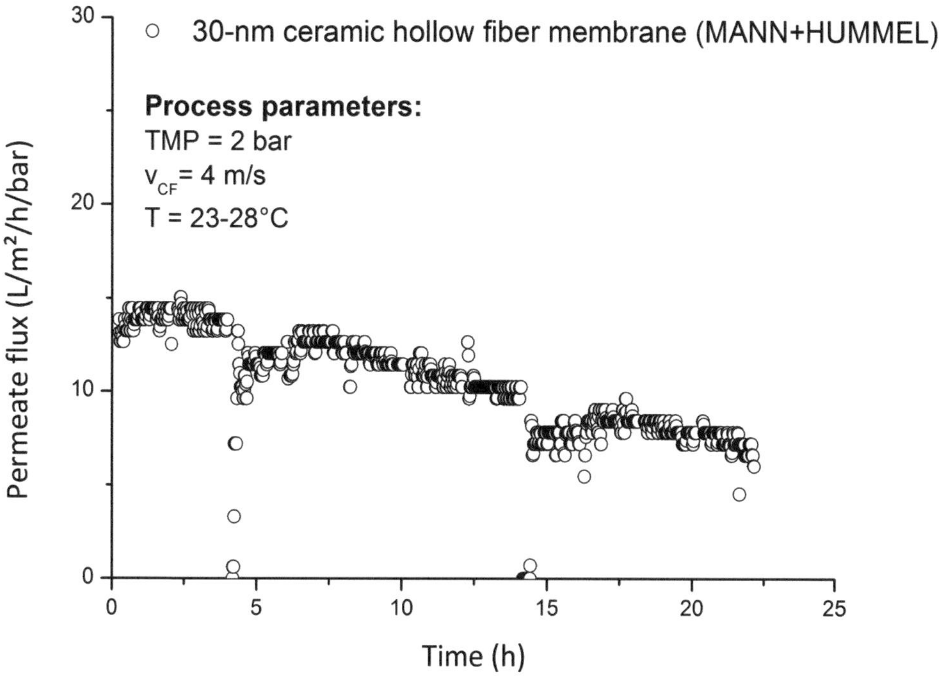

Figure 8. Flux profile during the batch filtration of oxidized lignosulfonate in methanol (TMP = 2 bar, v_{CF} = 4 m/s, T = 23–28°C, 30-nm MANN + HUMMEL ceramic hollow fiber membrane).

profile is shown in **Figure 8**. The interruptions in the flux curve at ~4 and ~14 h reflect safety-related overnight system shutdowns.

The anticipated mass reduction of 50% was achieved within 22 h after supplying 3 kg of feed. The flux decline from ~15 to ~7 L/m^2/h/bar probably reflected two effects: fouling on the membrane surface and/or membrane pores and a concentration-related increase in viscosity. The characteristics of these effects will be investigated in further studies to determine their individual contributions to filtration performance. The retention of the residual lignosulfonate fraction was calculated based on size-exclusion chromatography (SEC) analytics. The molecular weight distributions in the feed, retentate after 22 h and permeate after 22 h are compared in **Figure 9**.

The characteristics of the illustrated molecular weight distribution curves are summarized in **Table 3**. The presentation of the curves and the associated data indicate that a notable fraction of partially degraded or non-degraded lignosulfonate is retained by the 30-nm membrane. The number average molar mass (M_n) and the mass average molar mass (M_w) of the permeate were lower than those of the feed and retentate. The polydispersity index (PDI), which represents the width of a molecular weight distribution, was also significantly lower in the permeate than in the feed and retentate.

The retention of lignosulfonate fractions much smaller than the MWCO of the membrane (800 kDa) may reflect the filtration effects of a fouling layer. Further investigations will

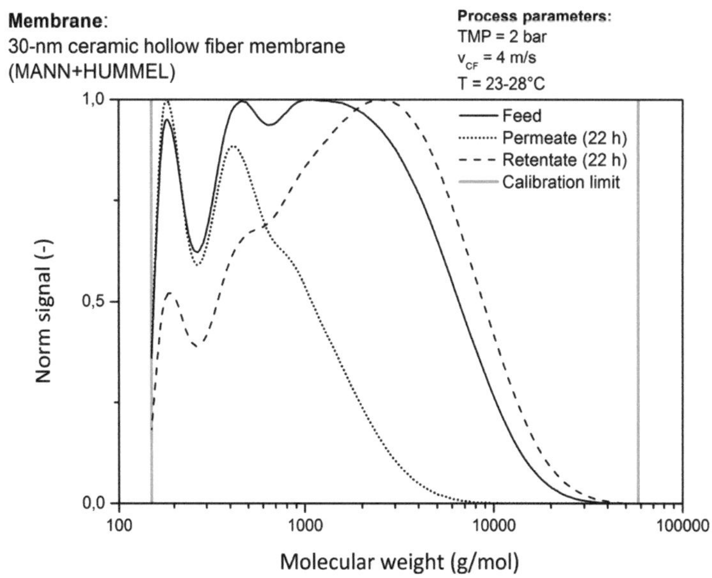

Figure 9. Molecular weight distribution of feed, retentate and permeate (TMP = 2 bar, v_{CF} = 4 m/s, T = 23–28°C, 30-nm ceramic hollow fiber membrane (MANN + HUMMEL)).

Sample name	M_n (g/mol)	M_w (g/mol)	PDI (–)
Feed	457	1768	3.87
Permeate (22 h)	305	559	1.83
Retentate (22 h)	614	2665	4.34

Table 3. Characteristics of the presented molecular weight distributions.

characterize the fouling effects observed with this setup and a potential retention of products such as vanillin and syringaldehyde.

2.3.5. Conclusion and outlook

Thermochemical lignin degradation has been investigated extensively, and each thermochemical degradation process generates characteristic degradation products. Hydrogenolysis creates simple phenolic products with low functionality, suitable as bulk chemicals, whereas oxidative conversion generates products with increased functionality, suitable for the production of aromatic aldehydes (e.g., vanillin). Lignin degradation is followed by the purification of the reaction mixtures, including the removal of residual partially degraded or non-degraded lignin which can be recycled into the reaction to improve lignin conversion rates and total process efficiency. However, the operability of the degradation process may be restricted by such recycling.

As we have shown, membrane filtration systems are suitable as a separation technology for the retention of partially degraded or non-degraded lignin fractions. Further research will characterize the fouling effects and will help to optimize filtration performance. It is clear that filtration efficiency depends on the feed characteristics, and our data presented herein are therefore relevant only for processes featuring the oxidative degradation of lignosulfonate with methanol as the solvent. Other lignin sources, degradation methods, solvents and catalysts are likely to influence the filtration performance, and membrane filtration processes must therefore be designed on a case-by-case basis. Intensive research to create a database covering all process parameters (particularly the lignin source, solvent and type of degradation process) could reduce the preliminary effort required to design optimal membrane-based filtration processes.

Acknowledgements

We acknowledge Dr Richard M. Twyman for editing the manuscript, MANN + HUMMEL GmbH for supplying the ceramic membranes and Sappi Europe, Stockstadt Mill, Germany and CMBlu Projekt AG, Alzenau, Germany for supplying the feed solutions. This work has been financed by *Fachagentur für nachwachsende Rohstoffe e.v. (FNR)* through funds of *Bundeministeriums für Ernährung und Landwirtschaft (BMEL)* out of the special asset *"Energie- und Klimafonds"* by a resolution of *Deutscher Bundestag (FKZ: 22402816).*

Author details

Stefan Schoenherr[1], Mehrdad Ebrahimi[1] and Peter Czermak[1,2,3]*

*Address all correspondence to: peter.czermak@kmub.thm.de

1 Institute of Bioprocess Engineering and Pharmaceutical Technology, University of Applied Sciences Mittelhessen, Giessen, Germany

2 Department of Chemical Engineering, Kansas State University, Manhattan, KS, USA

3 Faculty of Biology and Chemistry, Justus-Liebig University of Giessen, Germany

References

[1] Sun Y, Cheng J. Hydrolysis of lignocellulosic materials for ethanol production: A review. Bioresource Technology. 2002;**83**(1):1-11. DOI: https://doi.org/10.1016/S0960-8524(01)00212-7

[2] Zakzeski J, Bruijnincx PCA, Jongerius AL, Weckhuysen BM. The catalytic valorization of lignin for the production of renewable chemicals. Chemical Reviews. 2010;**110**(6):3552-3599. DOI: https://doi.org/10.1021/cr900354u 20218547

[3] Kirk TK, Farrell RL. Enzymatic "combustion": The microbial degradation of lignin. Annual Review of Microbiology. 1987;**41**:465-505. DOI: https://doi.org/10.1146/annurev.mi.41.100187.002341 3318677

[4] Hammel KE. Fungal degradation of lignin. In: Cadisch G, editor. Driven by Nature: Plant Litter Quality and Decomposition. Wallingford: CAB International. 1997;33-45

[5] Pandey MP, Kim CS. Lignin depolymerization and conversion: A review of thermochemical methods. Chemical Engineering and Technology. 2011;**34**(1):29-41. DOI: https://doi.org/10.1002/ceat.201000270

[6] Vanholme R, Demedts B, Morreel K, Ralph J, Boerjan W. Lignin biosynthesis and structure. Plant Physiology. 2010;**153**(3):895-905. DOI: https://doi.org/10.1104/pp.110.155119 20472751

[7] Faix O. Classification of lignins from different botanical origins by FT-IR spectroscopy. Holzforschung. 1991;**45**(s1):21-28. DOI: https://doi.org/10.1515/hfsg.1991.45.s1.21

[8] Pinto PC, Evtuguin DV, Neto CP. Effect of structural features of wood biopolymers on hardwood pulping and bleaching performance. Industrial and Engineering Chemistry Research. 2005;**44**(26):9777-9784. DOI: https://doi.org/10.1021/ie050760o

[9] Evtuguin DV, Neto CP, Silva AMS, et al. Comprehensive study on the chemical structure of dioxane lignin from plantation *Eucalyptus globulus* wood. Journal of Agricultural and Food Chemistry. 2001;**49**(9):4252-4261. DOI: https://doi.org/10.1021/jf010315d

[10] Calvo-Flores FG, Dobado JA. Lignin as renewable raw material. ChemSusChem. 2010; **3**(11):1227-1235. DOI: https://doi.org/10.1002/cssc.201000157 20839280

[11] Gonzalo G, de Colpa DI, Habib MHM, Fraaije MW. Bacterial enzymes involved in lignin degradation. Journal of Biotechnology 2016;**236**:110-119. DOI: https://doi.org/10.1016/j.jbiotec.2016.08.011. PMID: 27544286

[12] Faulon J-L, Hatcher PG. Is there any order in the structure of lignin? Energy & Fuels. 1994;**8**(2):402-407. DOI: https://doi.org/10.1021/ef00044a018

[13] Rodrigues Pinto PC, Borges da Silva EA, Rodrigues AE. Insights into oxidative conversion of lignin to high-added-value phenolic aldehydes. Industrial and Engineering Chemistry Research. 2011;**50**(2):741-748. DOI: https://doi.org/10.1021/ie102132a

[14] Awungacha Lekelefac C, Busse N, Herrenbauer M, Czermak P. Photocatalytic based degradation processes of lignin derivatives. International Journal of Photoenergy. 2015;**2015**(12):1-18. DOI: https://doi.org/10.1155/2015/137634

[15] Busch R, Hirth T, Liese A, et al. Nutzung nachwachsender Rohstoffe in der industriellen Stoffproduktion. Chemie Ingenieur Technik. 2006;**78**(3):219-228. DOI: https://doi.org/10.1002/cite.200500178

[16] Gustavsson L, Haus S, Lundblad M, et al. Climate change effects of forestry and substitution of carbon-intensive materials and fossil fuels. Renewable and Sustainable Energy Reviews. 2017;**67**:612-624. DOI: https://doi.org/10.1016/j.rser.2016.09.056

[17] Allison EH, Perry AL, Badjeck M-C, et al. Vulnerability of national economies to the impacts of climate change on fisheries. Fish and Fisheries. 2009;**10**(2):173-196. DOI: https://doi.org/10.1111/j.1467-2979.2008.00310.x

[18] Quadrelli R, Peterson S. The energy–climate challenge: Recent trends in CO_2 emissions from fuel combustion. Energy Policy. 2007;**35**(11):5938-5952. DOI: https://doi.org/10.1016/j.enpol.2007.07.001

[19] Bugg TDH, Rahmanpour R. Enzymatic conversion of lignin into renewable chemicals. Current Opinion in Chemical Biology. 2015;**29**:10-17. https://doi.org/10.1016/j.cbpa.2015.06.009 PMID: 26121945

[20] Saxena RC, Adhikari DK, Goyal HB. Biomass-based energy fuel through biochemical routes: A review. Renewable and Sustainable Energy Reviews. 2009;**13**(1):167-178. DOI: https://doi.org/10.1016/j.rser.2007.07.011

[21] Hileman JI, Donohoo PE, Stratton RW. Energy content and alternative jet fuel viability. Journal of Propulsion and Power. 2010;**26**(6):1184-1196. DOI: https://doi.org/10.2514/1.46232

[22] Mathews SL, Grunden AM, Pawlak J. Degradation of lignocellulose and lignin by Paenibacillus glucanolyticus. International Biodeterioration & Biodegradation. 2016;**110**: 79-86. DOI: https://doi.org/10.1016/j.ibiod.2016.02.012

[23] Kamm B, Kamm M. Principles of biorefineries. Applied Microbiology and Biotechnology. 2004;**64**(2):137-145. DOI: https://doi.org/10.1007/s00253-003-1537-7 14749903

[24] Strassberger Z, Tanase S, Rothenberg G. The pros and cons of lignin valorisation in an integrated biorefinery. RSC Advances. 2014;**4**(48):25310-25318. DOI: https://doi.org/10.1039/C4RA04747H

[25] JE Holladay, JF White, JJ Bozell, and D Johnson. Top Value-Added Chemicals from Biomass – Volume II – Results of Screening for Potential Candidates from Biorefinery Lignin. Prepared for the U.S. Department of Energy under Contract DE-AC05-76RL01830; 2007

[26] Gosselink R, Jong E. De, Guran B, Abächerli a. Co-ordination network for lignin—Standardisation, production and applications adapted to market requirements (EUROLIGNIN). Industrial Crops and Products. 2004;**20**(2):121-129. DOI: https://doi.org/10.1016/j.indcrop.2004.04.015

[27] Silva EB, Zabkova M, Araújo JD, et al. An integrated process to produce vanillin and lignin-based polyurethanes from Kraft lignin. Chemical Engineering Research and Design. 2009;**87**(9):1276-1292. DOI: https://doi.org/10.1016/j.cherd.2009.05.008

[28] Kuhad RC, Gupta R, Singh A. Microbial cellulases and their industrial applications. Enzyme Research. 2011;**2011**:280696. DOI: https://doi.org/10.4061/2011/280696. PMID: 21912738

[29] Saha BC. Hemicellulose bioconversion. Journal of Industrial Microbiology & Biotechnology. 2003;**30**(5):279-291. DOI: https://doi.org/10.1007/s10295-003-0049-x 12698321

[30] Leonowicz A, Matuszewska A, Luterek J, et al. Biodegradation of lignin by white rot fungi. Fungal Genet Biol. 1999;**27**(2-3):175-185. [https://doi.org/10.1006/fgbi.1999.1150] [PMID: 10441443]

[31] Sánchez C. Lignocellulosic residues: Biodegradation and bioconversion by fungi. Biotechnology Advances. 2009;**27**(2):185-194. DOI: https://doi.org/10.1016/j.biotechadv.2008.11.001 19100826

[32] Timell TE, Eriksson K-EL, Blanchette RA, Ander P. Microbial and Enzymatic Degradation of Wood and Wood Components. Berlin, Heidelberg: Springer; 1990

[33] Cowling EB. Comparative biochemistry of decay of sweetgum sapwood by white-rot and brown-rot fungi. USDA Technical Bulletin; 1961

[34] Wong DWS. Structure and action mechanism of ligninolytic enzymes. Applied Biochemistry and Biotechnology. 2009;**157**(2):174-209. DOI: https://doi.org/10.1007/s12010-008-8279-z 18581264

[35] Tuor U, Winterhalter K, Fiechter A. Enzymes of white-rot fungi involved in lignin degradation and ecological determinants for wood decay. Journal of Biotechnology. 1995;**41**(1):1-17. DOI: https://doi.org/10.1016/0168-1656(95)00042-O

[36] Monrroy M, Ortega I, Ramírez M, Baeza J, Freer J. Structural change in wood by brown rot fungi and effect on enzymatic hydrolysis. Enzyme and Microbial Technology. 2011;**49**(5):472-477. DOI: https://doi.org/10.1016/j.enzmictec.2011.08.004 22112620

[37] Jensen KA, Houtman CJ, Ryan ZC, Hammel KE. Pathways for extracellular Fenton chemistry in the brown rot basidiomycete Gloeophyllum trabeum. Applied and Environmental Microbiology. 2001;**67**(6):2705-2711. DOI: https://doi.org/10.1128/AEM.67.6.2705-2711.2001 11375184

[38] Green F, Highley TL. Mechanism of brown-rot decay: Paradigm or paradox. International Biodeterioration & Biodegradation. 1997;**39**(2–3):113-124. DOI: https://doi.org/10.1016/S0964-8305(96)00063-7

[39] Dashtban M, Schraft H, Syed TA, Qin W. Fungal biodegradation and enzymatic modification of lignin. International Journal of Biochemistry and Molecular Biology. 2010;**1**(1):36-50

[40] Findlay WPK, Savory JG. Breakdown of timber in water cooling towers. 7th International Botanical Congress, Stockholm; 1950. p. 1

[41] Levy J. The soft rot fungi: Their mode of action and significance in the degradation of wood. In: Advances in Botanical Research. Vol. 2. Elsevier; 1966. pp. 323-357

[42] Floudas D, Binder M, Riley R, et al. The Paleozoic origin of enzymatic lignin decomposition reconstructed from 31 fungal genomes. Science. 2012;**336**(6089):1715-1719. DOI: https://doi.org/10.1126/science.1221748 22745431

[43] Butler MJ, Day AW. Destruction of fungal melanins by ligninases of *Phanerochaete chrysosporium* and other white rot fungi. International Journal of Plant Sciences. 1998;**159**(6):989-995. DOI: https://doi.org/10.1086/314093

[44] Sugano Y. DyP-type peroxidases comprise a novel heme peroxidase family. Cellular and Molecular Life Sciences. 2009;**66**(8):1387-1403. DOI: https://doi.org/10.1007/s00018-008-8651-8 19099183

[45] Abdel-Hamid AM, Solbiati JO, Cann IKO. Insights into lignin degradation and its potential industrial applications. Advances in Applied Microbiology. 2013;**82**:1-28. DOI: https://doi.org/10.1016/B978-0-12-407679-2.00001-6 23415151

[46] Guillen F, AT M, MMJ. Substrate specificity and properties of the aryl-alcohol oxidase from the ligninolytic fungus *Pleurotus eryngii*. European Journal of Biochemistry. 1992;**209**(2):603-611. DOI: https://doi.org/10.1111/j.1432-1033.1992.tb17326.x

[47] Kersten PJ. Glyoxal oxidase of *Phanerochaete chrysosporium* its characterization and activation by lignin peroxidase. Proceedings of the National Academy of Sciences. USA 1990;(87):2936-2940

[48] Gutiérrez A, Caramelo L, Prieto A, Martínez MJ, Martínez AT. Anisaldehyde production and aryl-alcohol oxidase and dehydrogenase activities in ligninolytic fungi of the genus Pleurotus. Applied and Environmental Microbiology 1994;**60**(6):1783-1788. [PMID: 8031078]

[49] Guillén F, Martínez MJ, Muñoz C, Martínez AT. Quinone redox cycling in the ligninolytic fungus Pleurotus eryngii leading to extracellular production of superoxide anion radical. Archives of Biochemistry and Biophysics. 1997;**339**(1):190-199. DOI: https://doi.org/10.1006/abbi.1996.9834 9056249

[50] Henriksson G, Johansson G, Pettersson G. A critical review of cellobiose dehydrogenases. Journal of Biotechnology. 2000;**78**(2):93-113. DOI: https://doi.org/10.1016/S0168-1656(00)00206-6

[51] Thurston CF. The structure and function of fungal laccases. Microbiology. 1994;**140**(1):19-26. DOI: https://doi.org/10.1099/13500872-140-1-19

[52] Hofrichter M. Review: Lignin conversion by manganese peroxidase (MnP). Enzyme and Microbial Technology. 2002;**30**(4):454-466. DOI: https://doi.org/10.1016/S0141-0229(01)00528-2

[53] Martínez ÁT, Speranza M, Ruiz-Dueñas FJ, Ferreira P, Camarero S, Guillén F, Martínez MJ, Gutiérrez A. Biodegradation of lignocellulosics: Microbial, chemical, and enzymatic aspects of the fungal attack of lignin. International Microbiology. 2005;**8**:195-204

[54] Claus H. Laccases: Structure, reactions, distribution. Micron. 2004;**35**(1–2):93-96. DOI: https://doi.org/10.1016/j.micron.2003.10.029 15036303

[55] Bourbonnais R, Paice MG. Oxidation of non-phenolic substrates. FEBS Letters. 1990;**267**(1):99-102. DOI: https://doi.org/10.1016/0014-5793(90)80298-W

[56] Kawai S, Nakagawa M, Ohashi H. Degradation mechanisms of a nonphenolic β-O−4 lignin model dimer by Trametes versicolor laccase in the presence of 1-hydroxybenzotriazole. Enzyme and Microbial Technology. 2002;**30**(4):482-489. DOI: https://doi.org/10.1016/S0141-0229(01)00523-3

[57] Dedhia BS, Vetal MD, Rathod VK, Levente C. Xylanase and laccase aided bio-bleaching of wheat straw pulp. Canadian Journal of Chemical Engineering. 2014;**92**(1):131-138. DOI: https://doi.org/10.1002/cjce.21798

[58] María E. Eugenio, Manuel Hrenández, Raquel Moya, Raquel Martín-Sampedro, Juan C. Villar, Arias ME. Evaluation of a new laccase produced by *Streptomyces ipomoea* on biobleaching and ageing of kraft pulps. BioResources Aug 2011;**6**(3):3231-3241

[59] Valli K, Wariishi H, Gold MH. Oxidation of monomethoxylated aromatic compounds by lignin peroxidase: Role of veratryl alcohol in lignin biodegradation. Biochemistry. 2002;**29**(37):8535-8539. DOI: https://doi.org/10.1021/bi00489a005

[60] Lundell T, Wever R, Floris R, et al. Lignin peroxidase L3 from Phlebia rediata. Pre-steady-state and steady-state studies with veratryl alcohol and a non-phenolic lignin model compound 1-(3,4-dimethoxyphenyl)-2-(2-methoxyphenoxy)propane-1,3-diol. European Journal of Biochemistry. 1993;**211**(3):391-4020. DOI: https://doi.org/10.1111/j.1432-1033.1993.tb17562.x

[61] Miki K, Renganathan V, Gold MH. Mechanism of beta-aryl ether dimeric lignin model compound oxidation by lignin peroxidase by *Phanerochaete chrysosporium*. Biochemistry. 2002;**25**(17):4790-4796. DOI: https://doi.org/10.1021/bi00365a011

[62] Hammel KE, Moen MA. Depolymerization of a synthetic lignin in vitro by lignin peroxidase. Enzyme and Microbial Technology. 1991;**13**(1):15-18. DOI: https://doi.org/10.1016/0141-0229(91)90182-A

[63] Reddy GVB, Sridhar M, Gold MH. Cleavage of nonphenolic beta-1 diarylpropane lignin model dimers by manganese peroxidase from *Phanerochaete chrysosporium*. Evidence for a hydrogen abstraction mechanism. European Journal of Biochemistry. 2003;**270**(2):284-292. DOI: https://doi.org/10.1046/j.1432-1033.2003.03386.x

[64] Ruiz-Dueñas FJ, Camarero S, Pérez-Boada M, Martínez MJ, Martínez ATA. New versatile peroxidase from Pleurotus. Biochemical Society Transactions. 2001;**29**(2):116-122. DOI: https://doi.org/10.1042/bst0290116

[65] Camarero S, Sarkar S, Ruiz-Dueñas FJ, Martínez MJ, Martínez ÁT. Description of a versatile peroxidase involved in the natural degradation of lignin that has both manganese peroxidase and lignin peroxidase substrate interaction sites. The Journal of Biological Chemistry. 1999;**274**(15):10324-10330. DOI: https://doi.org/10.1074/jbc.274.15.10324

[66] Liers C, Bobeth C, Pecyna M, Ullrich R, Hofrichter M. DyP-like peroxidases of the jelly fungus *Auricularia auricula*-judae oxidize nonphenolic lignin model compounds and high-redox potential dyes. Applied Microbiology and Biotechnology. 2010;**85**(6):1869-1879. DOI: https://doi.org/10.1007/s00253-009-2173-7 19756587

[67] Colpa DI, Fraaije MW, van Bloois E. DyP-type peroxidases: A promising and versatile class of enzymes. Journal of Industrial Microbiology & Biotechnology. 2014;**41**(1):1-7. DOI: https://doi.org/10.1007/s10295-013-1371-6. PMID: 24212472

[68] Kim SJ, Shoda M. Purification and characterization of a novel peroxidase from Geotrichum candidum dec 1 involved in decolorization of dyes. Applied and Environmental Microbiology. 1999;**65**(3):1029-1035. [PMID: 10049859]

[69] Adav SS, Ng CS, Arulmani M, Sze SK. Quantitative iTRAQ secretome analysis of cellulolytic Thermobifida fusca. Journal of Proteome Research. 2010;**9**(6):3016-3024. DOI: https://doi.org/10.1021/pr901174z 20408570

[70] Vicuña R. Bacterial degradation of lignin. Enzyme and Microbial Technology. 1988;**10**(11): 646-655. DOI: https://doi.org/10.1016/0141-0229(88)90055-5

[71] Zimmermann W. Degradation of lignin by bacteria. Journal of Biotechnology. 1990;**13** (2–3): 119-130. DOI: https://doi.org/10.1016/0168-1656(90)90098-V

[72] Liese W, editor. Biological Transformation of Wood by Microorganisms. Berlin, Heidelberg: Springer; 1975

[73] Davis JR, Goodwin L, Teshima H, et al. Genome sequence of *Streptomyces viridosporus* strain T7A ATCC 39115, a lignin-degrading actinomycete. Genome Announcements. 2013;**1**(4). DOI: https://doi.org/10.1128/genomeA.00416-13 23833133

[74] Brown ME, Walker MC, Nakashige TG, Iavarone AT, Chang MCY. Discovery and characterization of heme enzymes from unsequenced bacteria: Application to microbial lignin degradation. Journal of the American Chemical Society. 2011;**133**(45):18006-18009. DOI: https://doi.org/10.1021/ja203972q 21671563

[75] Bugg TDH, Ahmad M, Hardiman EM, Singh R. The emerging role for bacteria in lignin degradation and bio-product formation. Current Opinion in Biotechnology. 2011;**22**(3): 394-400. DOI: https://doi.org/10.1016/j.copbio.2010.10.009 21071202

[76] Sharma P, Goel R, Capalash N. Bacterial laccases. World Journal of Microbiology and Biotechnology. 2007;**23**(6):823-832. DOI: https://doi.org/10.1007/s11274-006-9305-3

[77] Ahmad M, Roberts JN, Hardiman EM, Singh R, Eltis LD, Bugg TDH. Identification of DypB from *Rhodococcus jostii* RHA1 as a lignin peroxidase. Biochemistry. 2011;**50**(23): 5096-5107. DOI: https://doi.org/10.1021/bi101892z 21534568

[78] Kellner H, Luis P, Zimdars B, Kiesel B, Buscot F. Diversity of bacterial laccase-like multicopper oxidase genes in forest and grassland Cambisol soil samples. Soil Biology and Biochemistry. 2008;**40**(3):638-648. DOI: https://doi.org/10.1016/j.soilbio.2007.09.013

[79] Alexandre G, Zhulin IB. Laccases are widespread in bacteria. Trends in Biotechnology. 2000;**18**(2):41-42. DOI: https://doi.org/10.1016/S0167-7799(99)01406-7

[80] Fernandes TAR, Silveira WBd, Passos FML, Zucchi TD. Laccases from actinobacteria—What we have and what to expect. Archives of Iranian Medicine. 2014;**04**(06):285-296. DOI: https://doi.org/10.4236/aim.2014.46035

[81] Lambertz C, Ece S, Fischer R, Commandeur U. Progress and obstacles in the production and application of recombinant lignin-degrading peroxidases. Bioengineered. 2016;**7**(3): 145-154. DOI: https://doi.org/10.1080/21655979.2016.1191705 27295524

[82] Ihssen J, Reiss R, Luchsinger R, Thöny-Meyer L, Richter M. Biochemical properties and yields of diverse bacterial laccase-like multicopper oxidases expressed in *Escherichia coli*. Scientific Reports. 2015;**5**:10465. DOI: https://doi.org/10.1038/srep10465 26068013

[83] Sainsbury PD, Hardiman EM, Ahmad M, et al. Breaking down lignin to high-value chemicals: The conversion of lignocellulose to vanillin in a gene deletion mutant of *Rhodococcus jostii* RHA1. ACS Chemical Biology. 2013;**8**(10):2151-2156. DOI: https://doi.org/10.1021/cb400505a 23898824

[84] Giles RL, Zackeru JC, Elliott GD, Parrow MW. Fungal growth necessary but not sufficient for effective biopulping of wood for lignocellulosic ethanol applications. International Biodeterioration & Biodegradation. 2012;**67**:1-7. DOI: https://doi.org/10.1016/j.ibiod.2011.11.007

[85] Eriksson K-E. Biotechnology in the pulp and paper industry. Wood Science and Technology 1990;**24**(1):79-101. [https://doi.org/10.1007/BF00225309]

[86] Akhtar M, Scott GM, Aziz A, Lentz MJ, Horn EG, Inventors. Biopulping Industrial Wood Waste. US 6,402,887 B1. Nov 30, 1998

[87] Singh P, Sulaiman O, Hashim R, Rupani PF, Peng LC. Biopulping of lignocellulosic material using different fungal species: A review. Reviews in Environmental Science and Biotechnology. 2010;**9**(2):141-151. DOI: https://doi.org/10.1007/s11157-010-9200-0

[88] Breen A, Singleton FL. Fungi in lignocellulose breakdown and biopulping. Current Opinion in Biotechnology. 1999;**10**(3):252-258. DOI: https://doi.org/10.1016/S0958-1669(99)80044-5

[89] Ferraz A, Guerra A, Mendonça R, et al. Technological advances and mechanistic basis for fungal biopulping. Enzyme and Microbial Technology. 2008;**43**(2):178-185. DOI: https://doi.org/10.1016/j.enzmictec.2007.10.002

[90] Arias ME, Arenas M, Rodriguez J, Soliveri J, Ball AS, Hernandez M. Kraft pulp biobleaching and mediated oxidation of a nonphenolic substrate by Laccase from *Streptomyces cyaneus* CECT 3335. Applied and Environmental Microbiology. 2003;**69**(4):1953-1958. DOI: https://doi.org/10.1128/AEM.69.4.1953-1958.2003

[91] Busse N, Kraume M, Czermak P. Modeling the design and operational mode of a continuous membrane reactor for enzymatic lignin modification. Biochemical Engineering Journal. 2017;**124**:88-98. DOI: https://doi.org/10.1016/j.bej.2017.04.007

[92] Busse N, Fuchs F, Kraume M, Czermak P. Treatment of enzyme-initiated delignification reaction mixtures with ceramic ultrafiltration membranes: Experimental investigations and modeling approach. Separation Science and Technology. 2016;**76**(3):1-20. DOI: https://doi.org/10.1080/01496395.2016.1167739

[93] Taboada-Puig R, Lu-Chau TA, Eibes G, Feijoo G, Moreira MT, Lema JM. Continuous removal of endocrine disruptors by versatile peroxidase using a two-stage system. Biotechnology Progress. 2015;**31**(4):908-916. DOI: https://doi.org/10.1002/btpr.2116 26033915

[94] Méndez-Hernández JE, Eibes G, Arca-Ramos A, et al. Continuous removal of nonylphenol by versatile peroxidase in a two-stage membrane bioreactor. Applied Biochemistry and Biotechnology. 2015;**175**(6):3038-3047. DOI: https://doi.org/10.1007/s12010-014-1474-1 255 88527

[95] Busse N. Reaction kinetics of versatile peroxidase for the degradation of lignin compounds. American Journal of Biochemistry and Biotechnology 2013;**9**(4):365-394. DOI: https://doi.org/10.3844/ajbbsp.2013.365.394

[96] Busse N, Kraume M, Czermak P. Optimal permeate flux for an enzymatic oxidation of technical lignins in a membrane reactor. Separation Science and Technology. 2016;**52**(2): 374-380. DOI: https://doi.org/10.1080/01496395.2016.1213748

[97] Peña N, Gallego S, del Vigo F, Chesters SP. Evaluating impact of fouling on reverse osmosis membranes performance. Desalination and Water Treatment. 2013;**51**(4–6): 958-968. DOI: https://doi.org/10.1080/19443994.2012.699509

[98] Ebrahimi M, Busse N, Kerker S, Schmitz O, Hilpert M, Czermak P. Treatment of the bleaching effluent from sulfite pulp production by ceramic membrane filtration. Membranes (Basel). 2015;**6**(1). DOI: https://doi.org/10.3390/membranes6010007 26729180

[99] Arkell A, Olsson J, Wallberg O. Process performance in lignin separation from softwood black liquor by membrane filtration. Chemical Engineering Research and Design. 2014;**92**(9):1792-1800. DOI: https://doi.org/10.1016/j.cherd.2013.12.018

[100] Ebrahimi M, Kerker S, Daume S, et al. Innovative ceramic hollow fiber membranes for recycling/reuse of oilfield produced water. Desalination and Water Treatment. 2014;**55**(13): 3554-3567. DOI: https://doi.org/10.1080/19443994.2014.947780

[101] Gupta VK. Microbial Enzymes in Bioconversions of Biomass. Cham: Springer International Publishing; 2016

[102] Ayala M, Pickard MA, Vazquez-Duhalt R. Fungal enzymes for environmental purposes, a molecular biology challenge. Journal of Molecular Microbiology and Biotechnology. 2008;**15**(2–3):172-180. DOI: https://doi.org/10.1159/000121328 18685269

[103] Casella L, Monzani E, Nicolis S. Potential applications of peroxidases in the fine chemical industries. In: Torres E, Ayala M, editors. Biocatalysis Based on Heme Peroxidases: Peroxidases as Potential Industrial Biocatalysts. Berlin, Heidelberg: Imprint Springer; 2010. pp. 111-153

[104] Conesa A, van den Hondel CAMJJ, Punt PJ. Studies on the production of fungal peroxidases in Aspergillus niger. Applied and Environmental Microbiology. 2000;**66**(7):3016–3023. DOI: https://doi.org/10.1128/AEM.66.7.3016-3023.2000

[105] de Weert S, Lokman BC. Heterologous expression of peroxidases. In: Torres E, Ayala M, editors. Biocatalysis Based on Heme Peroxidases: Peroxidases as Potential Industrial Biocatalysts. Berlin, Heidelberg: Imprint Springer; 2010. pp. 315-333

[106] Andreozzi R. Advanced oxidation processes (AOP) for water purification and recovery. Catalysis Today. 1999;**53**(1):51-59. DOI: https://doi.org/10.1016/S0920-5861(99)00102-9

[107] Ross F, Ross AB. Selected Specific Rates of Reactions of Transients from Water in Aqueous Solution. Gaithersburg, MD: National Bureau of Standards; 1977

[108] Hoigné J, Bader H. Rate constants of reactions of ozone with organic and inorganic compounds in water—II. Water Research. 1983;**17**(2):185-194. DOI: https://doi.org/10.1016/0043-1354(83)90099-4

[109] Linsebigler AL, Lu G, Yates JT. Photocatalysis on TiO_2 surfaces: Principles, mechanisms, and selected results. Chemical Reviews. 1995;**95**(3):735-758. DOI: https://doi.org/10.1021/cr00035a013

[110] Nosaka Y, Fox MA. Kinetics for electron transfer from laser-pulse irradiated colloidal semiconductors to adsorbed methylviologen: Dependence of the quantum yield on incident pulse width. The Journal of Physical Chemistry. 1988;**92**(7):1893-1897. DOI: https://doi.org/10.1021/j100318a039

[111] Rajeshwar K. Photoelectrochemistry and the environment. Journal of Applied Electrochemistry. 1995;**25**(12):1067-1082. [https://doi.org/10.1007/BF00242533]

[112] Hashimoto K, Irie H, Fujishima A. TiO_2 Photocatalysis: A historical overview and future prospects. Japanese Journal of Applied Physics. 2005;**44**(12):8269-8285. DOI: https://doi.org/10.1143/JJAP.44.8269

[113] Fujishima A, Zhang X, Tryk D. TiO_2 photocatalysis and related surface phenomena. Surface Science Reports. 2008;**63**(12):515-582. DOI: https://doi.org/10.1016/j.surfrep.2008. 10.001

[114] Tang H, Prasad K, Sanjinès R, Schmid PE, Lévy F. Electrical and optical properties of TiO_2 anatase thin films. Journal of Applied Physics. 1994;**75**(4):2042-2047. DOI: https://doi.org/10.1063/1.356306

[115] Lekelefac CA, Czermak P. Degradation of lignin derivatives by Photocatalysts. In: Cao W, editor. Semiconductor Photocatalysis – Materials, Mechanisms and Applications. Rijeka, Croatia: InTech; 2016

[116] Kansal SK, Singh M, Sud D. Studies on TiO(2)/ZnO photocatalysed degradation of lignin. Journal of Hazardous Materials. 2008;**153**(1–2):412-417. DOI: https://doi.org/10.1016/j.jhazmat.2007.08.091 17936502

[117] Ohnishi H, Matsumura M, Tsubomura H, Iwasaki M. Bleaching of lignin solution by a photocatalyzed reaction on semiconductor photocatalysts. Industrial and Engineering Chemistry Research. 1989;**28**(6):719-724. DOI: https://doi.org/10.1021/ie00090a012

[118] Serpone N. Relative photonic efficiencies and quantum yields in heterogeneous Photocatalysis. Journal of Advanced Oxidation Technologies. 1997;**2**(1):1-12. [https://doi.org/10.1515/jaots-1997-0104]

[119] Hoffmann MR, Martin ST, Choi W, Bahnemann DW. Environmental applications of semiconductor photocatalysis. Chemical Reviews. 1995;**95**(1):69-96. DOI: https://doi.org/10.1021/cr00033a004

[120] Rothenberger G, Moser J, Graetzel M, Serpone N, Sharma DK. Charge carrier trapping and recombination dynamics in small semiconductor particles. Journal of the American Chemical Society. 1985;**107**(26):8054-8059. DOI: https://doi.org/10.1021/ja00312a043

[121] Kobayakawa K, Sato Y, Nakamura S, Fujishima A. Photodecomposition of kraft lignin catalyzed by titanium dioxide. Bulletin of the Chemical Society of Japan. 1989;**62**(11):3433-3436. DOI: https://doi.org/10.1246/bcsj.62.3433

[122] Cristina Yeber M, Rodríguez J, Freer J, Durán ND, Mansilla H. Photocatalytic degradation of cellulose bleaching effluent by supported TiO_2 and ZnO. Chemosphere 2000;**41**(8): 1193-1197. DOI: https://doi.org/10.1016/S0045-6535(99)00551-2

[123] Peralta-Zamora P, de Moraes SG Pelegrini R, et al. Evaluation of ZnO, TiO_2 and supported ZnO on the photoassisted remediation of black liquor, cellulose and textile mill effluents. Chemosphere. 1998;**36**(9):2119-2133. DOI: https://doi.org/10.1016/S0045-6535(97)10074-1

[124] Stephen Davidson R, Dunn LA, Castellan A, Nourmamode A. A study of the photobleaching and photoyellowing of paper containing lignin using fluorescence spectroscopy. Journal of Photochemistry and Photobiology A: Chemistry. 1991;**58**(3):349-359. DOI: https://doi.org/10.1016/1010-6030(91)87054-Y

[125] Gierer J. Formation and involvement of superoxide (O_{2-}/HO_2) and hydroxyl (OH·) radicals in TCF bleaching processes: A review. Holzforschung. 1997;**51**(1):34-46. DOI: https://doi.org/10.1515/hfsg.1997.51.1.34

[126] Catalkaya EC, Kargi F. Color, TOC and AOX removals from pulp mill effluent by advanced oxidation processes: A comparative study. Journal of Hazardous Materials. 2007;**139**(2):244-253. DOI: https://doi.org/10.1016/j.jhazmat.2006.06.023 16839682

[127] Jokela JK, Laine M, Ek M, Salkinoja-Salonen M. Effect of biological treatment on halogenated organics in bleached kraft pulp mill effluents studied by molecular weight distribution analysis. Environmental Science & Technology. 1993;**27**(3):547-557. DOI: https://doi.org/10.1021/es00040a014

[128] Balcioglu I, Arslan I. Application of photocatalytic oxidation treatment to pretreated and raw effluents from the Kraft bleaching process and textile industry. Environmental Pollution. 1998;**103**(2–3):261-268. DOI: https://doi.org/10.1016/S0269-7491(98)00112-2

[129] Tanaka K, Calanag R, Hisanaga T. Photocatalyzed degradation of lignin on TiO_2. Journal of Molecular Catalysis A: Chemical. 1999;**138**(2–3):287-294. DOI: https://doi.org/10.1016/S1381-1169(98)00161-7

[130] Turchi CS, Ollis DF. Comment. photocatalytic reactor design: An example of mass-transfer limitations with an immobilized catalyst. The Journal of Physical Chemistry. 1988;**92**(23):6852-6853. DOI: https://doi.org/10.1021/j100334a070

[131] Awungacha Lekelefac C, Czermak P, Herrenbauer M. Evaluation of Photocatalytic active coatings on sintered glass tubes by methylene blue. International Journal of Photoenergy. 2013;**2013**(4):1-9. DOI: https://doi.org/10.1155/2013/614567

[132] Miyata Y, Miyazaki K, Miura M, Shimotori Y, Aoyama M, Nakatani H. Solventless delignification of wood flour with TiO_2/poly(ethylene oxide) photocatalyst system. Journal of Polymers and the Environment. 2013;**21**(1):115-121. DOI: https://doi.org/10.1007/s10924-012-0465-y

[133] Huber GW, Iborra S, Corma A. Synthesis of transportation fuels from biomass: Chemistry, catalysts, and engineering. Chemical Reviews. 2006;**106**(9):4044-4098. DOI: https://doi.org/10.1021/cr068360d

[134] Ferdous D, Dalai AK, Bej SK, Thring RW. Pyrolysis of lignins: Experimental and kinetics studies. Energy & Fuels. 2002;**16**(6):1405-1412. DOI: https://doi.org/10.1021/ef0200323

[135] Bridgwater AV. Principles and practice of biomass fast pyrolysis processes for liquids. Journal of Analytical and Applied Pyrolysis. 1999;**51**(1–2):3-22. DOI: https://doi.org/10.1016/S0165-2370(99)00005-4

[136] Gardner DJ, Schultz TP, McGinnis GD. The pyrolytic behavior of selected lignin preparations. Journal of Wood Chemistry and Technology. 2007;**5**(1):85-110. DOI: https://doi.org/10.1080/02773818508085182

[137] Piskorz J, Majerski P, Radlein D, Scott DS. Conversion of lignins to hydrocarbon fuels. Energy & Fuels. 1989;**3**(6):723-726. DOI: https://doi.org/10.1021/ef00018a011

[138] Meier D, Ante R, Faix O. Catalytic hydropyrolysis of lignin: Influence of reaction conditions on the formation and composition of liquid products. Bioresource Technology. 1992;**40**(2):171-177. DOI: https://doi.org/10.1016/0960-8524(92)90205-C

[139] Thring RW, Chornet E, Overend RP. Thermolysis of glycol lignin in the presence of tetralin. Canadian Journal of Chemical Engineering. 1993;**71**(1):107-115. DOI: https://doi.org/10.1002/cjce.5450710115

[140] Kudsy M, Kumazawa H, Sada E. Pyrolysis of kraft lignin in molten ZNCL 2 -KCL media with tetralin vapor addition. Canadian Journal of Chemical Engineering. 1995;**73**(3):411-415. DOI: https://doi.org/10.1002/cjce.5450730319

[141] Song Q, Wang F, Cai J, et al. Lignin depolymerization (LDP) in alcohol over nickel-based catalysts via a fragmentation–hydrogenolysis process. Energy & Environmental Science. 2013;**6**(3):994. DOI: https://doi.org/10.1039/c2ee23741e

[142] Connors WJ, Johanson LN, Sarkanen KV, Winslow P. Thermal degradation of kraft lignin in tetralin. Holzforschung. 1980;**34**(1):29-37. DOI: https://doi.org/10.1515/hfsg.1980.34.1.29

[143] Anderson EM, Katahira R, Reed M, et al. Reductive catalytic fractionation of corn Stover lignin. ACS sustainable. Chemical Engineer. 2016;**4**(12):6940-6950. DOI: https://doi.org/10.1021/acssuschemeng.6b01858

[144] Pinto PCR, Costa CE, Rodrigues AE. Oxidation of lignin from *Eucalyptus globulus* pulping liquors to produce syringaldehyde and vanillin. Industrial and Engineering Chemistry Research. 2013;**52**(12):4421-4428. DOI: https://doi.org/10.1021/ie303349j

[145] Araújo JD, Grande CA, Rodrigues AE. Vanillin production from lignin oxidation in a batch reactor. Chemical Engineering Research and Design. 2010;**88**(8):1024-1032. DOI: https://doi.org/10.1016/j.cherd.2010.01.021

[146] Villar JC, Caperos A, García-Ochoa F. Oxidation of hardwood kraft-lignin to phenolic derivatives with oxygen as oxidant. Wood Science and Technology. 2001;**35**(3):245-255. DOI: https://doi.org/10.1007/s002260100089

[147] Bjørsvik H-R, Norman K. Fine chemicals from Lignosulfonates. 2. Synthesis of Veratric acid from Acetovanillon. Organic Process Research and Development. 1999;**3**(5):341-346. DOI: https://doi.org/10.1021/op9900030

[148] Humpert D, Ebrahimi M, Czermak P. Membrane Technology for the Recovery of lignin: A review. Membranes (Basel). 2016;**6**(3). DOI: https://doi.org/10.3390/membranes6030042 27608047

[149] Žabková M, da Silva EB, Rodrigues AE. Recovery of vanillin from lignin/vanillin mixture by using tubular ceramic ultrafiltration membranes. Journal of Membrane Science 2007;**301**(1–2):221-237. DOI: https://doi.org/10.1016/j.memsci.2007.06.025

Chapter 3

Compositional Variability of Lignin in Biomass

Ana Lourenço and Helena Pereira

Additional information is available at the end of the chapter

http://dx.doi.org/10.5772/intechopen.71208

Abstract

The objective of this chapter is to provide a concise overview of lignin composition and structure in different species and materials (wood, barks and nonwood plants). It includes a brief review on the lignin precursors and their polymerization as well as of the analytical tools used for lignin characterization from wet chemical to spectroscopic methods. Wood of gymnosperms is characterized by high lignin content (25–35%) and a HG-type of lignin with more guaiacyl (G) units and a small portion of *p*-hydroxyphenyl (H) units. Wood of angiosperms has a lignin content of 15–28%, with a GS-lignin having different proportions of syringyl (S) units. Nonwoody monocotyledon species have different lignin content (9–20%) and a HGS type of lignin, characterized by a high proportion of H units. Bark lignin content ranges from 13 to 43% and is of HGS-type with species-specific composition and different in the bark components, phloem and cork. Lignin composition and macromolecular structure are key issues to understand the properties of lignocellulosic materials and to design a lignin-based pathway within biomass biorefineries. The available information on lignin composition is still limited to a few species and plant components. This is certainly an area where more research is needed.

Keywords: analytical tools, biomass, lignin composition, monolignols, S/G ratio

1. Introduction

Lignin has been the subject of continuous and intensive research over the last century. The Web of Science shows that since 1908 more than 27,000 publications were published with the topic "lignin," including articles, reviews, book chapters, notes and proceedings, under different subject areas, e.g., plant science, biotechnology, applied microbiology, chemistry, wood, pulp & paper, materials, energy and fuels.

Lignin is the second most abundant biopolymer in nature and accounts for almost 30% of the plants [1]. Its deposition in the cell wall is of great importance for plant development: (i) it provides rigidity and strength to the cell wall, giving mechanical support for the plant organs; (ii) it presents hydrophobicity favoring the transport of water and solutes in the vascular system and (iii) it protects the cell against pathogens [1–5]. Lignin is linked to the other structural components of the cell wall—cellulose and hemicelluloses—by covalent linkages, forming lignin-carbohydrate complex (LCC) [6].

The first lignin studies were mainly driven by its importance for the pulp & paper industry, where the objective is to remove lignin from the wood cells to obtain a fibrous product rich in cellulose. Thus, the studies on lignin, whether related to content, composition or structure, were focused on pulpwoods [7–11]. Lignin was also studied in herbaceous plants [12–16] partly triggered by digestability and dietary conversion issues in animal feed [2]. More recently, other wood species as well as various lignocellulosic residues and wastes attracted attention within the biorefinery concept providing opportunities for production of green chemicals, bioproducts and energy, calling for the need to include lignin valorization.

The lignin content shows a large variability between species: in general, in monocotyledons, it ranges between 5 and 12%, in softwoods between 25 and 35% and in hardwoods between 15 and 30%. The structural arrangement of lignin also differs between these three groups. This chapter makes a review on the compositional variability of lignin in various species and biomass components after an introductory compilation of the macromolecular assembly and the analytical tools used in lignin research.

2. The molecular construction of lignin

2.1. Precursors and monomers

Lignin is a heterogeneous aromatic polymer mainly constituted by three precursors: *p*-coumaryl alcohol, coniferyl alcohol and sinapyl alcohol (**Figure 1**). These precursors are often mentioned as phenylpropane or C9 units, where the hydroxyl group is linked to the C4 and substitutions with one or two methoxyl groups may be present at the C3 and C5. Therefore, the aromatic ring of the three alcohols is called *p*-hydroxyphenyl (H), guaiacyl (G) and syringyl (S) if the ring is unmethoxylated, or has one or two methoxyl groups, respectively. The side-chain carbons are designated as α-, β- and γ-, with Cα attached to the aromatic C1.

Researchers have recently recognized that lignin polymerization may involve other monomers, such as hydroxycinnamic acids and aldehydes, coniferyl and sinapyl acetates and coumarates [17–19].

The deposition of lignin and cellulose in the cell wall proceeds in three phases and starts after the deposition of pectins and the formation of the secondary wall S1 layer has begun [3, 20]. The first phase starts by the lignification at cell corners and middle lamella; the second phase corresponds mainly to the deposition in the S2 layer of cellulose in microfibrils and of xylan and mannan, with lignin being slowly added; in the third phase, lignin deposition proceeds extensively across the cell wall after the deposition of cellulose in the microfibrils of the S3 layer.

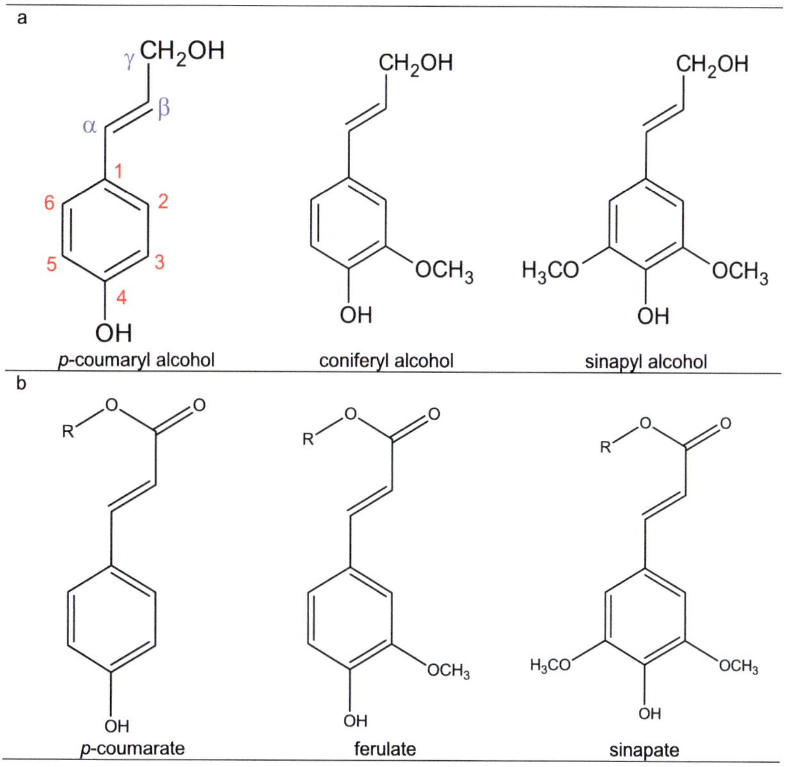

Figure 1. Chemical structures of the primary lignin precursors (a) and the hydroxycinnamates (b). R = H. Adapted from Ralph [19].

The composition of lignin changes during cell development. According to Terashima and Fukushima [20], in conifers, the middle lamella and cell corners are first enriched in *p*-hydroxyphenyl lignin, followed by the deposition of guaiacyl lignin in the middle lamella and in the secondary wall, while a small amount of syringyl lignin may be deposited later in the secondary wall. Under mechanical stress, gymnosperm trees form compression wood where the cells are characterized by a higher lignin content and an increased proportion of *p*-hydroxyphenyl lignin [2]. This differentiation of monolignol incorporation in the cell wall at different stages of xylem development and cell wall regions suggests a tightly regulated pathway for the lignin unit formation [21]. Lignin deposition and composition depend on the environmental conditions and are subject to modulation at different levels during normal development and in response to different stresses [1, 2]. It is still unclear whether lignin biosynthesis is explained by a single pathway in all species and tissues and under every environment conditions [2, 4].

Lignin heterogeneity may be related to enzyme diversity and specificity regarding substrates, thereby affecting the metabolite flux into diverse branches of the biosynthetic pathway [2]. A clear example is the difference in lignin type between softwoods and hardwood: softwood lignin is mainly constituted by G units and minor amounts of H units, whereas hardwood lignin has G and S units.

2.2. Polymerization and molecular assembly

The theory underlying lignification was presented by Freudenberg and Neish [22] based on chemical processes involving the oxidative coupling of phenols and addition of the available phenolic substrates to the polymer [23, 24]. The oxidation produces a phenolic radical with unpaired electron density delocalized at the C1, C3, C5 and O-4 positions of the aromatic ring and at the propanolic Cβ, forming resonance structures.

The lignin polymerization starts with the coupling of two monomeric radicals and continues by coupling of monomer radicals with phenoxy radicals formed on the growing polymer [25]. This concept explains some features of the lignin composition and structure, e.g., the evidences that other monomers such as coniferyl and sinapyl acetates and coumarates are also incorporated in the polymer [17–19]. Although a lignin polymerization model based on a protein-controlled radical coupling was proposed [26–29], the idea was not fully accepted by the scientific community given its flaws [30]. It was proved that lignification is malleable to plant needs and the polymer can be manipulated by changing the lignin-biosynthetic pathway genes [17, 31], and plants may incorporate other monomers into the lignin [32, 33].

The dehydrogenation of the lignin monomers is made by peroxidases (or peroxidase-H_2O_2 system) that are capable of removing a proton from the phenolic hydroxyl forming the resonance-stabilized free radicals, using the H_2O_2 produced by the peroxidase enzyme as an electron-acceptor substrate [34]. Laccase is a phenoloxidase also related to lignin biosynthesis [35].

After formation of the phenoxy radicals, the reaction is no longer controlled by enzymes but is a random radical polymerization process at the reactive sites [34]. The most reactive positions are the phenoxy oxygen and the Cβ that readily couple into aryl-ether linkages; the β-O-4′ linkage is predominant in lignin, e.g., almost 50% of all intermonomeric linkages in softwoods and 60% in hardwoods [36]. Overall, the coupling of the lignin monomers may be by ether bonds (β-O-4′, 4-O-5′, 1-O-4′) and by carbon-carbon bonds (5–5′, β-5′, β-β′, β-1′) is often called condensed bonds [25]. Some of these linkages are shown in **Figure 2**.

2.3. Analytical tools and lignin compositional indicators

Lignin quantification is usually made through wet chemistry by acid hydrolysis with sulfuric acid, using standard methods, e.g., TAPPI T222 om-11 and UM 205 om-83, respectively, for Klason lignin (obtained as a solid residue) and acid-soluble lignin (measured at 205 mm in the solution) that together make up the total lignin in the sample [37]. The procedure was optimized for wood and may lead to overestimation if the raw material is rich in ash and proteins or contains carbohydrate degradation products such as furfural and hydroxymethyl furfural. In spite of these shortcomings, most of the available data on lignin content of lignocellulosic materials refer to Klason lignin determinations and therefore establish a comparative reference. More recently, lignin content has been calculated from analytical pyrolysis or estimated using FTIR and NIR spectra modeling, as described subsequently.

As regards the study of lignin composition, the ideal would be to have an isolated pure lignin, e.g., recovered after removal of extractives, cellulose and hemicelluloses, without chemical

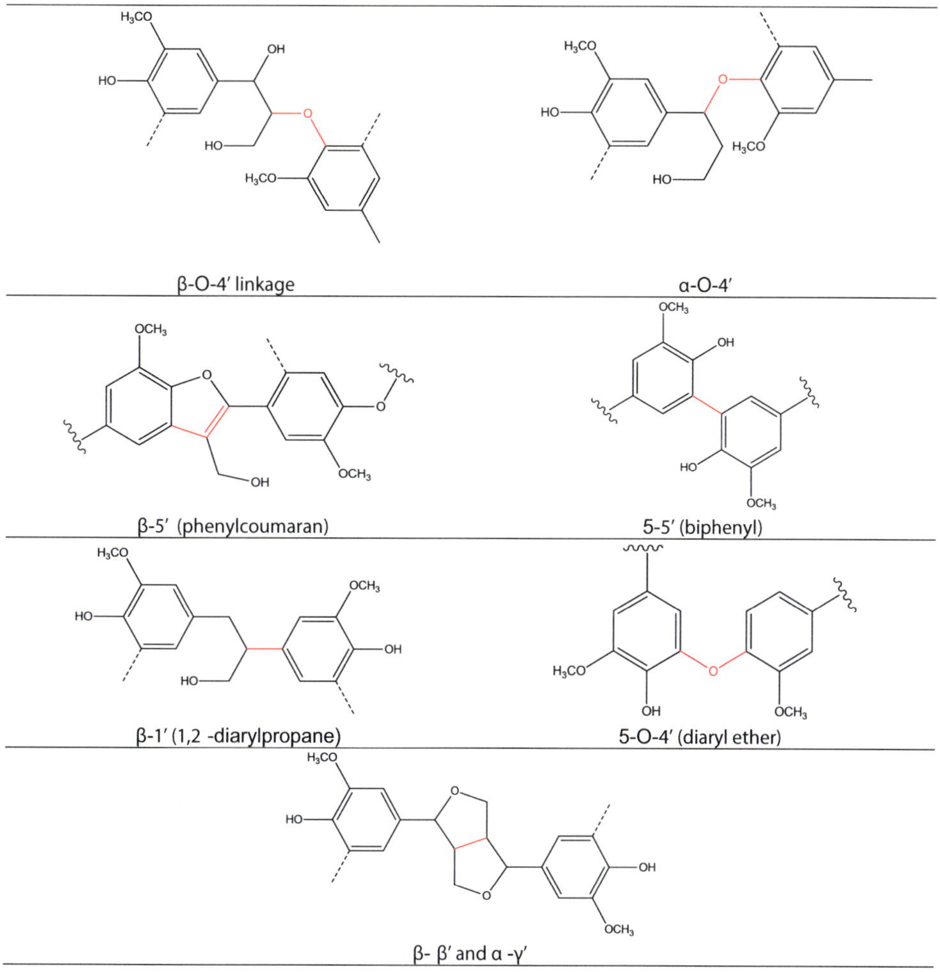

Figure 2. Linkages present in the lignin polymer. Adapted from Dimmel [25].

modification of the original lignin. This proved unfeasible although some procedures approach the requirements [38]. Lignin is frequently isolated by the classical Björkman method [39] and called milled wood lignin (MWL) if wood is the starting material. The lignin is obtained by milling the extracted sample in a planetary ball mill, followed by extraction with dioxane; after evaporation, the lignin is dissolved in acetic acid, precipitated into water, dried and dissolved in dichloroethane:ethanol solution and precipitated into ethyl ether [39]. MWL is still contaminated with carbohydrates [39] and represents only a part of the total cell wall lignin [40] whose structural features are correlated with the yield and, to a less degree, milling time [41]. Other procedures involving enzymes, e.g., cellulolytic enzymes, to remove the carbohydrates are used to increase the lignin yield [42].

Much of the present understanding of the composition and structure of lignin is based upon interpretation and extrapolation of data obtained from chemical degradative methods that include thioacidolysis, nitrobenzene oxidation, derivatization followed by reductive cleavage and analytical pyrolysis. Nondestructive methods, such as Fourier transform infrared (FTIR), near-infrared spectroscopy (NIRS) and nuclear magnetic resonance (NMR), are also used for lignin characterization.

2.3.1. Thioacidolysis

Thioacidolysis involves the solvolysis of the extractive-free material in dioxane/ethanediol (9/1) containing boron trifluoride etherate [43, 44]. The reaction depolymerizes part of the lignin and the released monomers can be analyzed, e.g., by GC-MS allowing to estimate the amount and composition of uncondensed alkyl ether structures [43]. Most studies were focused on wood but thioacidolysis was also applied to other materials, e.g., cork [45]. Calculation of the S/G ratio is usually made.

2.3.2. Nitrobenzene oxidation (NO)

Alkaline nitrobenzene oxidation leads to formation of aromatic aldehydes (*p*-hydroxybenzaldehyde, vanillin, syringaldehyde) as the major compounds, with benzoic acids (*p*-hydroxybenzoic, vanillic and syringic acids) and dehydrodivanillin or acetoguaiacone in minor amounts [46]. The yield of the compounds is related to the lignin structure in the raw material: vanillin is the major product from softwoods, syringaldehyde and vanillin are produced in great amounts from hardwoods and *p*-hydroxybenzaldehyde is produced from grasses [46, 47]. Nitrobenzene oxidation provides a satisfactory yield of *p*-hydroxybenzaldehyde, vanillin and syringaldehyde from the lignin H, G and S units [48]. The calculation of the S/G ratio is based on the obtained syringaldehyde and vanillin [49].

2.3.3. Derivatization followed by reductive cleavage (DFRC)

DFRC provides information on the occurrence of acylated γ-OH units [50–53]. The derivatization is made with acetyl bromide in acetic acid (called DFRC) or propionyl bromide in propionic acid (DFRC modified) at 50°C. The products are dissolved in dioxane/propionic acid/water (5:4:1, v/v/v), with the addition of zinc for the reductive cleavage, and the final derivatization step made with acetic anhydride or propionic anhydride depending of the method chosen (DFRC or DFRC modified) [53]. The lignin degradation compounds are collected after evaporation and analyzed by GC/MS.

2.3.4. Analytical pyrolysis

Pyrolysis transforms a nonvolatile compound into a volatile degradation mixture by heat in the absence of oxygen and by the breaking of chemical bonds using thermal energy [54, 55]. The ground biomass sample (particle sizes from mm to μm) is heated at temperatures from 400 to 1000°C, in the absence of oxygen, for 300 s to less than 0.5 s, producing charcoal (solid), bio-oil (liquid) and fuel gas products [56, 57]. Analytical pyrolysis was perfected for the pyrolysis of small samples for analytical purposes: a mixture of volatile compounds derived from the three macromolecular constituents of biomass (cellulose, hemicelluloses and lignin) is obtained and separated through a capillary column, identified by mass spectrometry (MS) and quantified by a flame ionization detector (FID). The MS identification of pyrolysis products was mostly made by Faix et al. [58–61] and by Ralph and Hatfield [62]. Nowadays, quantification is also made by Py-GC/MS [63]. Pyrolysis is particularly interesting to characterize the lignin monomeric composition into the phenolic S, G and H precursor monomers. The ratio of monomers is given in the form of H:G:S or, more frequently, as the S/G ratio.

2.3.5. Fourier transform infrared (FTIR)

FTIR spectroscopy is a rapid technique for lignin characterization that permits determination of the lignin monomeric composition, of methoxy and carbonyl groups, and the calculation of the ratio of phenolic to aliphatic groups [64]. It is based on the interaction of infrared radiation (4000–500 nm wavelength) with the sample and the fact that each molecule absorbs energy characteristic from its specific intramolecular bonds. Some of the bands related to lignin are found at 1600 and 1500 cm^{-1} due to the aromatic skeleton vibration of the benzene ring, at 1300 and 1200 cm^{-1} related to syringyl and guaiacyl lignin units and the bands at 1716 and 1711 cm^{-1} are attributed to phenol esterification and the alcohol of the propanoic chain ($C\alpha$ and $C\gamma$) [65, 66]. The spectral acquisition data can be made in transmission mode (TR), attenuated total reflectance (ATR), diffuse reflectance (DR) and photoacoustic (PA). Most lignin studies have used FTIR-TR but FTIR-DR is suitable to study the wood surface oxidation and weathering [64]. Watkins et al. [66] used FTIR-ATR to characterize organosolv lignins. FTIR spectral data were also used to model lignin content in wood of, e.g., *Picea sitchensis* [67], *Picea abies* [68] and *Eucalyptus globulus* [69].

2.3.6. Near-infrared spectroscopy (NIRS)

NIRS is a fast nondestructive technique that requires minimal or no sample preparation. The NIR region spans the wavelength range 780–2500 nm (12,821–4000 cm^{-1}), in which absorption bands correspond mainly to overtones and combinations of fundamental molecular vibrations, especially stretching and bending [70, 71]. The NIR spectrum is a superposition of scatter and light absorbance signals [72] and consequently contains information specific to the molecular vibrational aspects and their physical environments. NIRS methods require multivariate calibration algorithms (PCA, SIMCA, PCR or PLSR) usually referred to as chemometric methods to model spectral response to chemical or physical properties of a calibration sample set [71]. Spectra pre-processing is used to eliminate or minimize variability not related to the investigated property [72] and include normalization, derivatives (usually first or second), multiplicative scatter correction (MSC), standard normal variate (SNV), de-trending (DT) or a combination thereof [70]. NIRS has been successfully applied to evaluate different features of lignocellulosic materials [73, 74]. For instance, it was used to estimate Klason lignin content in *E. camaldulensis* and *L. leucocephala* wood [75], and lignin composition in maritime pine wood [76], and in *E. camaldulensis* and *E. urophylla* wood [44] by screening the H/G and S/G ratios, respectively.

2.3.7. Two-dimensional nuclear magnetic resonance (2D NMR)

Nuclear magnetic resonance spectrometry (NMR) is based on the measurement of absorption of electromagnetic radiation in the radiofrequency region of 4–900 MHz [77]. The NMR parameters, chemical shifts (δ), coupling constants (J), relaxation times and signal intensities, are related to the electronic structure and chemical environments of nuclei involved in the resonance phenomenon [78]. NMR spectroscopy can be proton (^1H) NMR or carbon (^{13}C), and a two-dimensional (2D) ^{13}C-^1H correlation spectrum has also been used. The acquisition can be made by: i) homonuclear correlation spectroscopy: COZY, TOCSY meaning that the protons correlate with the other protons; and ii) heteronuclear correlation (^{13}C-^1H): HMQC or HSQC where each carbon correlates with its attached proton; HMQC-TOCSY or HSQC-TOCSY where a carbon correlates with the proton attached to him and to other protons in the same coupling network [79]. The 2D HSQC NMR spectra are analyzed in the aromatic and the aliphatic regions. The aromatic region contains the correlations of C2-H2 at δ_C/δ_H 100–150/6.0–9.0 and includes the hydroxycinnamates and cinnamyl alcohol end-groups [49]. The H:G:S relation is quantified by volume integration of the contours. Also, the presence of acylated structures in lignin is identified in this region (especially in Cγ). The aliphatic-oxygenated region (around δ_C/δ_H 50–90/2.5–6.0) provides information on the inter-unit bonds in lignin. Signals from the β-O-4′ alkyl-aryl ethers, phenylcoumarans, resinols and dibenzodioxocins are a few examples of signals that are visualized in this region. The presence of polysaccharides signals is also detected here [49]. Overall, NMR is a powerful tool for structural investigation, since it allows to accurately assess the chemical structures, functionalities and nature of chemical bonds in the lignin macromolecule [78, 79], including an accurate measurement of the S/G ratio.

3. Lignin composition in biomass

Lignin differs naturally in content and composition between biomass materials at various levels, e.g., between species, within species and between components (such as wood and bark), and is influenced by plant growth stage and environmental stress [2, 15, 62, 65, 80].

Most studies on lignin composition were focused on the economic important pulpwoods, such as the pine and spruce softwoods, and the eucalyptus and birch hardwoods. The S/G ratio has been the mostly evaluated lignin composition parameter due to its importance in the pulping reactions; it is considered a pulpwood quality trait and a selection parameter in breeding programmes. Lignin composition has also been investigated in herbaceous plants in relation with their use in animal nutrition. More recently, and with the growing interest of biomass as a feedstock for biorefineries, lignin has increasingly been investigated in different species and components, namely in barks.

Table 1 makes a synthesis of the available information on lignin composition in wood of gymnosperms and angiosperms, in biomass from monocotyledons and barks. **Figure 3** compares the monomeric composition of lignin (S/G) in wood and bark for the species for which this determination was made.

3.1. Wood of gymnosperms

Softwoods are gymnosperms (mostly conifers), generally needle-leaved evergreen trees, e.g., pines (*Pinus* spp.) and spruces (*Picea* spp.). The lignin content in softwoods varies between 25 and 35% and consists almost exclusively of guaiacyl units with a low proportion of *p*-hydroxyphenyl

	Total lignin (% o.d.)	H:G:S	S/G or H/G*	Linkages (%) β–O–4′, phenylcoumarans, resinols	Acylation (%)	Ref.
Gymnosperms						
Pice abies						
wood	25-29%	-	-	69%, 18%, 10%	no	[37, 68, 80, 93, 84]
Pinus pinaster						
wood	23-30%	-	0.041	-	-	[91–94]
bark	33%	20:80:0	0.25*	-	-	[167]
Pinus taeda						
wood	28%	-	0.01*	-	-	[95]
bark	33-44%	-	0.59*	-	-	[95, 166]
Pinus sylvestris						
wood	19-25%	-	0.048*	-	-	[96–98]
Angiosperms Dicotyledons						
E. globulus						
wood	15-28%	-	1.5-2.9	76%, 2%, 17%	no	[105–113]
younger trees	-	1:4:6	1.3	68%, 5%, 18%	-	[108]
older trees	-	1:10:39	3.8	69%, 1%, 19%	-	[108]
bark	-	1:29:23	0.80	-	-	[179]
Betula pendula						
wood	22-25%	-	3.0-7.0	-	-	[116, 117]
bark (cork)	14%	1:43:6	0.1			[176, 177]
Fagus sylvatica						
wood	20%	1:35:26	0.75	-	-	[118, 119]
Acacia mangium						
wood	28%	1:1:0.1	-	40%, 43%,	-	[122]
Quercus suber						
wood	24%	1:44:55	1.2	77%, 9%, 8%	-	[125]
phloem	38%	1:58:41	0.7	71%, 13%, 7%	-	[125]
bark (cork)	27%	1:43:7	0.1	68%, 20%, 4%	48%, at G-units	[125, 170, 173]
Tectona grandis	32-43%					
sapwood	32%	1:34:24;	0.7	-	-	[126]

	Total lignin (% o.d.)	H:G:S	S/G or H/G*	Linkages (%) β–O–4′, phenylcoumarans, resinols	Acylation (%)	Ref.
heartwood	32%	1:29:23	0.8	-	-	[126]
bark	22%	1:11:9	0.8	-	-	[126, 178]
Cynara cardunculus						
stalks	16-19%	0:58:42	0.7	70%, 14%, 7%	12%, acetates (S-units)	[131, 134–136]
depithed stalks	23-26%	1:6:8	1.3	-	-	[135]
pith	19-25%	1:4:9	2.1	-	-	[135]
Monocotyledons						
Arundo donax						
stalks	20%	1:61:38	0.62	79%, 8%, 10%	43%	[139, 142]
internode	16-22%	1:2:0.5	1.2	49%, -, -	-	[140, 143]
node	15-20%	1:2:0.6	1.1	-	-	[140, 143]
Miscanthus **x** *giganteus*						
stalks	13%	1:13:11	0.7	93%, -, -	46% acetate or *p*-coumarate	[144–146]
Triticum						
stalks	5-16%	1:15:5		75%, 11%, -	10%, acetates in G-units (12%), in S (1%)	[14, 15, 149, 154]
Musa acuminata						
rachis	11%	1:0.7:1	1.4	$0.32/C_6$, -, -	-	[159, 160]
leaf sheaths	13%	1:2:0.5	0.25	-	-	[159, 160]
floral stalks	11%	1:1.6:1	0.63	$0.12/C_6$, -, -	-	[159, 160]
petioles/midrib	18%	1:1.9:2.1	1.1	-	-	[159, 160]
leaf blades	24%	1:9.3:6.3	0.60	-	-	[159, 160]

Table 1. Lignin composition in wood of gymnosperms, angiosperms, in biomass from monocotyledons and barks.

units (HG-lignin) with a methoxyl content of 15–16% [81]. In the compression wood of softwoods, the lignin is enriched in condensed structures such as H units [4] and G units [82] depending on the species.

3.1.1. *Picea*

Picea abies (Norway spruce) is a fast-growing conifer that is mostly planted in the North of Europe for timber and pulp production. Spruce lignin ranges from 25.3 to 28.6% [37, 80] and was estimated as 28.5% using FTIR modeling [68]. It is composed predominantly of G units with minor amounts of H units [83, 84]. Wood and MWL were analyzed by 2D NMR: the main linkage was the β-O-4′ aryl ether bond (69 and 65% of the total intermonomeric bonds), followed by phenylcoumarans (18%), resinols (10 and 11%), dibenzodioxocins (3 and 5%), with no acylation [84]. Similar results were presented by Capanema et al. [85] who compared different structural models for spruce MWL. Lawoko et al. [86] showed the presence of covalent linkages between spruce lignin and carbohydrates. The potential of spruce lignin genetic modification was evaluated by Wadenbäck [87] who could reduce the lignin content and *p*-hydroxyphenyl fraction by 5 and 23%, respectively. The developmental lignification in Norway spruce was studied by Koutaniemi et al. [88, 89] who were able to identify a group of potential genes for the monolignol polymerization. Recently, the spectra of spruce MWL, black liquor and lignin models were studied by ^{13}C-NMR and it was found that the signals of some carbons can shift if using different solvents and ionization processes [90].

3.1.2. *Pinus*

Maritime pine (*Pinus pinaster*) is a fast-growing pine adapted to Mediterranean climates and used for construction and pulping. A total lignin content of 23% was reported by Esteves et al. [91] and a Klason lignin content between 25.8 and 35.3% by Alves et al. [92], whereas lignin content determined by Py-GC/FID was between 23.0 and 29.6% [92]. Pine lignin is a HG-lignin, with H/G ratio values from 0.041 to 0.113 [76, 93]. The prediction of lignin content and composition by NIRS was possible [76, 93]. Baptista et al. [94] using permanganate analysis showed that noncondensed structures predominate in MWL comparatively to lignins isolated from black liquors: the ratio condensed/noncondensed structures was 0.17 in MWL and increased to 1.06 in one of the isolated lignins. The NMR analysis revealed that the delignification induced the formation of new condensed structures (α-5, β-6 or 5–6) and arabinoglucoronoxylans were the hemicelluloses preferentially linked to the lignin.

Loblolly pine (*P. taeda*) has a total lignin content of 28%, and a H/G ratio of 0.01, i.e., guaiacyl units largely predominate over *p*-hydroxyphenyl units [95].

Lignin content of *Pinus sylvestris* ranges between 18.8 and 24.5% [96, 97]. The H/G ratio ranges from 0.042 to 0.048 (NIR) in trees with a Klason lignin content ranging from 26.9 and 27.8% [98]. The lignin deposition in early and latewood xylem cells was studied by Antanova et al. [99]: the lignification of earlywood cells occurs faster, decreasing the intensity towards the end of latewood cell differentiation.

3.1.3. *Pseudotsuga menziesii*

Douglas-fir wood is used for timber and pulping, mainly in North America and also in Europe. The lignin content can range from 19.7 to 32.8% [100, 101]. The MWL was characterized as a HG-type, presenting an total amount of β-O-4′ aryl ether bonds of approximately 1700 μmol/g and a total amount of phenolic hydroxyl groups of 1500 μmol/g, with 40% of condensed structures, and the average molecular weight was 7400 g/mol [42].

3.2. Wood of angiosperms

Hardwoods belong to angiosperms, typically broadleaf deciduous trees. Lignin in hardwoods is constituted mainly by guaiacyl and syringyl units (GS-lignin), with a methoxyl content of 21%, and the β-O-4′ as the most common linkage, with a proportion of 71% or higher of the intermonomeric linkages [81, 102, 103].

3.2.1. *Eucalyptus*

Eucalyptus globulus is a fast-growing hardwood species that is highly appreciated for pulping [104]. The lignin content determined by wet chemical analysis ranges from 15 to 28% [105–107]. Using the acetyl bromide method, Rodrigues et al. [69] obtained a lignin content of 23–34% (in extractive-free base), and a similar range by FTIR modeling. The monomeric composition analysis shows that it is a GS-lignin with minor amounts of H units. The monomeric composition varies with tree age (1-month, 18-months, 9 years) with H units decreasing and S units increasing: the H:G:S was 1:4:6 (youngest trees) and 1:10:39 (oldest trees) [108]. Sapwood and heartwood present a H:G:S relation of, respectively, 1:8:29 and 1:11:39 [109]. The S/G ratio ranged from 1.5 to 6.4 determined by analytical pyrolysis [109–112] and 2.8 by 2D NMR [84]. Eucalypt MWL was characterized by a predominance of β-O-4′ aryl ether bond (76%), followed by resinol structures (17%) and in minor amounts of phenylcoumarans (2%), spirodienones (2%) and cinnamyl alcohol end groups (3%) [84]. The presence of phenylcoumarans and cinnamyl end groups decreased with age [108]. Eucalypt lignin presents no acylation [84]. Evtuguin et al. [113] obtained similar results with eucalypt lignin isolated by a modified mild acidolysis: high abundance of β-O-4 (0.56/C6) structures, units linked by α-O-4 bonds (0.23/C6), low presence of phenylcoumaran structures (0.03/C6) and slightly higher amounts of β-β substructures (0.13/C6).

MWL from *E. nitens* and *E. grandis* revealed similar characteristics of those of *E. globulus* lignin: predominance of syringyl units, β-O-4′ aryl ether bonds (83 and 77%, respectively), resinols (9 and 10%) and phenylcoumarans (5 and 8%) [63].

Lignin content differs between eucalyptus species: Neiva et al. [114] reported values of 21.6% (*E. maculata*), 24.0% (*E. camaldulensis*), 24.8% (*E. globulus*), 26.0% (*E. ovata*), 26.6% (*E. sideroxylon* and *E. saligna*), 26.8% (*E. rudis* and *E. viminalis*), 27.1% (*E. botryoides*), 27.8% (*E. grandis*), 29.9% (*E. propinqua*) and 30.8% (*E. resinifera*). The lignin content in *E. urophylla* was 29.9% and presents a S/G of 2.4 [115].

3.2.2. *Betula pendula*

B. pendula (silver birch) is a deciduous tree used for timber and pulping. Total lignin content ranges from 22 to 25% [116, 117], and its monomeric composition showed that it is a GS-lignin as determined by thioacidolysis and by oxidation with copper with a S/G ratio ranging from 3 to 7 depending on the tissue, e.g., in lignified xylem, the ratio reached 7 [116].

3.2.3. *Fagus sylvatica*

The European beech (*F. sylvatica*) wood is one of the most important and wide spread trees in Europe with a wide range of uses from furniture, musical instruments and pulp. The lignin content in sapwood and heartwood was, respectively, 22.9 and 24.9% [118]. The monomeric composition showed that beech lignin is a HGS-type with a H:G:S of 1:35:26, corresponding to a S/G of 0.75 [119].

3.2.4. *Acacia*

Acacia species are fast-growing trees used for timber and pulp production. Total lignin in *A. melanoxylon* ranged from 21.0 to 28.2% [120, 121], whereas in *A. mangium* the value was 28% [122]. Overall, the lignin is of the HGS-type with proportions differing between species. Lignin from *A. mangium* presented a H:G:S relation of 1:16:16 (by NMR) and 1:21:12 (by permanganate oxidation), showing a very low content of H units, with a corresponding S/G of 0.98 and 0.56 [122]. The lignin presented a high degree of condensation and low content of β-O-4′ structures that are associated to the low reactivity during pulping [122]. *A. mearnsii, A. mangium, A. auriculiformis* and hybrids were studied by Nawawi et al. [123] to relate lignin chemical characteristics (content and proportion of aromatic ring types) with pulpability; woods with higher syringyl ratios were easier to delignifiy due to the higher reactivity of the β-O-4′ structures.

3.2.5. *Quercus*

Quercus suber (cork oak), native to southwestern Europe and northwestern Africa, has a great economic importance as a producer of cork, the raw material used for the world-known wine cork stoppers [124]. The cork oak wood has a lignin content of 23.6%, and its MWL is characterized by a H:G:S molar ratio of 1:44:55, a S/G ratio of 1.2 (from NMR data), predominance of alkyl-aryl ethers (β-O-4′) (77%), lower amounts of condensed linkages and is scarcely acetylated, mainly over S units [125].

3.2.6. *Tectona*

Tectona grandis (teak), a tropical hardwood that grows naturally in Southeast Asia and has been planted in other countries, is one of the commercially most valuable timber species. Lignin content ranges from 32 to 43% [126–128]. Teak lignin composition was studied in sapwood and heartwood by Lourenço et al. [126]: teak has a GS-lignin, with monomeric composition determined by Py-GC/MS(FID) of H:G:S 1:34:24 (sapwood) and 1:29:23 (heartwood) and S/G ratio of 0.71 and 0.81, respectively.

3.2.7. *Cynara cardunculus*

Cynara cardunculus (cardoon) is an example of an angiosperm plant that is nonwoody (i.e., it forms only primary xylem) and has high productivities in Mediterranean countries [129, 130]. Cardoon is traditionally used for cheese making, fodder and human food [129] and has been researched as a multipurpose resource, e.g., for pulping and energy [131–133]. Lignin content in the stalks varies from 16.4 to 19.2% [131, 134, 135]. The lignin monomeric composition of cardoon stalks separated in depithed and pith regions shows a H:G:S relation of 1:6:8 and 1:4:9, respectively, determined by Py-GC/MS (FID), corresponding to a S/G ratio of 1.3 and 2.2, respectively [136]. Whole stalks lignin was studied after isolation by the Björkman method and characterized by Py-GC/MS, 2D NMR and DFRC′ [136]. The 2D NMR analysis showed that the isolated lignin has a H:G:S relation of 0:58:42, i.e., a S/G ratio of 0.7 and a predominance of β-O-4′ alkyl-aryl ether structures (70% of all inter-unit linkages), followed by a considerable amount of condensed structures: phenylcoumarans (β-5′, 14%), resinols (β-β′, 7%), spirodienones (β-1′, 5%) and dibenzodioxocins (5-5′, 4%). The lignin was partially acylated (12%) at the carbon γ-OH by acetate groups that were preferentially attached over S units (32% of S units and only 1% of G units). Lourenço et al. [136] showed that sinapyl acetate acts as a real monolignol involved in the lignification of cardoon stalks.

3.3. Biomass of monocotyledons

Monocotyledons are angiosperm flowering plants with seeds typically containing only one cotyledon that include the families *Poacea* (grasses, bamboos and sugar cane), *Arecaceae* (palms) and *Musaceae* (bananas) with several species having a great economic importance. Lignin in monocotyledon species is structurally different from softwood and hardwood lignin: it is a HGS-lignin with a higher proportion of *p*-hydroxyphenyl units, a significant amount of hydroxycinnamate esters and some acylation by *p*-coumarates in the S units [4]. The structure and biogenesis of the cell walls of grasses were reviewed by Carpita [137], who concluded that in spite of the differences between grasses and other plants (e.g., in wall composition), the grass cells respond similarly to environmental signals and growth regulators. A review of herbaceous lignin was also published by Buranov and Mazza [15] describing the lignin in different straws (wheat, rice, flax and corn).

3.3.1. Arundo donax

Arundo donax (giant reed) is a perennial plant that grows in Mediterranean climates with high biomass production and potential for pulp, chemicals and energy [138]. The lignin content in the whole stalk is 21.1% [139] with 16–22% in the internodes and 15–20% in nodes [140]. The lignin is of HGS-type mainly constituted by guaiacyl and syringyl units with some *p*-hydroxyphenyl units [141]. The lignin in the whole stalk has a S/G ratio of 0.62 and a high content of β-O-4′ alkyl-aryl ether structures (79%) and less of other structures, such as resinol (9%), phenylcoumaran structures (8%) and spirodienone (3%), and is highly acylated in the Cγ (43%) [142]. In internodes and nodes, the H:G:S relation was 1:2:0.5 and 1:2:0.6, respectively, corresponding to S/G ratios of 0.25 and 0.3 [140]. Higher values of S/G ratio of 1.23 (internodes) and 1.13 (nodes) were also reported [143]. Internode lignin presented in relation to node lignin more β-O-4′ linkages (0.49 *vs.* 0.32 per aromatic unit), suggesting that internode lignin is of a less condensed nature [143].

3.3.2. Miscanthus

Miscanthus sinensis, M. sacchariflorus, M. tinctorius and the hybrid *Miscanthus x giganteus* (*M. sinensis x M. sacchariflorus*) show potential as energy crops [144]. Lignin content varies from 9.2% in *M. sinensis* to 12.6% in *M. x giganteus* [144]. *M. x giganteus* lignin (MWL) is of HGS-type with a relation of 1:13:11 [145], a S/G ratio of 0.7 and is highly acylated at the Cγ side chain (46%) with acetate or p-coumarate groups [146] and has a predominance of β-O-4′ linkages (93% of all linkages, Brosse et al. [144]. Lignins isolated by mild formosolv (AL), alkaline organosolv (BL) and cellulolytic enzymes (CL) showed differences: CL lignin contained more carbohydrates (12.8%), more β-O-4′ linkages (82%) and the lowest S/G ratio of 0.7 [147]. Organosolv lignins isolated using different ethanol concentrations (65–95%) showed less carbohydrate content (3.6–1.1%) and molecular weight (2.72–2.25 Ka) with the increase of ethanol concentration, but an increase of p-coumaric and ferulic acids while no effect was found in the S/G ratio (0.63) [148].

3.3.3. *Triticum*

Wheat is extensively cultivated for seed production, leaving the straw as a widely available residue with great potential for bioenergy, including bioethanol [15, 149]. The lignin content in wheat straw ranges from 5 to 16% [15, 149]. It is a HGS-type of lignin with a proportion of 1:11:5 [14] and 1:10:9 [15], corresponding to S/G ratios of 0.45 and 0.9, respectively. Milled straw lignin presents *p*-coumarates and ferulates, and the flavone tricin was also found to be incorporated into the polymer [14]. The *p*-coumaric acid is ester linked to lignin [150], whereas ferulic acid is linked by ether bonds (52–68%) and ester bonds (32–48%) forming ester-ether bridges in the lignin fragments [151, 152]. Ferulic acid is also ester linked to polysaccharides [15, 153]. The main structures in wheat straw lignin are β-O-4′-ethers (~75%), followed by phenylcoumarans (~11%) among others such as pinoresinol [14, 154]. The lignin is partially acylated (around 10%) at the Cγ, predominantly with acetates in the guaiacyl units (12%) and in minor amounts in syringyl units (1%) [14].

3.3.4. *Bamboo*

Phyllostachys pubescens is one of the bamboo species that is used for different purposes, as a material, e.g., flooring, furniture and mats, and as a fiber source. Its lignin content can vary from 14.6 to 18.3% (2-month-old bamboo) and 25.4 to 27.1% (3-year-old bamboo) [155]. Bamboo culm is divided into green, yellow, timber and pith. Milled wood lignins were isolated from green (MWLg) and yellow bamboo (MWLy) and the lignin found to be of a HGS-type with values of 1:9:7 (MWLg) and 1:9:10 (MWLy) and a S/G of, respectively, 0.74 and 1.16 [156]. The main substructures in MWLg and MWLy were β-O-4 alkyl-aryl ether (38.2 and 39.8%, respectively), resinol (6.9 and 6.3%), phenylcoumaran (3.8 and 2.9%), spirodienone (1.7 and 2.1%) and α,β-diaryl ether (0.4 and 0.3%). The lignin of green bamboo presented lower acylation degree (17.2 vs. 21.1%), and in both lignins, tricin was detected [156]. Wen et al. [157] studied MWL of *P. pubescens* of control and torrefied samples. Torrefication promoted depolymerization, demethoxylation, bond cleavage and condensation reactions. The torrefied samples at 275 and 275°C were enriched in lignin content but dramatically different from the original lignin with almost no β-O-4 linkages, resinol and phenylcoumaran that were not detected by 2D NMR. Also, the presence of H units increased; in the starting material, the H:G:S relation was 1:15:34 (S/G of 2.26) and in the torrefied sample at 300°C, the relation was 1:0.3:0.8 (S/G of 2.37). Shao et al. [158] treated bamboo by steam explosion and its MWL presented a reduction of the β-aryl ether linkages and a reduction of ester bonds between lignin and *p*-coumaric acid.

3.3.5. *Banana*

Musa acuminata Colla var. cavendish (Dwarf Cavendish) is one of the banana varieties under commercial production [159]. The banana plant has a HGS-type of lignin but its content and composition differ in the different morphological parts [160]. The lignin content varied from 10.5 to 24.3%: in rachis, 10.5%; floral stalks, 10.7%; leaf sheaths, 13.3%; petioles/midrib, 18.0% and leaf blades, 24.3%. By nitrobenzene oxidation, the lignin composition determined as follows: the G units predominated in leaf blades (with a H:G:S of 1:9.3:6.3), in floral stalk (1:1.6:1) and leaf sheaths (1:2:0.5), and the S/G ratios were 0.68, 0.63 and 0.25, respectively. The S units slightly predominated in petioles/midrib (1:1.9:2.1) and rachis (1:0.7:1.0) with S/G ratio values of 1.10 and 1.42. Oliveira et al. [159] further studied the lignin composition in floral stalks (LFS) and raquis (LR) after lignin isolation by dioxane method. LR presented significantly higher amount of β-O-4′ structures (0.32/C6 vs. 0.12/C6) and higher molecular weight (5400 Da vs. 3750 Da) comparatively to LFS. LFS had more condensed structures (~72%) such as β-5′ and 5-5′ types, whereas in LR the more abundant were the 4-O-5′-diaryl ether structures. Both lignin presented H units linked to coumarates by ester bonds, and both were structurally associated with suberin-like components.

3.4. Barks

Barks are complex and heterogeneous components of plants that include phloem and periderm and eventually rhytidome (periderms interspersed by phloem), as schematically represented in Sen et al. [161]. Phloem is produced by the cambium and the periderm by the phellogen [162]. Barks are a largely available residue from the timber and pulp industries mostly used for energy but increasingly considered as potential feedstocks for biorefineries given their chemical and structural diversity [163, 164]. Cork is one component of bark periderms that may attain considerable proportions in some species [165]. The cork from *Quercus suber* (cork oak) is the basis of an important industrial chain and therefore has been extensively studied (as reviewed by Pereira [124]). But overall the literature on barks is limited and little information is available on lignin composition and structural features except for a few cases.

3.4.1. *Pinus barks*

In *Pinus taeda*, bark lignin ranged from 32.9 to 43.5% [95, 166]. It is mainly constituted by guaiacyl and *p*-hydroxyphenyl units, with reported H/G ratio values of 0.59 (the value for wood lignin is 0.01) [95] and 0.28 [166]. Compared to wood lignin, bark lignin contains more condensed structures (5-5′ or β-5′), less dibenzodioxocin and β-O-4′ structures and fewer methoxyl groups [95].

P. pinaster bark has a Klason lignin content of 33.2% (% o.d. initial bark) determined after extraction of the polyphenolic tannin compounds; it is a HG-lignin composed of *p*-hydroxyphenyl and guaiacyl units with a H/G ratio of 0.25 [167].

3.4.2. *Quercus barks*

Quercus suber cork has a total lignin content ranging from 17.1 to 36.4% [168, 169]. The lignin monomeric composition of cork was assessed after isolation by the Björkman method by Marques [170–173] and Lourenço et al. [125]. The S/G ratio of cork was 0.029 determined by Py-GC/FID and 0.1 by NMR, revealing a strong predominance of guaiacyl units that was also confirmed by FTIR analysis [170]. Cork lignin in young trees was characterized by 2D NMR as a HGS-lignin constituted mainly by guaiacyl units, with a H:G:S of 1:43:7 and a S/G ratio of 0.16. The alkyl-aryl ethers (β-O-4′) were the predominant structures (68–77%), but it also contained condensed structures of phenylcoumarans (β-5′, 18–20%) and dibenzodioxocins (5-5′, 1–5%) [125, 173]. The cork lignin is partially acylated at the $C\gamma$ of the side chain (48–50%), mainly over G units [125, 173]. Ferulic acid was present in cork lignin (6%) and considered to participate in the crosslinking between lignin and suberin [173]. Lopes et al. [174] isolated the cork lignin by an organosolv protocol and characterized it by permanganate oxidation. The lignin presented a H:G:S relation of 1:47:3 with a S/G ratio of 0.05.

Lourenço et al. [125] also studied the lignin content and composition in the bark phloem after isolation by Björkman method. The lignin content was 38.4% (% o.d material), with a H:G:S ratio of 1:58:41 and a S/G ratio of 0.7 (from NMR). The lignin was characterized mainly by β-O-4′ alkyl-aryl ethers (71%), with low amounts of condensed linkages, and was scarcely acetylated, mainly over S units.

Cork from *Quercus cerris* bark has a lignin content of 28.1% [175], and the monomeric composition is mainly of G units, with a low proportion of S and H units, respectively, 93.7, 2.7 and 3.6%, that correspond to a S/G ratio of 0.03 [176].

3.4.3. *Betula pendula barks*

Cork from *Betula pendula* presents a lignin content of 14.3% [177], characterized predominantly by guaiacyl units (85.7%) with a minor proportion of syringyl units (11.9%) and *p*-hydroxyphenyl (2.4%), which correspond to a S/G of 0.14, determined by Py-GC-MS/FID [176].

3.4.4. *Tectona grandis barks*

Teak bark has a lignin content of 22.4% determined by wet chemistry [178]. It has a GS-lignin with a composition in H:G:S of 1:11:9, and a S/G ratio of 0.8 [126].

3.4.5. *Eucalyptus globulus barks*

Lignin from eucalypt bark was isolated by mild acidolysis and characterized by nitrobenzene oxidation (NO) and NMR (^{13}C and ^{31}P NMR) [179]. Bark lignin was of HGS-type, with a H:G:S relation of 1:6:18 (^{13}C NMR). The S/G ratio values differed between techniques, with the higher values attained by the nitrobenzene oxidation since it only quantifies the noncondensed structures: 1.5 (NMR), 1.5 (^{31}P NM), 3.17 (^{13}C NMR) and 5.9 (NO).

3.5. Variability of lignin composition at tissue and cell levels

Few studies have compared lignin composition in different biomass components of the same plant (**Figure 3**). For *Quercus suber*, lignin was characterized in wood, phloem and cork, revealing a distribution of H:G:S quite distinct. Wood and phloem were enriched in S-lignin with H:G:S values of 1:45:55 and 1:58:41, respectively, whereas cork was considerably enriched in G units and with minor proportions of H and S units (2:85:13) [125]. Similar differences were found in the proportions of interunit linkages: wood and phloem presented more alkyl-aryl ethers (β-O-4′) representing 77 and 71% in accordance to the predominance of S units in these tissues, whereas lower value was reported in cork (68%). Therefore, the main condensed structures were found in cork: 20% phenylcoumarans and 5% dibenzodioxocins.

Variation of lignin composition also occurs at cellular level. Lignin formation and composition are cell specific, e.g., lignin differs between tracheary elements, vessels and sclerenchyma cells, and presents a distinctive feature at subcellular localization [2, 180]. During the early phases of xylem lignification, the H units are incorporated in the cell and G units are present in the middle lamella and cell corners, whereas in the next phase, the lignification of the cell primary wall and outer layers of secondary wall is mainly by G units [181–183]. In Arabidopsis, xylem vessels have a predominance of H units and cell corners and middle lamella have a G-lignin, while the fibers are rich in S units [184]. In white birch, the vessels secondary wall has a G:S relation of 88:12, the fibers 12:88 and the ray parenchyma 49:51 [185]. In *Acer* species, Watanabe and Fukazawa [186] observed several patterns: in some, vessels and fibers were richer in S units, whereas in others, vessels and fibers were richer in G units; but it was also observed that vessels were richer in G units and fibers richer in S units. Saito et al. [187] confirmed this last observation in *A. micranthum*, as well as Wu et al. [188] when they analyzed 25 Chinese hardwoods species.

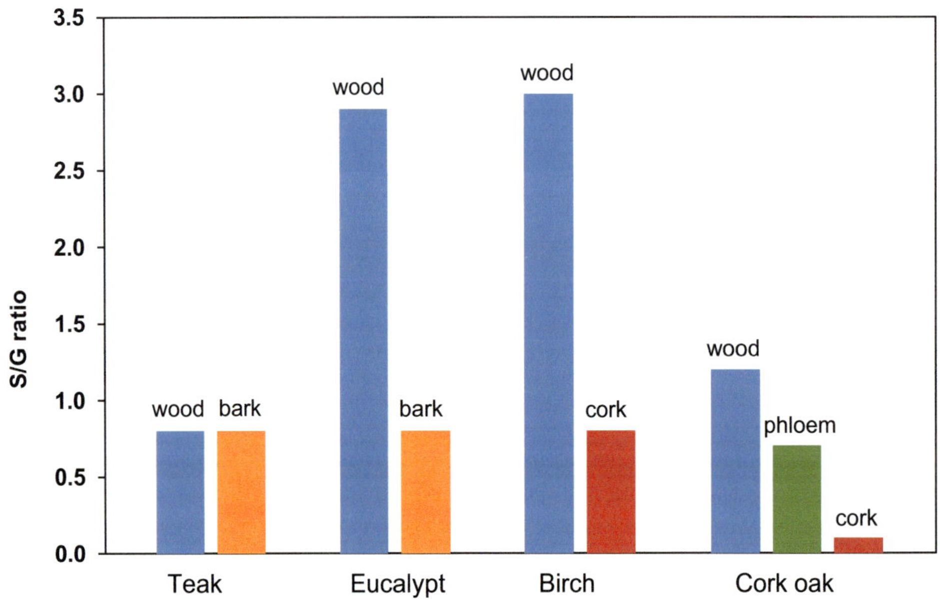

Figure 3. Variation of S/G in wood and barks of *Tectona grandis* (teak) and *Eucalyptus globulus* (eucalypt), wood and cork of *Betula pendula* (birch) and wood, phloem and cork of *Quercus suber* (cork oak).

4. Concluding remarks

This chapter provides an overview on lignin content and composition, showing its complexity as a polymer and variability between and within plant species. The lignin polymer is built with three monomers (*p*-coumaryl alcohol, coniferyl alcohol and sinapyl alcohol), which are known since the early lignin research studies, to which other monomers were more recently added (hydroxycinnamic acids and aldehydes, coniferyl and sinapyl acetates). However, the biosynthetic pathway of monomer formation and lignin construction are still open research areas seeking to explain the variability that is found as well as the flexibility in lignin construction.

Lignin diversity is well demonstrated when analyzing the available data on the lignin monomeric composition of various species and biomass components. The usual classification of lignin in three types—HG-lignin, GS-lignin and HGS-lignin—broadly assigned to softwoods, hardwoods and monocots, respectively, is a crude generalization and do not encompass the diversity found within each group. It is true that the knowledge on lignin composition and structure is still restricted to a limited number of species and plant components. This is clearly an area in which more research is needed. The various analytical tools that have been developed, including wet chemistry, spectroscopic, magnetic and pyrolytic methodologies allow a better insight into lignin structure and the possibility of making a much more extensive coverage of biomass materials.

Lignin plays an important role in plant cell walls providing support and protection, and it is the second most abundant polymer in nature after cellulose. Increased knowledge on lignin will therefore contribute to our understanding of plant physiology and adaptation, as well as support a lignin platform within future biorefineries providing combined valorization routes for chemicals, materials and energy.

Acknowledgements

The support for this work was provided by Fundação para a Ciência e a Tecnologia (FCT) through funding of the Forest Research Center (UID/AGR/00239/2013). Ana Lourenço acknowledges support from FCT through a postdoctoral grant (SFRH/BPD/95385/2013). A word of appreciation to Vanda Oliveira, Duarte M. Neiva and Jorge Gominho for their help.

Author details

Ana Lourenço* and Helena Pereira

*Address all correspondence to: analourenco@isa.ulisboa.pt

Forest Research Center, School of Agriculture, University of Lisbon, Lisboa, Portugal

References

[1] Boudet A. Lignins and lignification: Selected issues. Plant Physiology and Biochemistry. 2000;**38**:81-96. DOI: 10.1016/S0981-9428(00)00166-2

[2] Campbell MM, Sederoff RR. Variation in lignin content and composition. Mechanism of control and implications for the genetic improvements of plants. Plant Physiology. 1996; **110**:3-13

[3] Donaldson LA. Lignification and lignin topochemistry – An ultrastructural view. Phytochemistry. 2001;**57**:859-873. DOI: 10.1016/S0031-9422(01)00049-8

[4] Boerjan W, Ralph J, Baucher M. Lignin biosynthesis. Annual Review of Plant Biology. 2003;**54**:519-546. DOI: 10.1146/annurev.arplant.54.031902.134938

[5] Vanholme R, Morreel K, Ralph J, Boerjan W. Lignin engineering. Current Opinion in Plant Biology. 2008;**11**:278-285. DOI: 10.1016/j.pbi.2008.03.005

[6] Henriksson G. Lignin. In: Ek M, Gellerstedt G, Henriksson G, editors. Pulp and Paper Chemistry Technology. Wood Chemistry Wood Biotechnology. Vol. 1. Berlin: Walter de Gruyter GmbH Co; 2009. p. 121-145

[7] Rydholm SA. Pulping Process. Hertfordshire: Interscience Publishers, John Wiley Sons; 1965

[8] Biermann CJ. Essentials of Pulping and Papermaking. San Diego: Academic Press; 1983

[9] Biermann CJ. Handbook of Pulping and Papermaking. 2nd ed. San Diego: Academic Press Limited; 1996

[10] Gellersted G, Lindfors E. Structural changes in lignin during kraft pulping. Holzforschung. 1984;**38**:151-158. DOI: 10.1515/hfsg.1984.38.3.151

[11] Gierer J. Chemistry of delignification. Part 1: General concept and reactions during pulping. Wood Science and Technology. 1985;**19**:289-312. DOI: 10.1007/BF00350807

[12] del Río JC, Gutiérrez A, Rodríguez IM, Ibarra D, Martínez AT .Composition of non-woody plant lignins and cinnamic acids by Py-GC/MS, Py/TMAH and FT-IR. Journal of Analytical and Applied Pyrolysis 2007;**79**:39-46. doi:10.1016/j.jaap.2006.09.003

[13] del Río JC, Prinsen P, Rencoret J, Nieto L, Jiménez-Barbero J, Ralph J, Martínez AT, Gutiérrez A Structural characterization of the lignin in the cortex and pith of elephant grass (*Pennisetum purpureum*) stems. Journal of Agricultural and Food Chemistry 2012;**60**:3619-3634. DOI: 10.1021/jf300099g

[14] del Río JC, Rencoret J, Prinsen P, Martínez AT, Ralph J, Gutiérrez A Structural characterization of wheat straw lignin as revealed by analytical pyrolysis, 2D-NMR, and reductive cleavage methods. Journal of Agricultural and Food Chemistry 2012;**60**:5922-5935. DOI: 10.1021/jf301002n

[15] Buranov AU, Mazza G. Review: Lignin in straw of herbaceous crops. Industrial Crops and Products. 2008;**28**:237-259. DOI: 10.1016/j.indcrop.2008.03.008

[16] Marques G, Rencoret J, Gutiérrez A, del Río JC Evaluation of the chemical composition of different non-woody plant fibers used for pulp and paper manufacturing. The Open Agriculture Journal. 2010;**3**:93-101. DOI: 10.2174/1874331501004010093

[17] Ralph J, Lundquist K, Brunow G, Lu F, Kim H, Schatz PF, Marita JM, Hatfield RD, Ralph SA, Christensen JH, Boerjan W. Lignins: Natural polymers from oxidative coupling of 4-hydroxyphenylpropanoids. Phytochemistry Reviews. 2004;**3**:29-60

[18] Grabber JH, Schatz PF, Kim H, Lu F, Ralph J. Identifying new lignin bioengineering targets: 1. Monolignol-substitute impacts on lignin formation and cell wall fermentability. BMC Plant Biology. 2010;**10**:114-127. DOI: 10.1186/1471-2229-10-114

[19] Ralph J. Hydroxycinnamates in lignification. Phytochemistry Reviews. 2010;**9**(1):65-83. DOI: 10.1007/s11101-009-9141-9

[20] Terashima N, Fukushima K. Heterogeneity in formation of lignin – XI: An autoradiographic study of the heterogeneous formation and structure of pine lignin. Wood Science and Technology. 1988;**22**:259-270

[21] Dixon RA, Chen F, Guo D, Parvathi K. The biosynthesis of monolignols: A "metabolic grid", or independent pathways to guaiacyl and syringyl units? Phytochem. 2001;**57**:1069-1084. DOI: 10.1016/S0031-9422(01)00092-9

[22] Freudenberg K, Neish AC. Constitution and Biosynthesis of Lignin. Berlin, Germany: Springer-Verlag; 1968. p. 132. ISBN 978-3-540-04274-7

[23] Sarkanen KV, Ludwig CH, editors. Lignins, occurrence, formation, structure and reactions. California: Wiley-Interscience; 1971. p. 916. ISBN 0471754226

[24] Brunow G, Lundquist K, Gellerstedt G. Lignin. In: Sjöström E, Alén R, editors. Analytical Methods in Wood Chemistry, Pulping and Papermaking. Berlin, Germany: Springer-Verlag; 1999. p. 77-124

[25] Dimmel D. Overview. In: Heitner C, Dimmel DR, Schidt JA editors. Lignin and lignans. Advances in Chemistry. Taylor and Francis Group. New York, USA: CRC Press; 2010. p. 1-10

[26] Lewis NG, Davin LB. The biochemical control of monolignol coupling and structure during lignan and lignin biosynthesis. In: Lewis NG, Sarkanen S, editors. Lignin and Lignan Biosynthesis, Vol 697. American Chemistry Society Symposium Series. Washington, USA. 1998. p. 334-361

[27] Gang DR, Costa MA, Fujita M, Dinkova-Kostova AT, Wang HB, Burlat V, Martin W, Sarkanen S, Davin LB, Lewis NG. Regiochemical control of monolignol radical coupling: A new paradigm for lignin and lignin biosynthesis. Chemistry and Biology. 1999;**6**:143-151. DOI: 10.1016/S1074-5521(99)89006-1

[28] Davin LB, Lewis NG. Dirigent proteins and dirigent sites explain the mystery of specificity of radical precursor coupling in lignin and lignin biosynthesis. Plant Physiology. 2000;**123**:453-461. DOI: 10.1104/pp.123.2.453

[29] Davin LB, Lewis NG. Lignin primary structures and dirigent sites. Current Opinion in Biotechnology. 2005;**16**:407-415. DOI: 10.1016/j.copbio.2005.06.011

[30] Ralph J, Brunow G, Harris PJ, Dixon RA, Schatz PF, Boerjan W. Cap 2. Lignification: are lignins biosynthesized via simple combinatorial chemistry or via proteinaceous control and template replication? In: Daayf F, Lattanzio V, editors. Recent Advances in Polyphenol Research. Oxford, UK: Wiley-Blackwell; 2008. p. 36-99

[31] Sederoff RR, MacKay JJ, Ralph J, Hatfield RD. Unexpected variation in lignin. Current Opinion in Plant Biology. 1999;**2**:145-152

[32] Ralph J, MacKay JJ, Hatfield RD, O'Malley DM, Whetten RW, Sederoff RR. Abnormal lignin in a loblolly pine mutant. Science. 1997;**277**:235-239. DOI: 10.1126/science.277.5323.235

[33] Ralph J, Hatfield RD, Piquemal J, Yahiaoui N, Pean M, Lapierre C, Boudet AM. NMR characterization of altered lignins extracted from tobacco plants down-regulated for lignification enzymes cinnamyl alcohol dehydrogenase and cinnamoyl-CoA reductase. Proceedings of the National Academy of Sciences. 1998;**95**:12803-12808. DOI: 10.1073/pnas.95.22.12803

[34] Argyropoulos DS, Menachem SB. Lignin. In: Scheper T, editor. Advances in Biochemical Engineering/Biotechnology. Berlin, Germany: Springer-Verlag; 1997:**57**:127-158

[35] O'Malley DM, Whetten R, Bao W, Chen C, Sederoff R. The role of laccase in lignification. The Plant Journal. 1993;**4**(5):751-757. DOI: 10.1046/j.1365-313X.1993.04050751.x.

[36] Adler E. Lignin chemistry – Past, present and future. Wood Science and Technology. 1977; **11**:169-218

[37] Dence CW. The determination of lignin. In: Lin SY, Dence CW, editors. Methods in Lignin Chemistry. Berlin: Springer-Verlag; 1992. p. 33-62

[38] Dence CW, Lin SY. Introduction. In: Lin SY, Dence CW, editors. Methods in Lignin Chemistry. Berlin: Springer-Verlag; 1992. p. 3-21

[39] Björkman A. Studies on finely divided wood. Part I. Extraction of lignin with neutral solvents. Sven Papperstidn. 1956;**13**:477-485

[40] Fujimoto A, Matsumoto Y, Chang H, Meshitsuka G. Quantitative evaluation of milling effects on lignin structure during the isolation process of milled wood lignin. Journal of Wood Science. 2005;**51**:89-91. DOI: 10.1007/s10086-004-0682-7

[41] Capanema E, Balakshin M, Katahira R, Chang H, Jameel H. How well do MWL and CEL preparations represent the whole hardwood lignin? Journal of Wood Chemistry and Technology. 2015;**35**:17-26. DOI: 10.1080/02773813.2014.892993

[42] Guerra A, Filpponen I, Lucia LA, Argyropoulos DS. Comparative evaluation of three lignin isolation protocols for various wood species. Journal of Agricultural and Food Chemistry. 2006;**54**:9696-9705. DOI: 10.1021/jf062433c

[43] Roland C, Monties B, Lapierre C. Thioacidolysis. In: Lin S Y, Dence CW editors. Methods in Lignin Chemistry, Springer-Verlag, Berlin; 1992. p. 334-349

[44] Ramadevi P, Hedge DV, Varghese M, Kamalakannan R, Ganapathy SP, Gurumurth DS. Evaluation of lignin syringyl/guaiacyl ratio in *Eucalyptus camaldulensis* across three diverse sites based on near infrared spectroscopic calibration modelling with five eucalyptus species and its impact on kraft pulp yield. Journal of Near Infrared Spectroscopy. 2016;**24**:529-536

[45] Marques AV, Pereira H, Meier D, Faix O. Structural characterization of cork lignin by thioacidolysis and permanganate oxidation. Holzforschung. 1999;**53**:167-174. DOI: 10.1515/HF.1999.028

[46] Chen CL. Nitrobenzene and cupric oxide oxidations. In: Lin SY, Dence CW, editors. Methods in Lignin Chemistry. Springer-Verlag; 1992. pp. 301-321. ISBN 3-540-50295-5

[47] Min D, Xiang Z, Liu J, Jamaeel H, Chiang V, Jin Y, Chang H. Improved protocol for alkaline nitrobenzene oxidation of woody and non-woody biomass. Journal of Wood Chemistry and Technology. 2015;**35**:52-61. DOI: 10.1080/02773813.2014.902965

[48] Lapierre C. Determinating lignin structure by chemical degradations. In: Heitner C, Dimmel DR, Schmidt JA, editors. Lignin and Lignans. USA: CRC Press; 2010. pp. 11-48

[49] Mansfield SD, Kim H, Lu F, Ralph J. Whole plant cell wall characterization using solution-state 2D NMR. Nature Protocols. 2012;**7**(9):1579-1589. DOI: 10.1038/nprot.2012.064

[50] Lu F, Ralph J. The DFRC method for lignin analysis. Part 1. A new method for β-aryl ether cleavage: lignin model studies. Journal of Agricultural and Food Chemistry 1997;**45**:4655-4660. DOI: S0021-561(97)00539-6

[51] Lu F, Ralph J. Derivatization followed by reductive cleavage (DFRC method), a new method for lignin analysis: Protocol for analysis of DFRC monomers. Journal of Agricultural and Food Chemistry 1997;**45**:2590-2592. DOI: S0021-8561(97)00258-6

[52] Lu F, Ralph J. Efficient ether cleavage in lignins: the derivatization followed by reductive cleavage procedure as a basis for new analytical methods. In: Lewis NG, Sarkanen S. editors. Lignin and Lignan Biosynthesis. New Orleans, USA: American Chemical Society; 1998. p. 294-322

[53] Ralph J, Lu F. The DFRC method for lignin analysis. 6. A simple modification for identifying natural acetates in lignin. Journal of Agricultural and Food Chemistry. 1998;**46**:4616-4619. DOI: 10.1021/jf980680d

[54] Meier D, Faix O. Pyrolysis-gas-chromatography-mass spectroscopy. In: Lin SY, Dence CW, editors. Methods in Lignin Chemistry. New York: Springer Series in Wood Science; 1992. p. 177-199

[55] Wampler TP. Analytical pyrolysis: An overview. In: Wampler TP, editor. Applied Pyrolysis Handbook. 2nd ed. New York: Taylor Francis Group; 2007

[56] Bahng M, Mukarakate C, Robichaud DJ, Nimlos MR. Current technologies for analysis of biomass thermochemical processing: A review. Analytica Chimica Acta. 2009;**651**:117-138. DOI: 10.1016/j.aca.2009.08.016

[57] Demirbas A, Arin G. An overview of biomass pyrolysis. Energy Sources. 2002;**24**:471-482. DOI: 10.1080/00908310252889979

[58] Faix O, Meier D, Fortman I. Thermal degradation products of wood. A collection of electron-impact (EI) mass spectra of monomeric lignin derived products. Holz als Roh-und Werkstoff. 1990;**48**:351-354

[59] Faix O, Meier D, Fortman I. Thermal degradation products of wood. Gas chromatographic separation and mass spectrometric characterization of monomeric lignin derived products. Holz als Roh-und Werkstoff. 1990;**48**:281-285

[60] Faix O, Fortman I, Bremer J, Meier D. Thermal degradation products of wood. Gas chromatographic separation and mass spectrometric characterization of polysaccharide derived products. Holz als Roh-und Werkstoff. 1991;**49**:213-219

[61] Faix O, Fortman I, Bremer J, Meier D. Thermal degradation products of wood. A collection of electron-impact (EI) mass spectra of polysaccharide derived products. Holz als Roh-und Werkstoff. 1991;**49**:299-304

[62] Ralph J, Hatfield RD. Pyrolysis-GC-MS characterization of forage materials. Journal of Agricultural and Food Chemistry. 1991;**39**:1426-1437. DOI: 10.1021/jf00008a014

[63] Rencoret J, Marques G, Gutierrez A, Ibarra D, Li J, Gellersted G, Santos I, Jimenez-Barbero J, Martinez A, del Rio JC. Structural characterization of milled wood lignins from different eucalypt species. Holzforschung 2008;**62**: 514-526. DOI: 10.1515/HF.2008.096

[64] Faix O. Fourier transform infrared spectroscopy. In: Lin SY, Dence CW, editors. Methods in Lignin chemistry. Berlin, Germany: Springer-Verlag; 1992. p. 83-109. ISBN 3-540-50295-5

[65] Faix O. Classification of lignins from different botanical origins by FTIR. Holzforschung. 1991;**45**:21-27

[66] Watkins D, Nuruddin M, Hosur M, Tcherbi-Narteh A, Jeelani S. Extraction and characterization of lignin from different biomass resources. Journal of Materials Research and Technology. 2015;**4**(1):26-32. DOI: 10.1016/j.jmrt.2014.10.009

[67] Costa-e-Silva J, Nielsen BH, Rodrigues JC, Pereira H, Wellendorf U. Rapid determination of the lignin content in Sitka spruce (*Picea sitchensis* (bong.) Carr.) wood by Fourier transform infrared spectrometry. Holzforschung. 1999;**53**(6):597-602

[68] Hannrup B, Cahalan C, Chantre G, Grabner M, Karlsson B, Le Bayon I, Jones GL, Muller U, Pereira H, Rodrigues JC, Rosner S, Rozenberg P, Wilhelmsson L, Wimmer R. Genetic parameters of wood properties in *Picea abies*. Scandinavian Journal of Forest Research. 2004;**19**(1):14-29. DOI: 10.1080/02827580310019536

[69] Rodrigues JC, Faix O, Pereira H. Determination of lignin content of *Eucalyptus globulus* wood using FTIR spectroscopy. Holzforschung. 1998;**52**(1):46-50. DOI: 10.1515/hfsg.1998.52.1.46

[70] Blanco M, Villarroya I. NIR spectroscopy: A rapid-response analytical tool. Trends in Analytical Chemistry. 2002;**21**(4):240-250

[71] Workman J Jr., Weyer L. Practical Guide to Interpretive Near-Infrared Spectroscopy. Taylor & Francis Group. New York, USA: CRC Press; 2008. p. 344

[72] Sandak J, Sandak A, Meder R. Assessing trees, wood and derived products with near infrared spectroscopy: Hints and tips. Journal of Near Infrared Spectroscopy. 2016;**24**:485-505

[73] Schwanninger M, Rodrigues JC, Fackler K. A review of band assignments in near infrared spectra of wood and wood components. Journal of Near Infrared Spectroscopy. 2011;**19**:287-308. DOI: https://doi.org/10.1255/jnirs.955

[74] Tsuchikawa S, Schwanninger M. A review of recent near-infrared research for wood and paper. Applied Spectroscopy Reviews. 2013;**48**:560-587. DOI: 10.1080/05704920601036707

[75] Ramadevi P, Meder R, Varghese M. Rapid estimation of kraft pulp yield and lignin in *Eucalyptus camaldulensis* and *Leucaena leucocephala* by diffuse reflectance near-infrared spectroscopy (NIRS). Southern Forests: A Journal of Forest Science. 2010;**72**(2):107-111. DOI: 10.2989/20702620.2010.507462

[76] Alves A, Schwanninger M, Pereira H, Rodrigues J. Calibration of NIR to assess lignin composition (H/G ratio) in maritime pine wood using analytical pyrolysis as the reference method. Holzforschung. 2006;**60**:29-31. DOI: 10.1515/HF.2006.006

[77] Skoog DA, Holler FJ, Crouch SR. Principles of instrumentation analysis. Belmont, USA: Harris D publisher; 2007. ISBN-10:0-495-01201-7

[78] Robert, D. Characterization in solution: Spectroscopic methods. 5.4. Carbon-13 nuclear magnetic resonance spectrometry. In: Lin SY, Dence CW, editors. Methods in Lignin Chemistry, Berlin: Springer-Verlag; 1992. p. 250-273

[79] Ralph J, Landucci LL. NMR of lignins. Heitner C, Dimmel DR, Schmidt JA, editors. Lignin and lignans: advances in chemistry. New York, USA: CRC Taylor & Francis Group; 2010. ISBN 978-1-57444-486-5

[80] Cristiernin M. Structure of lignins in developing xylem of Norway spruce. Plant Physiology and Biochemistry. 2006;**44**:693-699. DOI: 10.1016/j.plaphy.2006.10.015

[81] Rowell RM, Pettersen R, Han JS, Rowell JS, Tshabalala MA. Cell wall chemistry. Part 1. Structure and chemistry. In: Rowell RM, editors. Handbook of Chemistry and Wood Composites. Florida: Taylor Francis; 2005

[82] Kutsuki H, Higuchi T. Activities of some enzymes of lignin formation in reaction wood of *Thuja orientalis, Metasequoia glyptostroboides* and *Robinia pseudoacacia*. Planta. 1981;**152**:365-368

[83] Lundquist K. 1H NMR spectral studies of lignins. Nordic Pulp & Paper Research Journal. 1992;**1**:4-16

[84] Rencoret J, Marques G, Gutiérrez A, Nieto L, Santos JI, Jiménez-Barbero J, Martínez AT, del Rio JC. 2-HSQC-NMR analysis of lignin in wood (*Eucalyptus globulus* and *Picea abies*) and non-woody (*Agave sisalana*) ball-milled plant materials at the gel state. Holzforchung. 2009; **63**:691-698. DOI: 10.1515/HF.2009.070

[85] Capanema EA, Balakhian MY, Kadla JF. A comprehensive approach for quantitative lignin characterization by NMR. Journal of Agricultural and Food Chemistry. 2004;**52**:1850-1860. DOI: 10.1021/jf035282b

[86] Lawoko M, Henriksson G, Gellerstedt G. Characterisation of lignin-carbohydrate complexes (LCCs) of spruce wood (*Picea abies* L.) isolated with two methods. Holzforschung. 2006;**60**(2):151-161. DOI: 10.1515/HF.2006.025

[87] Wadenbäck J. 2006. Lignin studies of transgenic Norway spruce. [Doctoral thesis] Uppsala. ISBN 91-576-7113-3 (pdf), Available from: https://pub.epsilon.slu.se/1237/

[88] Koutaniemi S. Lignin biosynthesis in Norway spruce: From model system to the tree. Academic dissertation; 2007. ISBN 9789521042928 (pdf), Available from: http://ethesis.helsinki.fi.

[89] Koutaniemi S, Warinowski T, Kärkönen A, Alatalo E, Fossdal CG, Saranpää P, Laakso T, Fagerstedt KV, Simola LK, Paulin L, Rudd S, Teeri TH. Expression profiling of the lignin biosynthetic pathway in Norway spruce using EST sequencing and real-time RT-PCR. Plant Molecular Biology. 2007;**65**:311-328. DOI: 10.1007/s11103-007-9220-5

[90] Fedulina TG, Kiryushina MF, Shevchenko SM, Pranovich AV. Effects of solvonation and ionization on 13C NMR spectra of lignin model compounds, spruce milled wood lignin and kraft lignin. Journal of Wood Chemistry and Technology. 2017;**37**(4):241-250. DOI: 10.1080/02773813.2016.1272126

[91] Esteves B, Gominho J, Rodrigues JC, Miranda I, Pereira H. Pulping yield and delignification kinetics of heartwood and sapwood of maritime pine. Wood Chemistry and Technology. 2005;**25**(4):217-230. DOI: 10.1080/02773810500366656

[92] Alves A, Schwanninger M, Pereira H, Rodrigues J. Analytical pyrolysis as a direct method to determine the lignin content in wood. Part 1: Comparison of pyrolysis lignin with Klason lignin. Journal of Analytical and Applied Pyrolysis. 2006;**76**:209-213. DOI: 10.1016/j.jaap.2005.11.004

[93] Alves A, Rodrigues J, Wimmer R, Schwanninger M. Analytical pyrolysis as a direct method to determine the lignin content in wood. Part 2: Evaluation of the common model and the influence of compression wood. Journal of Analytical and Applied Pyrolysis. 2008;**81**:167-172. DOI: 10.1016/j.jaap.2005.11.004

[94] Baptista C, Robert D, Duarte AP. Relationship between lignin structure and delignification degree in *Pinus pinaster* kraft pulps. Bioresource Technology. 2008;**99**:2349-2356. DOI: 10.1016/j.biortech.2007.05.012

[95] Huang F, Singh PM, Ragauskas AJ. Characterization of milled wood lignin (MWL) in loblolly pine stem wood, residue and bark. Journal of Agricultural and Food Chemistry. 2011;**59**:12910-12916. DOI: 10.1021/jf202701b

[96] Toivanen T, Alén R. Variations in the chemical composition within pine (*Pinus sylvestris*) trunks determined by diffuse reflectance infrared spectroscopy and chemometrics. Cellulose. 2006;**13**:53-61. DOI: 10.1007/s10570-005-9016-1

[97] Zaman A, Alén R, Kotilainen R. Thermal behavior of scots pine (*Pinus sylvestris*) and silver birch (*Betula pendula*) at 200-230°C. Wood and Fiber Science. 2000;**32**(2):138-143

[98] Fernandes C, Gaspar MJ, Pires J, Alves A, Simões R, Rodrigues JC, Silva ME, Carvalho A, Brito JE, Lousada JL. Physical, chemical and mechanical properties of *Pinus sylvestris* wood at five sites in Portugal. iForest. 2017;**10**:669-679. DOI: 10.3832/ifor2254-010

[99] Antanova GF, Varksina TN, Zheleznichenko TV, Stasova VV. Lignin deposition during earlywood and latewood formation in scots pine stems. Wood Science and Technology. 2014;**48**:919-936. DOI: 10.1007/s00226-014-0650-3

[100] Socha AM, Plummer SP, Stavila V, Simmons BA, Singh S. Comparison of sugar contente for ionic liquid pretreated Douglas-fir woodchips and forest residues. Biotechnology for Biofuels. 2013;**6**:61-71. DOI: 10.1186/1754-6834-6-61

[101] Alvarez-Vasco C, Ma R, Quintero M, Gio M, Geleynse S, Ramasamy KK, Wolcott M, Zhang X. Unique low-molecular-weight lignin with high purity extracted from wood by deep eutectic solventes (DES): A source of lignin for valorization. Green Chemistry. 2016;**18**:5133-5141. DOI: 10.1039/c6gc01007e

[102] Cristiernin M, Ohlsson AB, Berglund T, Henriksson G. Lignin isolated from primary walls of hybrid aspen cell cultures indicates significant differences in lignin structure between primary and secondary cell wall. Plant Physiology and Biochemistry. 2005;**43**: 777-785. DOI: 10.1016/j.plaphy.2005.07.007

[103] Abreu HS, Latorraca JVF, Pereira RPW, Monteiro MBO, Abreu FA, Amparado KF. A supramolecular proposal of lignin structure and its relation with the wood properties. Anais da Academia Brasileira de Ciências. 2009;**81**(1):137-142

[104] Pereira H, Miranda I, Gominho J, Tavares F, Quilhó T, Graça J, Rodrigues J, Shatalov A, Knapic S. Qualidade Tecnológica Do Eucalipto *Eucalyptus globulus*. Lisboa: Centro de Estudos Florestais ed., Instituto Superior de Agronomia, Universidade Técnica de Lisboa; 2010. 377 p

[105] Kojima Y, Isaji S, Yoon S, Ona T. Selection criteria of *Eucalyptus globulus* Labill. For production of chemithermomechanical pulps (CTMP). Holzforschung. 2008;**62**:71-76. DOI: 10.1515/HF.2008.010

[106] Miranda I, Pereira H. The variation of chemical composition and pulping yield with age and growth factors in young *Eucalyptus globulus*. Wood and Fiber Science. 2002; **34**(1):140-145

[107] Pereira H. Variability in the chemical composition of plantation eucalypts (*Eucalyptus globulus* Labill.). Wood and Fiber Science. 1988;**20**(1):82-90

[108] Rencoret J, Gutíerrez A, Neto L, Jiménez-Barbero J, Faulds CB, Kim H, Ralph J, Martínez AT, del Río JC. Lignin composition and structure in young versus adult *Eucalyptus globulus* plants. Plant Physiology 2011;**155**:667-682

[109] Lourenço A, Gominho J, Marques AV, Pereira H. Variation of lignin monomeric composition during kraft pulping of *Eucalyptus globulus* heartwood and sapwood. Journal of Wood Chemistry and Technology. 2013;**33**:1-18. DOI: 10.1080/02773813.2012.703284

[110] Rodrigues J, Meier D, Faix O, Pereira H. Determination of tree to tree variation in syringyl/guaiacyl ratio of *Eucalyptus globulus* wood lignin by analytical pyrolysis. Journal of Analytical and Applied Pyrolysis. 1999b;**48**:121-128. DOI: 10.1016/S0165-2370(98)00134-X

[111] del Río JC, Gutiérrez A, Hernando M, Landín P, Romero J, Martínez AT. Determining the influence of eucalypt lignin composition in paper pulp yield using Py-GC/MS. Journal of Analytical and Applied Pyrolysis 2005;**74**:110-115. DOI: 10.1016/j.jaap.2004.10.010

[112] Alves A, Simoes R, Stackpole DJ, Villancourt RE, Potts BM, Schwanninger M, Rodrigues J. Determination of syringyl/guaiacyl ratio of *Eucalyptus globulus* wood lignin by near infrared based partial least squares regression models using analytical pyrolysis as the reference method. Journal of Near Infrared Spectroscopy. 2011;**19**(5):343. DOI: 10.1255/jnirs.946

[113] Evtuguin DV, Neto CP, Silva AMS, Domingues PM, Amado FML, Robert D, Faix O. Comprehensive study on the chemical structure of dioxane lignin from plantation *Eucalyptus globulus* wood. Journal of Agricultural and Food Chemistry. 2001;**49**:4252-4261. DOI: 10.1021/jf010315d

[114] Neiva DM, Araújo S, Lourenço A, Gominho J, Fernandes L, Simões R, Pereira H. Chemical composition and kraft pulping potential of 12 *Eucalypt* species. Industrial Crops and Products. 2015;**66**:89-95. DOI: 10.1016/j.indcrop.2014.12.016

[115] Hein PRG, Lima JT, Chaix G. Effects of sample preparation on NIR spectroscopic estimation of chemical properties of *E. urophylla* S.T. Blake wood. Holzforschung. 2010;**64**:45-54. DOI: 10.1515/HF.2010.011

[116] Fagerstedt KV, Saranpää Tapanila T, Immanen J, Serra JAA, Nieminen K. Determining the composition of lignins in different tissues of silver birch. Plants 2015;**4**:183-195. DOI: 10.3390/plants4020183

[117] Chen C, Alén R, Lehto J, Pakkanen H. Combustion properties of birch (*Betula pendula*) black liquors from sulfur-free pulping. Journal of Wood Chemistry and Technology. 2016;**36**(6):401-411. DOI: 10.1080/02773813.2016.1203945

[118] Szczepkowski A, Nicewicz D, Koczon P. The relationship between tree health and chemical composition of beech (*Fagus sylvatica* L.) and oak (*Quercus robur* L.) wood and polish provenances. Acta Scientiarum Polonorum Silvarum Colendarum Ratio et Industria Lignaria. 2007;**6**(3):77-88

[119] Choi JW, faix O, Meier D. Characterization of residual lignins from chemical pulps of spruce (*Picea abies* L.) and beech (*Fagus sylvatica* L.) by analytical pyrolysis-gas chromatography/mass spectrometry. Holzforschung. 2001;**55**:185-192. DOI: 10.1515/HF.2001.031

[120] Lourenço A, Baptista I, Gominho J, Pereira H. The influence of heartwood in the pulping properties of *Acacia melanoxylon* wood. Journal of Wood Science. 2008;**54**:464-469. DOI: 10.1007/s10086-008-0972-6

[121] Santos AJA, Anjos O, Pereira H. Prediction of blackwood kraft pulps yields with wood NIR–PLSR models. Wood Science and Technology. 2016;**50**(6):1307-1322. DOI: 10.1007/s00226-016-0837-x

[122] Pinto PC, Evtuguin DV, Neto CP. Chemical composition and structural features of the macromolecular components of plantation *Acaia mangium* wood. Journal of Agricultural and Food Chemistry. 2005;**53**:7856-7862. DOI: 10.1021/jf058081b

[123] Nawawi DS, Syafii W, Tomoda I, Uchida Y, Akiyama T, Yokoyama T, Matsumoto Y. Characteristics and reactivity of lignin in *Acacia* and *Eucalyptus* woods. Journal of Wood Chemistry and Technology. 2017;**37**(4):273-282. DOI: 10.1080/02773813.2017.1291684

[124] Pereira H. Cork: Biology, Production and Uses. 1st ed. Elsevier; 2007. 336 p. ISBN-13: 978-0-444-5296-1

[125] Lourenço A, Rencoret J, Chematova C, Gominho J, Gutiérrez A, del Río JC, Pereira H. Lignin composition and structure differs between xylem, phloem and phellem in *Quercus suber* L. Frontiers in Plant Science 2016;**7**:1612. DOI: 10.3389/fpls.2016.01612

[126] Lourenço A, Neiva D, Gominho J, Marques AV, Pereira H. Characterization of lignin in heartwood, sapwood and bark from *Tectona grandis* using Py-GC-MS/FID. Wood Science and Technology. 2015;**49**(1):159-175. DOI: 10.1007/s00226-014-0684-6

[127] Miranda I, Sousa V, Pereira H. Wood properties of teak (*Tectona grandis*) from a mature unmanaged stand in East Timor. Journal of Wood Science. 2011;**57**:171-178. DOI: 10.1007/s10086-010-1164-8

[128] Windeisen E, Klassen A, Wegener G. On the chemical characterization of plantation teakwood from Panama. Holz als Roh- und Werkstoff. 2003;**61**:416-418. DOI: 10.1007/s00107-003-0425-2

[129] Fernández J, Curt MD, Aguado PL. Industrial applications of *Cynara cardunculus* L. for energy and other uses. Industrial Crops and Products. 2006;**24**:222-229. DOI: 10.1016/j.indcrop.2006.06.010

[130] Gominho J, Lourenço A, Curt MD, Férnandez J, Pereira H. *Cynara cardunculus* in large scale cultivation. A case study in Portugal. Chemical Engineering Transactions. 2014;**37**: 529-534

[131] Gominho J, Fernandez J, Pereira H. *Cynara cardunculus* L. – A new fibre crop for pulp and paper production. Industrial Crops and Products. 2001;**13**:1-10. DOI: 10.1016/S0926-6690(00)00044-3

[132] Sengo I, Gominho J, d'Orey L, Martins M, d'Almeida-Duarte E, Pereira H, Ferreira-Dias S. Response surface modeling and optimization of biodiesel production from *Cynara cardunculus* oil. European Journal of Lipid Science and Technology. 2010;**112**:310-320. DOI: 10.1002/ejlt.200900135

[133] Abelha P, Franco C, Pinto F, Lopes H, Gulyurtlu I, Gominho J, Lourenço A, Pereira H. Thermal conversion of *Cynara cardunculus* L. and mixtures with *Eucalyptus globulus* by fluidized bed combustion and gasification. Energy & Fuels. 2013;**27**(11):6725-6737. DOI: 10.1021/ef401246p

[134] Ballesteros M, Negro MJ, Manzanares P, Ballestros I, Sáez F, Oliva JM. Fractionation of *Cynara cardunculus* (cardoon) biomass by dilute-acid pretreatment. Applied Biochemistry and Biotechnology. 2007;**136-140**:239-252. DOI: 10.1007/s12010-007-9055-1

[135] Lourenço A, Neiva DM, Gominho J, Curt MD, Fernández J, Marques AV, Pereira H. Biomass production of four *Cynara cardunculus* clones and lignin composition analysis. Biomass and Bioenergy. 2015;**76**:86-95. DOI: 10.1016/j.biombioe.2015.03.009

[136] Lourenço A, Rencoret J, Chemetova C, Gominho J, Gutiérrez A, Pereira H, del Río JC. Isolation and structural characterization of lignin from cardoon (*Cynara cardunculus* L.) stalks. Bioenergy Research. 2015;**8**(4):1946-1955. DOI: 10.1007/s12155-015-9647-5

[137] Carpita NC. Structure and biogenesis of cell walls of grasses. Annual Review of Plant Physiology and Plant Molecular Biology. 1996;**47**:445-476. DOI: 10.1146/annurev.arplant.47.1.445

[138] Borin M, Barbera AC, Milani M, Molari G, Zimbone SM, Toscano A. Biomass production and N balance of giante reed (*Arundo donax* L.) under high water and N input in Mediterranean environments. European Journal of Agronomy. 2013;**51**:117-119. DOI: 10.1016/j.eja.2013.07.005

[139] Shatalov AA, Pereira H. Kinetics of organosolv delignification of fibre crop *Arundo donax* L. Industrial Crops and Products. 2005;**21**:203-210. DOI: 10.1016/j.indcrop.2004.04.010

[140] Neto CP, seca A, Nunes AM, Coimbra MA, Domingues F, Evtuguin D, Silvestre A, Cavaleiro JAS. Variations in chemical composition and structure of macromolecular components in different morphological regions and maturity stages of *Arundo donax*. Industrial Crops and Products. 1997;**6**(1):51-58. DOI: 10.1016/S0926-6690(96)00205-1

[141] Faix O, Meier D, Beinhoff O. Analysis of lignocelluloses and lignins from *Arundo donax* L. and *Miscanthus sinensis* Anderss, and hydroliquefaction of *Miscanthus*. Biomass. 1989;**18**:109-126. DOI: 10.1016/0144-4565(89)90088-7

[142] You T, Mao J, Yuan T, Wen J, Xu F. Structural elucidation of the lignins from stems and foliage of *Arundo donax* Linn. Journal of Agricultural and Food Chemistry. 2013;**61**:5361-5370. DOI: 10.1021/jf401277v

[143] Seca AML, Cavaleiro JAS, Domingues FMJ, Silvestre AJD, Evtuguin D, Neto CP. Structural characterisation of the lignin from the nodes and internodes of *Arundo donax* reed. Journal of Agricultural and Food Chemistry. 2000;**48**:817-824. DOI: 10.1021/jf9910988

[144] Brosse N, Dufour A, Meng X, Sun Q, Ragauskas A. *Miscanthus*: A fast-growing crop for biofuels and chemicals production. Biofuels, Bioproducts and Biorefining. 2012;**6**(5):580-598. DOI: 10.1002/bbb.1353

[145] Hage RE, Brosse N, Chrusciel L, Sanchez C, Sannigrahi P, Ragauskas A. Characterization of milled lignin and ethanol organosolv lignin from *Miscanthus*. Polymer Degradation and Stability. 2009;**94**:1632-1638. DOI: 10.1016/j.polymdegradstab.2009.07.007

[146] Villaverde JJ, Li J, Ek M, Ligero P, Veja A. Native lignin structure of *Miscanthus x giganteus* and its changes during acetic acid and formic acid fractionation. Journal of Agricultural and Food Chemistry. 2009;**57**:6262-6270. DOI: 10.1021/jf900483t

[147] Wang K, Bauer S, Sun R. Structural transformation of *Miscanthus* × *giganteus* lignin fractionated under mild formosolv, basic organosolv, and cellulolytic enzyme conditions. Journal of Agricultural and Food Chemistry. 2012;**60**:144-152. DOI: 10.1021/jf2037399

[148] Bauer S, Sorek H, Mitchell VD, Ibáñez AB, Wemmer DE. Characterization of *Miscanthus giganteus* lignin isolated by ethanol organosolv process under reflux condition. Journal of Agricultural and Food Chemistry. 2012;**60**:8203-8212. DOI: 10.1021/jf302409d

[149] Kim S, Dale BE. Global potential bioethanol production from wasted crops and crop residues. Biomass and Bioenergy. 2004;**4**:361-375. DOI: 10.1016/j.biombioe.2003.08.002

[150] Crestini C. Argyropoulos DS. Structural analysis of wheat straw lignin by quantitative ^{31}P and 2D NMR spectroscopy. The occurrence of ester bonds and α-O-4 substructures. Journal of Agricultural and Food Chemistry. 1997;**45**(4):1212-1219. DOI: S0021-8561(96) 00568-7

[151] Sun R, Sun X, Zhang S. Quantitative determination of hydroxycinnamic acids in wheat, rice, rye, and barley straws, maize stems, oil palm frond fiber, and fast-growing poplar wood. Journal of Agricultural and Food Chemistry. 2001;**49**:5122-5129. DOI: 10.1021/jf010500r

[152] Sun RC, Sun XF, Wang SQ, Zhu W, Wang XY. Ester and ether linkages between hydroxycinnamic acids and lignins from wheat, rice, rye, and barley straws, maize stems, and fast-growing poplar wood. Industrial Crops and Products. 2002;**15**(3):179-188. DOI: 10.1016/S0926-6690(01)00112-1

[153] Sun R, Lawther JM, Banks WBA. Tentative chemical structure of wheat straw lignin. Industrial Crops and Products. 1997;**6**:1-8. DOI: 10.1016/S0926-6690(96)00170-7

[154] Sun X, Sun R, Fowler P, Baird MS. Extraction and characterization of original lignin and hemicelluloses from wheat straw. Journal of Agricultural and Food Chemistry. 2005;**53**:860-870. DOI: 10.1021/jf040456q

[155] Wi SG, Lee D, Nguyen QA, Bae H. Evaluation of biomass quality in short-rotation bamboo (*Phyllostachys pubescens*) for bioenergy products. Biotechnology for Biofuels. 2017;**10**:127. DOI: 10.1186/s13068-017-0818-9

[156] Huang C, He J, Du L, Min D, Yong Q. Structural characterization of the lignins from the green and yellow bamboo of bamboo culm (*Phyllostachys pubescens*). Journal of Wood Chemistry and Technology. 2016;**36**:157-172. DOI: 10.1080/02773813.2015.1104544

[157] Wen J, Sun S, Yuan T, Xu F, Sun R. Understanding the chemical and structural transformations of lignin macromolecule during torrefaction. Applied Energy. 2014;**121**:1-9. DOI: 10.1016/j.apenergy.2014.02.001

[158] Shao S, Jin Z, Wen G, Iiyama K. Thermo characteristics of steam-exploded bamboo (*Phyllostachys pubescens*) lignin. Wood Science and Technology. 2009;**43**:643-652. DOI: 10.1007/s00226-009-0252-7

[159] Oliveira L, Evtuguin D, Cordeiro N, Silvestre AJD. Structural characterization of stalks lignin from banana plant. Industrial Crops and Products. 2009;**29**:86-95. DOI: 10.1016/j.indcrop.2008.04.012

[160] Oliveira L, Cordeiro N, Evtuguin DV, Torres IC, Silvestre AJD. Chemical composition of different morphological parts from 'Dwarf Cavendish' banana plant and their potential as a non-wood renewable source of natural products. Industrial Crops and Products. 2007;**26**:163-172. DOI: 10.1016/j.indcrop.2007.03.002

[161] Sen A, Quilhó T, Pereira H. Bark anatomy of *Quercus cerris* L. Var. Cerris from Turkey. Turkish Journal of Botany. 2011;**35**:45-55

[162] Esau K. Anatomy of Seed Plants. 2nd ed. New York: John Wiley & Sons; 1960. p 1-4

[163] Miranda I, Gominho J, Mirra I, Pereira H. Chemical characterisation of barks from *Picea abies* and *Pinus sylvestris* after fractioning into different particle sizes. Industrial Crops and Products. 2012;**36**:395-400. DOI: 10.1016/j.indcrop.2011.10.035

[164] Miranda I, Gominho J, Mirra I, Pereira H. Fractioning and chemical characterisation of barks of *Betula pendula* and *Eucalyptus globulus*. Industrial Crops and Products. 2013;**41**:299-305. DOI: 10.1016/j.indcrop.2012.04.024

[165] Leite C, Pereira H. Cork-containing barks – A review. Frontiers in Materials. 2017;**3**:63. DOI: 10.3389/fmats.2016.00063

[166] Pan S, Pu Y, Foston M, Ragauskas AJ. Compositional characterization and pyrolysis of loblolly pine and Douglas-fir bark. Bioenergy Research. 2013;**6**:24-34. DOI: 10.1007/s12155-012-9223-1

[167] Fradinho DM, Neto CP, Evtuguin D, Jorge FC, Irle MA, Gil MH, Jesus JP. Chemical characterisation of bark and of alkaline bark extracts from maritime pine grown in Portugal. Industrial Crops and Products. 2002;**16**:23-32. DOI: 10.1016/S0926-6690(02)00004-3

[168] Pereira H. Chemical composition and variability of cork from *Quercus suber* L. Wood Science and Technology. 1988;**22**:211-218. DOI: 10.1007/BF00386015

[169] Pereira H. Variability of the chemical composition of cork. BioResources. 2013;**8**(2): 2246-2256

[170] Marques AV, Pereira H, Meier D, Faix O. Quantitative analysis of cork (*Quercus suber* L.) and milled cork lignin by FTIR spectroscopy, analytical pyrolysis, and total hydrolysis. Holzforschung. 1994;**48**:43-50. DOI: 10.1515/hfsg.1994.48.s1.43

[171] Marques AV, Pereira H, Meier D, Faix O. Isolation and characterization of a guaiacyl lignin from saponified cork of *Quercus suber* L. Holzforschung. 1996;**50**:393-400. DOI: 10.1515/hfsg.1996.50.5.393

[172] Marques AV, Pereira H, Rodrigues J, Meier D, Faix O. Isolation and comparative characterization of a Björkman lignin from the saponified cork of Douglas-fir bark. Journal of Analytical and Applied Pyrolysis. 2006;**77**:169-176. DOI: 10.1016/j.jaap.2006.03.003

[173] Marques AV, Rencoret J, Gutiérrez A, del Río JC, Pereira H. Ferulates and lignin structural composition in cork. Holzforschung 2015;**70**(4):275-289. DOI: 10.1515/hf-2015-0014

[174] Lopes M, Neto CP, Evtuguin D, Silvestre AJD, Gil A, Cordeiro N, Gandini A. Products of the permanganate oxidation of cork, desuberized cork, suberin and lignin from *Quercus suber* L. Holzforschung. 1998;**52**:146-148. DOI: 10.1515/hfsg.1998.52.2.146

[175] Sen A, Miranda I, Santos S, Graça J, Pereira H. The chemical composition of cork and phloem in the rhytidome of Quercus Cerris bark. Industrial Crops and Products. 2010;**31**:417-422. DOI: 10.1016/j.indcrop.2010.01.002

[176] Marques AV, Pereira H. Lignin monomeric composition of corks from the barks of *Betula pendula, Quercus suber* and *Quercus cerris* determined by Py-GC-MS/FID. Journal of Analytical and Applied Pyrolysis. 2013;**100**:88-94. DOI: 10.1016/j.jaap.2012.12.001

[177] Ferreira JPA, Quilhó T, Pereira H. Characterization of *Betula pendula* outer bark regarding cork and phloem components at chemical and structural levels in view of biorefinery integration. Journal of Wood Chemistry and Technology. 2017;**37**:10-25. DOI: 10.1080/02773813.2016.1224248

[178] Baptista I, Miranda I, Quilhó T, Gominho J, Pereira H. Characterisation and fractioning of *Tectona grandis* bark in view of its valorisation as a biorefinery raw-material. Industrial Crops and Products. 2013;**50**:166-175. DOI: 10.1016/j.indcrop.2013.07.004

[179] Costa CAE, Pinto PCR, Rodrigues AR. Evaluation of chemical processing impact on *E. globulus* wood lignin and comparison with bark lignin. Industrial Crops and Products. 2014;**61**:479-491. DOI: 10.1016/j.indcrop.2014.07.045

[180] Barros J, Serk H, Granlund I, Pesquet E. The cell biology of lignification in higher plants. Annals of Botany. 2015;**115**:1053-1074. DOI: 10.1093/aob/mcv046

[181] Takabe K, Fujita H, Harada H, Saiki H. Lignification process of Japanese black pine (*Pinus thunbergii* Parl.) tracheids. Mokuzai Gakkaishi. 1981;**27**:813-820

[182] Takabe K, Fujita M, Harada H, Saiki H. Autoradiographic investigation of lignification in the cell-walls of Cryptomeria (*Cryptomeria japonica* D don). Mokuzai Gakkaishi. 1985;**31**:613-619

[183] Terashima N, Fukushima K. Biogenesis and structure of macromolecular lignin in cell wall of tree xylem as studied by microautoradiography. In: Lewis N, Paice M editors. Plant Cell Wall Polymers: Biogenesis and Biodegradation, Vol. 399, Chap 11. Washington, USA: American Chemistry Society Symposium Series; 1989. p. 160-168

[184] Schuetz M, Smith R, Ellis B. Xylem tissue specification, patterning, and differentiation mechanisms. Journal of Experimental Botany. 2012;**64**(1):11-31. DOI: 10.1093/jxb/ers287

[185] Saka S, Goring DAI. The distribution of lignin in white birch wood as determined by bromination with TEM-EDXA. Holzforschung. 1988;**42**:149-153. DOI: 10.1515/hfsg.1988.42.3.149

[186] Watanabe Y, Fukazawa K. Lignin heterogeneity of the cell walls on the genus Acer. Research Bulletins of the Hokkaido University Forests. 1993;**50**:349-389

[187] Saito K, Watanabe Y, Shirakawa M, Matsushita Y, Imai T, Koike T, Sano Y, Funada R, Fukazawa K, Fukushima K. Direct mapping of morphological distribution of syringyl and guaiacyl lignin in the xylem of maple by time-of-flight secondary ion mass spectrometry. The Plant Journal. 2012;**69**:542-552. DOI: 10.1111/j.1365-313X.2011.04811.x

[188] Wu J, Fukazawa K, Ohtani J. Distribution of syringyl and guaiacyl lignins in hardwoods in relation to habitat and porosity form in wood. Holzforschung. 1992;**46**:181-185. DOI: 10.1515/hfsg.1992.46.3.181

Chapter 4

Base-Catalyzed Depolymerization of Lignin: History, Challenges and Perspectives

Björn Rößiger, Gerd Unkelbach and
Daniela Pufky-Heinrich

Additional information is available at the end of the chapter

http://dx.doi.org/10.5772/intechopen.72964

Abstract

Bio-based phenolic compounds available from lignin are promising candidates for industrial application, e.g., within polymer resins or as biogenic fuel substitutes. Among numerous conversion methods for the valorization of lignin, the base-catalyzed depolymerization (BCD) has considerable advantages with respect to other processes. By this method, lignin and lignin-containing biorefinery streams can be catalytically transferred to valuable, defined products with tailored specifications. Continuous process operation allows conversions at short residence times and, thus, enables its industrial implementation more easily due to economic reasons. This review reflects the development in the field of BCD on various types of lignin. A historical overview will be given and the principal application of the method is shown. Challenges for operations are addressed, mainly to the development of efficient and selective methods for product separation and purification of the alkylphenolic moieties and the reduction of char formation during the process. An outlook will be given by showing trends and perspectives, especially in the field of industrial applications. Here, hydrotreatment methods for refining BCD intermediates for fuel and platform chemical production are shown. Furthermore, the application of BCD for the conversion of woody biomass and black liquor is discussed.

Keywords: lignin, base-catalyzed depolymerization BCD, bio-based aromatic compounds, bio-based phenols, downstream processing, hydrogenolysis, hydrodeoxygenation

1. Introduction

Lignin is one of the major components in lignocellulosic biomass together with cellulose and hemicellulose and the world's only naturally occurring polymers, which is available as a renewable resource in large quantities. Lignin is obtained as a major by-product during second-generation bio-ethanol production and also isolated as Kraft black liquor or lignosulfonates during pulp production. Also new lignocellulose bio-refinery processes for the production of chemicals or fuels will release an enormous amount of different kinds of lignin. Apart from its use in the form of wood, lignin is mainly used as energy source as a by-product of the paper and pulp industry. Every year, around 50 million tons of lignin are currently produced in this way worldwide. Conventionally, lignin is employed for the generation of heat and power in these industry processes [1]. Utilizations of lignin with regard to material application are only established to a limited extent. Increase in application depends solely on the availability of valuable chemicals synthesized from lignin that have a corresponding market value higher than their fuel value.

However, valorization of lignin is one of the most important challenges for the development of sustainable and cost-effective biorefinery processes based on lignocellulosic biomass. The alkylphenolic structure of the lignin molecule can be cleaved into low molecular weight compounds such as phenols, alkylphenols and phenol resins and replacing those obtained from fossil resources [2, 3]. The products are highly promising fuels or fuel additives [4, 5] or antifungal components [6] or can be used for the preparation of polyols and polyurethane resins [7]. Yet, the natural complexity and high stability of intramolecular bonds make lignin depolymerization a challenging task. Moreover, effective technologies for cleavage product separation from the reaction solutions are still subject of developments and have not been satisfactorily resolved. A prerequisite for industrial implementation are techno-economic feasible processes, which can be integrated into existing process chains. So far, low selectivity to defined products, subsequent separation and purification and char formation are the main hurdles for effective conversion of lignins. Among different methods, the so-called base-catalyzed depolymerization (BCD) is one option to transfer lignin and lignin-containing streams catalytically into valuable products even to those with narrow product specification. It can be operated in continuous process mode allowing conversions at short residence times (≤1 h) and low char formation. The latter is as phenomena, which might occur during hydrothermal conversion of lignin feedstock, and is an effect of concurrent repolymerization processes of lignin fragments during processing under the harsh reaction conditions needed [8].

This chapter comprehensively reviews the state-of-the-art applications and challenges for BCD processes of different kinds from solid lignin and lignin-containing feedstocks. It also points out future trends and perspectives in this field of research and development. Besides scientific approaches, methods and concepts for industrial implementation of this technology, e.g., in the pulp and paper industry or the chemical and process industry, are touched and discussed. In this context, patent applications on this subject are also mentioned, as well as processes and applications are discussed. In order to increase the technology readiness level of the BCD process and, thus, to develop an industry-relevant process, it is necessary to establish an overall approach regarding material and energy efficiency as well as to examine its technical and economic feasibility.

2. Methods on lignin depolymerization

Research in the field of lignin depolymerization is of strong interest. The number of scientific publications has developed exponentially in recent years and increased a hundredfold worldwide since 1980 starting with 2 to 1998 with 25 and until 2016 with 221 publications annually [9]. From this, it is evident that the viability of recovering of hydrocarbons and aromatics from lignins has been under intensive investigation over the past years. Literature and publications have been reviewed and extensively discussed in terms of the generation of valuable chemicals and fuels [1, 4, 5, 10–12] as well as derivatives for polymer resins [7]. Along with this and beyond, mechanistical aspects of chemical, biological and biotechnological depolymerization strategies are compared in detail and advantages and disadvantages as well as limits in their applications are presented [3, 13, 14]. The focus is primarily on the presentation of scientific work and innovations. Trends and new methods are partially presented in detail and summarized clearly. These include, above all, the current developments in transition-metal-catalyzed conversion of lignins for catalytic cracking, oxidation and/or hydrogenolysis for the utilization of lignin feed streams as fuels or fuel additives and bulk chemicals, respectively [2, 4, 13, 15].

Similar development can be overserved in the status on patent publications. Over the last decades, IP applications increased strongly in the last few years, whereas approximately 80% of patent applications have been made within the last 10 years [16]. The first application of BCD was published in 1983 by Stake Technology LTD as "process for depolymerization and extraction of lignin utilizing steam explosion technology" [17]. Technology applications might be found in the production of bioaromatic compounds for platform and fine chemicals [18, 19], as gasoline and biofuels [20, 21] or blending components [22].

2.1. Lignin structure

Lignin is a cross-linked amorphous three-dimensional copolymer synthesized from radical, random polymerization of the three primary phenylpropene units: coumaryl-, coniferyl-, and sinapyl-alcohols, joined by C—O—C (ether) and C—C bonds and collectively called monolignins (**Figure 1**). Structure and, thus, ratio of these three primary monomer units vary among different plants and species. In native lignin, the most abundant linkage is β—O—4 ether bond, which comprises around 45–60% of all linkages within lignin, whereas hardwood lignin contains roughly 1.5 times more compared to that of softwood. In total, approximately two third of linkages are ether bonds, while the others are C—C bonds [4, 11, 23]. The polyphenolic aromatic structure of lignin is ideally suited to obtain aromatic molecules, either as oligomeric derivatives or as low molecular weight monomeric compounds.

2.2. Strategies for lignin conversion

Chemical and enzymatic conversion strategies have been developed in recent years aiming to synthesize aromatic substructures or valorized chemicals from lignin. Lignin can be cleaved

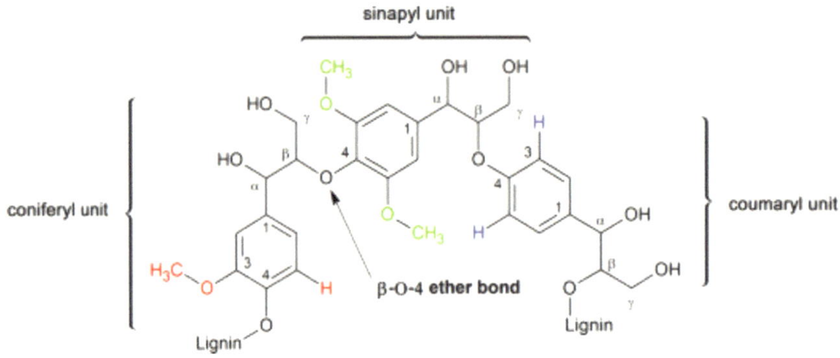

Figure 1. Details of the lignin structure with its varying phenylpropene units connected by β–O–4 ether bond.

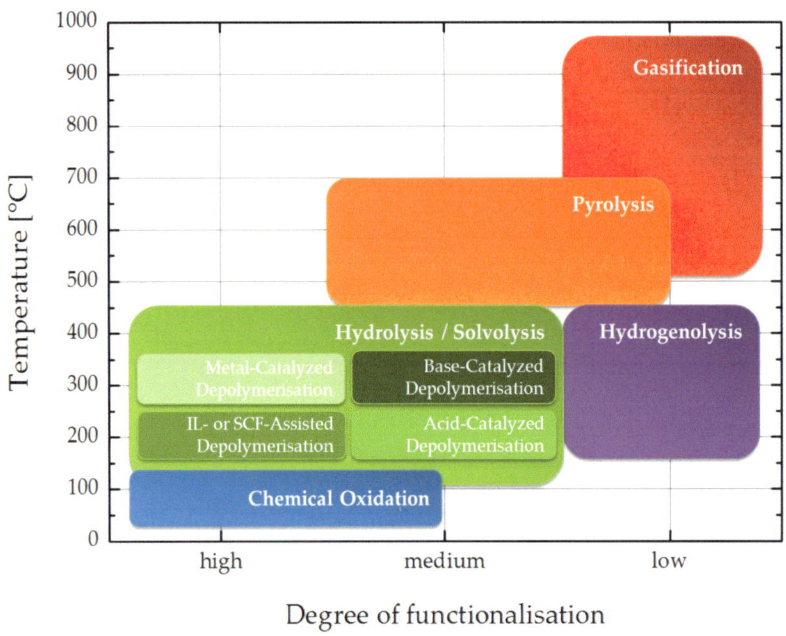

Figure 2. Summary of thermochemical processes for conversion of lignin; temperature applied vs. degree of functionalization within the product species.

by thermochemical processes that means by thermal treatment in the presence of various solvents, chemical additives or catalysts. Alternatively, there are approaches for enzymatic depolymerization using laccases or manganese peroxidases [24, 25]. A simplified summary of processes for chemical lignin conversion is given in **Figure 2**. Here, applied process temperatures are compared to the degree of functionalization within the product lignin species. Essential processes are pyrolysis, gasification, hydrogenolysis, chemical oxidation and hydrolysis. Depending on the process, the products of the depolymerization possess different proportions of gaseous (gas), liquid (oil) and/or solid (oligomers and/or char) products with

http://dx.doi.org/10.5772/intechopen.72964

specific compositions. With pyrolysis, the thermal treatment at 300–600°C with exclusion of oxygen, mainly gaseous and volatile hydrocarbons, such as methane, ethane, acetone, methanol, acetaldehyde, phenol and carbon mono- and carbon dioxide, is formed. Gasification is a process in which lignin is converted at temperatures between 700 and 1000°C into a gas mixture of hydrogen, carbon monoxide, methane and carbon dioxide (syngas) [26]. The reductive cleavage with hydrogen (hydrogenolysis) leads to a mixture of differently substituted phenols and benzenes, whereas products of oxidative cleavage are mostly phenolic aldehydes such as vanillin and syringyl alcohol [11]. With hydrolysis, the ether bonds in the lignin are cleaved by a homogeneous or heterogeneously catalyzed aqueous reaction. The hydrolytic cleavage can be further categorized according to different chemicals applied in the depolymerization process, namely base-catalyzed, acid-catalyzed and metal-catalyzed or with the aid of ionic liquids or supercritical fluids. Wang et al. summarized recent scientific developments with regard to this classification [14]. Depolymerization methods have also been reviewed according to the product specification, which is either gaseous as methane, carbon dioxide and formaldehyde or liquid as a so-called bio-oil [27]. This view is very product-driven and above all illustrates the scope of application of lignin-derived products. From the point of view of chemists and process engineers, classification is also carried out according to the mode of chemical conversion. Homogeneous and heterogeneous catalytic conversions are in this case considered separately and thus condition the process design and process implementation [5, 12].

3. Base-catalyzed depolymerization of lignin

3.1. Nature of base-catalyzed depolymerization of lignin

All degradation strategies have the objective of reducing the complexities of the natural lignin molecule, lowering its molecular weight and, moreover, increasing the chemical reactivity of the degradation products. Three fractions are formed: liquid oil; oligomers, often called tar fraction; and higher polymeric lignin species called char. Side products are formic acid, acetic acid, methanol and carbon dioxide. This is compared to degradation processes such as pyrolysis and hydropyrolysis, which lead to a mixture of oligomer and polymer oxyaromatic fractions. The later is caused by the initiation of radical formation inside the lignin molecule during the cleavage of weak phenolic bonds. Subsequently, recombination of radical moieties may lead to the formation of new kinds of carbon-carbon bonds and ultimately to oligomers and higher condensed structures named as char. The BCD of lignin is carried out in dilute alkaline solution at temperatures between 250 and 350° C, high pressures (150–300 bar) and short residence times (5–15 min), preferably performed in a continuous flow tube reactor to ensure short residence times and, thus, to avoid repolymerization.

In principle, lignin is cleaved into a mixture of aliphatic degradation products (methanol, formic acid and acetic acid), phenolic mono- (e.g., guaiacol, syringol and catechol), di- and oligomers and carbon dioxide. This reaction mode is simplified in **Figure 3**.

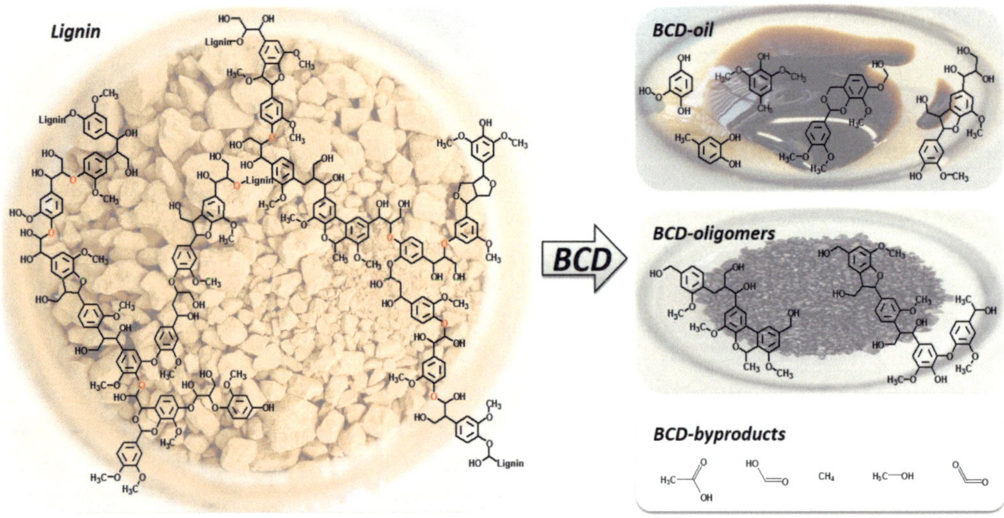

Figure 3. Reaction scheme for cleavage via base-catalyzed depolymerization (BCD) at β—O—4 ether bonds (red) with structures of lignin, BCD-oil (phenolic mono- and dimers), BCD-oligomers (phenolic oligomers) and BCD-by-products (organic acids, alcohols and gases).

Mechanism of bond-breaking and product formation: Lignin is solved in alkaline water, whereas the alkaline or earth alkaline metal ions polarize the ether bond. At reaction conditions, bond-breaking occurs mainly at the β—O—4 bond (aryl-glyceryl-β-aryl ether bond) and the 4—O—5 bond (diaryl ether bond) as aryl-alkyl-ether bonds are the weakest bonds in lignin. According to the structure of the lignin molecule, up to 25% of monomer units, i.e., phenol, guaiacol, syringol and catechol derivatives, are formed. The ideal process would be a reaction affording high yields in monomers and, thus, a nearly entire cleavage of all aryl-ether bonds within the lignin molecule [28].

A detailed description and analytical characterization of the monomeric aromatic compounds is given in [28–31]. In addition to these monomers, a large proportion of dimeric and oligomeric structures are obtained having the following types of bonds: 5-5 (biphenyl bond), β-5 (phenylcoumaran), β-β (THF or resinol type) and β-1 (1, 2-diarylpropane). These compounds accumulate in the liquid product (BCD-oil) depending on the processing method for separation. The BCD-oligomer fraction contains oligomeric polyphenols, higher condensed structures and unreacted lignin. Short-chain acids formed during the reaction neutralize the base and can hinder the hydrolysis process. At worst, the pH in the reactor drops to such an extent that dissolved phenolates precipitate in the reactor, which can lead to clogging and blockages of the reactor. For this reason, the pH value, due to a sufficiently high base concentration, must be at least pH 12 at all times [32].

Demethoxylation and demethylation reaction also occur at more harsh process parameters resulting in the formation of catechol-type molecules with simultaneous formation of methanol, formic acid and acetic acid. A shift in the yield to catechols with simultaneous decrease in syringol concentration was clearly demonstrated by BCD of beech wood organosolv lignin and kraft lignin operating above 300°C. Here, the formation of catechol-type monomers by BCD, in the absence of a H_2 donor, mainly depends on retention in the reactor and the process temperature [28]. Investigations on hardwood organosolv lignin showed demethoxylation processes starting at 280°C by decreasing syringol concentration in the monomeric product phase [33]. Kinetic studies on lignin model compounds mimicking the alkyl-aryl or aryl-aryl-ether linkages have shown that BCD is a rapid reaction that occurs within minutes under appropriate conditions [2, 32].

In general, the presence of alkali and earth alkali in base media modifies the reaction routes, facilitates bond-breakings and in some cases enhances the formation of formic or acetic acid during the depolymerization process. Moreover and as already reported earlier, base catalysts are required in water to solubilize the lignin feedstock, to avoid coke formation and to increase liquid product yield. Condensation reaction is conspicuously suppressed during BCD processes compared to process in acid or neutral media [34].

3.2. State of the art and overview

Bond-breaking and recombination are strongly defined by process conditions, namely temperature, pressure, residence time and base concentration along with the type of feedstock used. Equally important is the chemical nature of the base. **Table 1** shows selected examples of BCD strategies and compares striking process parameters. In this case, only reactions in liquid homogeneous phase without the use of additional catalyst systems or other reaction agents were considered.

3.2.1. Lignin nature

The catalytic reagents are cheap and commercially available bases such as LiOH, NaOH and KOH. Its nature is important for the oil yield and product composition. Usually, stronger base gives higher conversion since the polarization of the base governs the kinetics and the mechanism of the depolymerization reaction. Highest BCD-oil yield could be observed with NaOH. Utilizing $Ca(OH)_2$ results in low BCD-oil formation when processing olive tree pruning lignin at 300°C [34]. Another example also describes that using strong bases (KOH and NaOH) converts more of the lignin to low molecular weight products than weak bases (LiOH, $Ca(OH)_2$ and Na_2CO_3). Here, maximum conversion was achieved at a NaOH/lignin ratio of 1.5–2. Higher ratios increased the conversion rate but not the conversion degree to lower molecular weights and higher BCD-oil yields. The optimum was given with 8 wt.% of insoluble material at a reaction time of 1 h at 330°C in batch microreactor system using up to 10 wt.% of lignin in solution. Further, depolymerization is more efficient and less dependent on temperature at higher base concentration [35]. This observation was also reported by Miller et al. and Roberts et al. who had found less insoluble products (unconverted lignin) with increasing NaOH/lignin ratios [32, 33]. In general, base concentration has to be sufficient in order to fully polarize and deprotonate the phenolic and catecholic species to alcoholates. This must be taken into account, in particular, in that the base concentration decrease during the cleavage process, as new hydroxyl groups, reactive organic acids and carbon dioxide are formed.

Feedstock	Base catalyst	Reaction parameters		Products and composition	Reference
		Reaction conditions	Reaction system		
Alcell/indulin lignin	10 wt.% lignin in MeOH or EtOH/ CsOH, NaOH, KOH, LiOH $Ca(OH)_2$ or Na_2CO_3- sol.	290°C, 60 min	Batchwise	Ether-soluble products ≤93 wt.%	[32]
Beech wood organosolv lignin and lignoboost kraft lignin	5 wt.% lignin in 1–5 wt.% NaOH-sol.	250–340°C, 250 bar; 5, 10, 15 min	Continuous mode, plug-flow reactor	Oil content ≤23 wt.%	[28, 39]
Organosolv lignin	2.5–10 wt.%, 2.5 wt.% NaOH-sol.	240–340°C, 250–315 bar, retention time 2–15 min	Continuous mode, plug-flow reactor	~ 15 wt.% oligomers, ~ 22 wt.% oil	[33]
Steam explosion hemp-lignin	5 wt.-% lignin in NaOH-sol.	300–330°C, 90–130 bar	Batchwise	Up to 11 wt.% monomer species	[29]
Softwood indulin lignin	10 wt.% lignin in 5 wt.% NaOH sol.	270–315°C, LHSV 1.4–4 h^{-1}	Continuous mode, plug-flow reactor	Small organic compounds ≤19 wt. %, solid products ≤70 wt.%	[37]
Organosolv lignin from olive tree pruning	4 wt.% lignin in NaOH, KOH, LiOH or K_2CO_3 – sol.	300 °C, 900 bar	Batchwise	Oil (monomers and dimers), yield 5–20 wt.%	[34]
Olive tree pruning organosolv lignin	5 wt.% lignin, 4 wt.% NaOH-sol.	300 °C, 90 bar, 80 min	Batchwise	Oil components ≤18.5 wt.%	[36]
Lignin-rich residue from corn stover	10 wt.% lignin, 2–4 wt. % NaOH sol.	270–300°C	Batchwise	Aqueous soluble fraction ≤78 wt.%	[31]
Gray ironbark organosolv lignin	5 wt.% lignin, 4 wt.% NaOH sol.	300 °C, 40 min	Batchwise, microreactor	Oil components ≤21.9 wt.%, solid products ≤58.3 wt.%	[6]
Beech wood organosolv lignin and kraft lignin	2.5–10 wt.% lignin in 2.5–7.5 wt.% NaOH-sol.	300–340°C, 250 bar; retention time 450–900 s	Continuous mode, plug-flow reactor	Oil content ≤14.5 wt.%	[8, 38]
Lignin-rich residue from corn stover	10 wt.% lignin, 1–2 wt. % NaOH sol.	60–240°C, 0.7–20 bar	Batchwise	Aromatic acids	[35]

Table 1. Selected examples and comparison of BCD of different kinds of lignin at various process conditions.

3.2.2. *Nature of the catalyst*

Similarly, it can be said that, besides, the base concentration temperature is the most important factor and has the strongest influence on the yield of oil and solid fractions. By cleaving organosolv lignin and sulfur-containing kraft lignin, it was shown that higher process intensity (temperature, pressure and residence time) results in higher depolymerization degree and, thus, in growing formation of monomeric and dimeric phenolic products and also of gaseous and liquid degradation products. Temperature influences not only the yield but also the molecular composition caused by the already mentioned successive reactions such as decarboxylation and demethoxylation at the substituent and side chains of the aromatic ring in the lignin molecule [8, 38].

Usually, the ether hydrolysis starts at 250°C. Ideal process conditions are described between 300 and 340°C in order to achieve complete cleavage of the ether bonds and to obtain a high yield of liquid products (see **Table 1**). Investigation on BCD at lower temperatures down to 60°C showed a poor yield on liquid products. On the other hand, at very mild reaction temperatures of up to 140°C, the exclusive formation of coumaric and ferulic acid can be observed [35]. Low yields are not surprising due to the manufacturing process of the lignin. Nearly all lignin preparation methods are thermochemical processes using temperatures between 150 and 180°C (e.g., kraft-, organosolv- and steam-explosion lignin).

3.2.3. Influence of the processing conditions

The composition of softwood and hardwood lignin is varying in the relative abundance of coumaryl, coniferyl and sinapyl alcohol units. Coniferyl alcohols constitute approximately 90% of softwood lignin, whereas roughly equal proportions of coniferyl alcohol and sinapyl alcohol appear in hardwood lignin [2]. These structural differences significantly influence the product composition. This was clearly showed in the comparison of BCD of hardwood organosolv lignin with softwood kraft lignin. Syringol and dimethoxyphenols are found almost exclusively in the BCD-oil of organosolv lignin with a significant amount in the monomer phase [8, 28, 30]. Other studies focused on the investigation of one type of wood but with different pretreatment methods applied. Erdocia et al. investigated the BCD of lignin from olive tree pruning fractioned by different organosolv processes, namely acetosolv, formosolv and acetosolv/formosolv. It could be summed up that low molecular weight lignin leads to more monomeric phenolic compounds, whereas the amount of obtained residual lignin is independent of the lignin nature [36]. The pretreatment method of the lignin stream has a significant influence on the product yields and nature since successability to BCD is strongly influenced by lignin structure and the amount of impurities, like carbohydrates, in the feedstock. Different mechanically refined substrates and acid-pretreated lignin-rich feedstock have been examined by Katahira et al. and the product dependence on the pretreatment has been demonstrated [31].

3.3. Challenges

3.3.1. Repolymerization and char formation

It is also proposed that in early stage of the hydrothermal treatment, lignin was decomposed to water-soluble fragments, and with the increase of the reaction time, the soluble products were transformed to insoluble products by recondensation reactions. Investigations on model compounds confirmed the theory that the two processes of lignin depolymerization and repolymerization are concurrent. The latter often leads to products that are more resilient than the initial lignin due to C—C bond formation [40, 41]. In the case of lignoboost kraft lignin processed at 350°C and 250 bar, high molecular weight fractions were found consisting of repolymerized macromolecules with new structural networks based on guaiacol, disubstituted aromatic ethers and polyaromatic hydrocarbon structures bound tightly together [42].

In order to overcome this phenomenon, the reaction parameters that determine the amount of repolymerization need to be optimized and the amount of lignin in the mixture should be carefully balanced [8, 12]. In addition, there is a need to scavenge and deactivate the reactive species possessing phenol, carbonyl or alkene functionalities. Boric acid acts as an excellent protecting or capping agent by forming esters with phenolic hydroxyl groups and, thus, reducing concurrent reaction of polymerization. The highest oil yield of 52 wt.% was found at an NaOH/boric acid ratio of 0.75 at 300°C and boric acid/lignin weight ratio of 2. Oil yield could be doubled relative to the standard NaOH-catalyzed process [33]. Similar results have been found utilizing phenol as capping agent. Interestingly, this method was proven to favor phenolic compound production (cresol, catechol and ferulic acid) and, at the same time, to avoid repolymerization [43].

3.3.2. Separation, purification and analytical characterization strategies

The assessment of process efficiency and selectivity requires the application of separation and purification methods for sample preparation and accurate analytical protocols for the specification and characterization of main and by-products. The methods described in literature are just as multivarious as depolymerization strategies. Kozliak et al. recently reviewed the efficiency, selectivity and product analysis of thermal liquefaction processes of lignin to aromatics [41]. Overall objectives are the detection and quantification of specific structural changes altered by depolymerization reaction, the separation and identification of individual phenolic compounds/fractions and attempts to close the mass balance.

Solid product fraction: Conventionally, downstream processing of the reaction outlet starts by precipitating the solid BCD fraction by adding a strong acid, e.g., sulfuric or hydrochloric acid. If possible, this quench must be carried out immediately after BCD in order to prevent further reaction and repolymerization of the phenolic compounds. The product slurry is subsequently treaded to separate water-soluble and water-insoluble products by centrifugation and/or filtration. Heating the precipitated suspension to 70°C for 15 min can significantly improve the filterability of the suspension [29]. This strategy is especially recommended for gravimetric determination of the BCD-oligomer fraction, since minor yield losses, water and ash contents compared to centrifugation. Analytical methods for structure identification include GPC, FT-IR, TGA, ^{1}H-NMR, ^{13}C-NMR and elemental analysis. Furthermore, an organic solvent solubilization is often used for further fractionation of the solid fraction. Toledano et al. used THF in order to separate lignin-derived oligomers (THF-soluble fraction) and coke (THF-insoluble fraction) [34].

Liquid product fraction: Strategies for the separation of value-added monomers from the BCD cleavage have been done by Vigneault et al. utilizing a steam-exploded aspen lignin. In addition to the described downstream strategy, a concept for the isolation of 12 monomeric lignin derivatives was shown. The strategy combines liquid-liquid extraction with subsequent vacuum distillation of monomers and liquid chromatography and crystallization for purification of the products like catechol, phenol, guaiacol derivatives, syringol derivatives and vanillin [44]. Unkelbach et al. evaluated organic solvents of different polarity (DCM, EtOAc, MIBK and toluene) for the extraction of different phenols from acidified/filtered BCD reactor water (pH value = 3). The extracts were analyzed gravimetrically and by GC- and LC-MS. The highest amount of oil was achieved with MIBK (28 wt.% of lignin). Extraction with EtOAc, DCM and toluene led to lower oil amounts (23, 12 and 9 wt.%) [39]. The results of Greminger et al. from measurements of equilibrium distribution coefficients for phenol, dihydroxybenzenes and trihydroxybenzenes also lead strongly toward the use of MIBK as a preferred solvent, especially for polyhydric phenols [45].

Extraction methods are particularly suitable for identification and quantification of individual phenolic compounds by gas chromatography and gravimetrical determination of selectivities referring to the used organic solvent. Additional characterization methods of the phenolic fraction are thin-layer chromatography, IR, ^{1}H- and ^{13}C-NMR. Total carbon (TC) measurement of BCD reactor water is recommended for mass balance, including all inorganic and organic compounds.

Gaseous product fraction: Joffres et al. characterized liquid, solid and gaseous products after batch catalytic (NiMo/Ag_2O_3) hydroconversion of a wheat straw soda lignin (Protobind 1000) for 5 h at 350°C, 8 MPa in tetralin solvent. Gaseous products were carbon dioxide (major gas), carbon monoxide, methane and light hydrocarbons (ethylene, ethane and propane), identified by GC-TCD/MS [12]. Long et al. analyzed gaseous products after batch depolymerization of pine lignin (220–280°C, 15–60 min) in MgO/THF solvent but did not take them into mass balance account, as the total amount was less than 1% of raw lignin [46]. BCD of commercial kraft lignin (indulin AT) was performed at 270–315°C, 130 bar and 15–45 min residence time (flow reactor) by Beauchet et al. quantifying gases (13,2–25,5%) by difference [37].

Generally, the chemical composition of the BCD reaction product mixture depends on the type and concentration of the lignin feedstock, the solvent and catalyst and the reaction conditions (temperature, pressure and retention time). A wide spectrum of gaseous (mainly CO_2), liquid (monomeric/dimeric oxyaromatics and low MW compounds, e.g., formic acid, acetic acid and methanol) and solid products (oligomeric oxyaromatics) formed during BCD degradation. Commonly used and promising separation, purification and analytical strategies are summarized at **Figure 4**.

3.3.3. *Increasing the technology readiness level for industrial implementation*

Industrial implementation of the processes requires fast processing in continuous operated reactors and plant equipment that can be implemented in existing infrastructures. Here, an economically sensible mode of operation is a prerequisite for the application of BCD in the process industry. A full commercial application where the technology is proven in operational environment and available for customers has not been launched so far (equal to technology readiness level TRL 9) [47]. Research activities in the field of BCD of lignin mostly range from applied research activities to small-scale prototypes in laboratory environment (TRL 3-4).

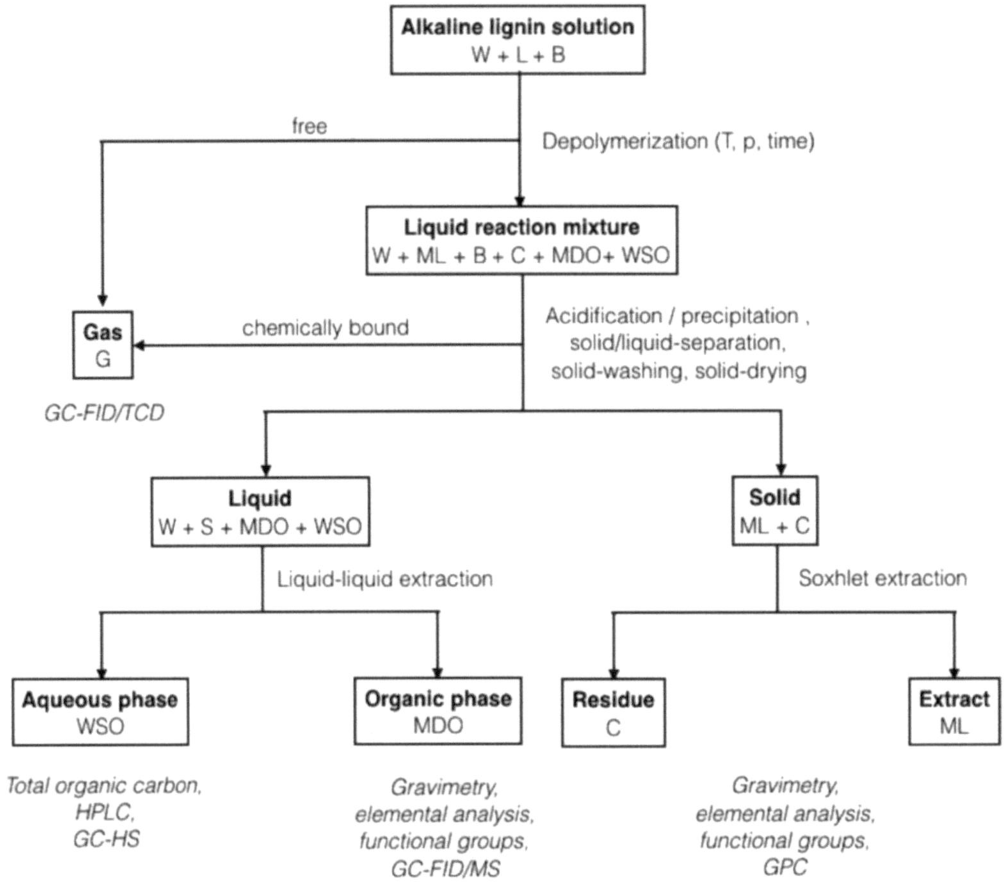

Figure 4. Strategies of separation, purification and analytical characterization of BCD cleavage products (W, water; L, lignin; B, base solvent; ML, modified lignin; C, coke; MDO, phenolic mono-,di-, and oligomers; WSO, water-soluble organics; S, salts).

Large-scale prototypes for plant equipment tested in intended industrially relevant environment close to expected performance (TRL 5-6) have been reported only by Rößiger et al. so far [8]. This includes not only the cleavage but also the corresponding downstream processing for the isolation and purification of the two product fractions BCD-oil and BCD-oligomers.

Commonly used batch reactors have inherent disadvantages including the alternation of heating and cooling, pressurization and depressurization, with inevitable energy losses [41]. Moreover, it can be pointed out in literature that char formation caused by repolymerization can be minimized by continuous operation of the BCD process due to a precise setting of the reaction parameter temperature and residence time. A customized engineering design of the reactor results in precise heating and cooling rates, residence time distribution and adjusted material design. Dedicated reactor systems have been described at laboratory and technical lab scale [18, 28, 33, 48] and also at a more upscale and direct approach [37] and in pilot scale [8]. The comparisons of the individual technologies are shown in **Table 2**.

Feedstock	BCD process	Reaction and reactor parameters		Reference
		Reaction conditions	Reaction system	
Hardwood organosolv lignin	In NaOH/aqueous solution with and without boric acid as capping agent, homogeneous phase	240–340°C, 250–315 bar, retention time 2–15 min	Plug-flow continuous mode, lab scale*,**	[33]
Beech wood organosolv lignin and kraft lignin	In NaOH/aqueous solution with and without formic acid as hydrogen donor, homogeneous phase	300–340°C, 250 bar, retention time 5, 10, 15 min	Plug-flow, continuous reactors, 0.25 L reactor volume**	[28]
Softwood indulin lignin	In NaOH/aqueous solution, homogeneous phase	270–315°C, LHSV 1.4–4 h-1	Plug-flow, continuous reactors, 1.0 L reactor volume	[37]
Lignoboost kraft lignin	In NaOH/aqueous solution with ZrO_2 catalyst with and without phenol as accepting agent, heterogeneous phase	350 °C, 250 bar, feed rate 1 kg/h	Fixed-bed continuous batch reactor, reactor volume 0.5 L	[48]
Alkali lignin and ligno-sulfonic acid	In base/water system and additional reagents, e.g., oxygen, peroxides and copper nitrate, homogeneous phase	200–380°C, 150–350 bar, residence times to 90 s	Plug-flow continuous reactors sequentially connected in series, total volume 9.8 ml**	[18]
Beech wood organosolv lignin and lignoboost kraft lignin	In NaOH/aqueous solution, homogeneous phase	250–340°C, 250 bar, retention time 450–900 s	Plug-flow, continuous reactors, 2.2 L reactor volume	[8]

*Size and flow rate not depicted.

**Quenching of the lignin solution with hot aqueous solution.

Table 2. Compilation of publications on BCD in continuous operation.

Heterogeneous catalysis has been performed in a continuous batch reactor applying a fixed bed of ZrO_2 catalyst [48]. For homogeneous base/aqueous feed streams, plug-flow reactor models are used. The heat input is predominantly realized via electrical preheaters and electrically operated heating jackets, whereby the lignin solution might be fed directly into the reactor. **Figure 5** shows the flow sheet of a pilot plant (feed rate up to 20 kg/h) for BCD cleavage operating with lignin feed streams up to 10 wt.% and subsequent isolation of the liquid-base product solution [8]. Cofeeding of the lignin/base solution to a water stream heated to operational temperature is an alternative approach that has been described several times in technical lab scale design [18, 28, 33]. Here, the two streams, water and lignin/base solution, are mixed right before entering the reactor. This procedure allows a precise heating of the lignin solution to the reaction temperature, thus, leading to a more exact and precise temperature control and also avoiding overheating in the system caused by preheating the lignin solution. As a consequence, the enhanced formation of higher molecular weight lignin fragments with a broad molar mass distribution is reduced, but the whole reaction solution is diluted due to the necessary secondary water stream. However, downstream-processing might be more consuming and expensive.

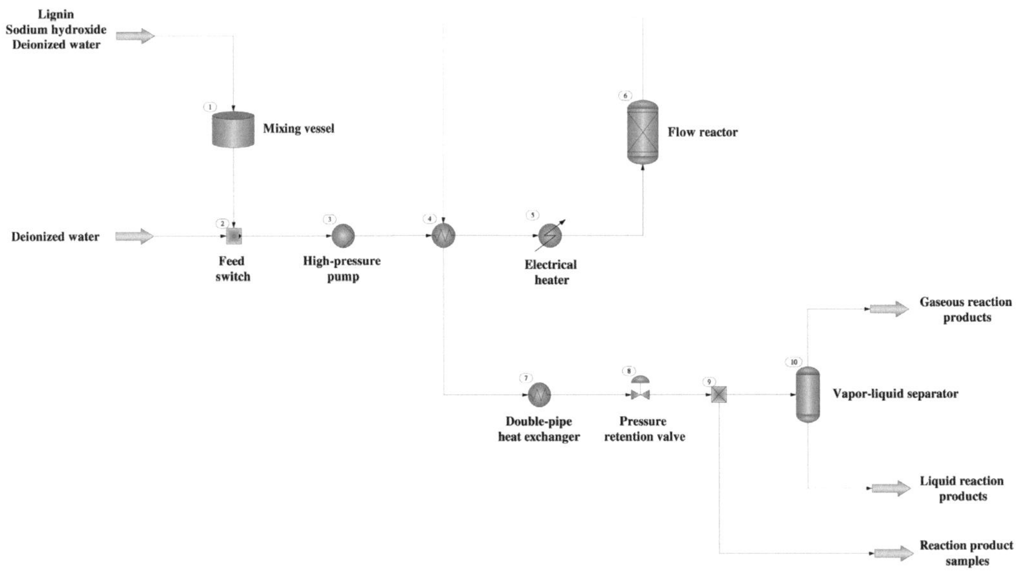

Figure 5. Schematic flow sheet of the plant for BCD depolymerization in pilot scale [8].

4. Trends and perspectives

The following chapter deals with developments beyond the original BCD in a homogeneous liquid phase. Innovative designs and processes include novel catalyst systems, subsequent and coupled reaction modes and the use of innovative reaction media. The depolymerization of lignocellulose-containing feedstocks by BCD has already been investigated and will also be discussed here exemplarily.

4.1. Solid-phase catalysts for improved lignin degradation

An efficient and economical catalytic system using the available industrial solid-phase catalyst MgO has been described by Long et al. [46]. Conversion rates in water system operating at 250°C show satisfactory results in conversion of lignin. Due to its excellent dissolution capabilities and the promotion effect for the catalyst organic solvent, in particular THF, it is found to be an efficient reaction media and leads to a significant increase in conversion rate. It was also shown that zeolite catalysts (NaX, NaY and NaP) lead to improved yield on monomeric products compared to other catalysts like MgO or CaO. A maximum of 51 wt.% of low molecular weight products was achieved in an ethanol/water medium at 250°C [23]. Layered double hydroxides (LDH) as recyclable, heterogeneous catalysts have been introduced recently for BCD of lignin. Such solid-base catalysts avoid the cost of liquid phase, nonrecyclable base and downstream processing steps like neutralization. LDH are prepared from Al and Mg salts and are stable in water and organic solvents at relatively high working temperature [49, 50].

4.2. Upgrading of BCD products by catalytic refinement

Lignin-derived fractions cannot be utilized directly as fuels or fuel additives due to their high oxygen content, acidity, instability, high viscosity and complexity of constituent compounds. It should be mentioned at this point that O/C ratio and H/C ratio of fossil species are 0 to 0.03 and from 1.6 to 2.1, respectively. In contrast, the O/C and H/C ratios of the BCD bio-oils range from 0.6 to 0.7 and from 1.0 to 1.35, respectively. Upgrading the cleavage products should therefore target oxygen removal, carbon preservation and hydrogenation of unsaturated carbon bonds. Thus, catalytic upgrading processes to convert the alkoxyphenol species are required for further chemical transformation to the improved and enhanced products. Procedures typically focus on further reduction in the oxygen content via hydrodeoxygenation (HDO) and/or further cleavage with larger oligomeric fragments of the oil, as so-called hydrocracking (HC). Solvolysis of lignin-utilizing reductive agents is summarized as hydrogenolysis (HGL) [3, 40, 51].

Hydrogenolysis of lignin: HGL has been investigated for decades and combines C—O bond-breaking with simultaneous hydrogenation at C and O moieties. Hydrogen-donating solvents such as tetralin, formiate or anthracene species have been successfully tested. Also alcohols have been widely used for *in situ* generation of hydrogen [5, 52]. These are highly attractive methods, especially considering that methanol or ethanol can be potentially derived from lignocellulose or other renewable resources. Hydrogen transfer from methanol over basic copper catalyst under supercritical or milder conditions reveals the formation of aromatic and aliphatic alcohols. The role of the multifunctional catalyst is to promote reforming of the methanol solvent to CO and H_2 that serve thereby as a "liquid syngas" and to catalyze diverse hydrogenolysis and hydrogenation processes. The latter consumes the produced hydrogen equivalents [13]. Besides, hydrotreating and hydrothermal treatment of alkaline lignin in methanol over Ru/Al_2O_3 have been found to improve product yield and selectivity and reduce char formation at lower working temperatures and pressures [53].

Another approach is introducing hydrogen H_2 directly into the reaction media within pressurized systems. Activation of H_2 is mandatory and realized by using various heterogeneous catalyst systems. Supported Ni (on Al_2O_3) or Ru (on carbon) materials have been used for example to break down lignin into monomeric units by introducing hydrogen at pressure up to 70 bar [5, 54]. To mention at this point is also the use of ionic liquids as a solvent and in parallel as an acid catalyst for improved product refinement. The amount of total hydroxyl groups could be increased in the final cleavage product using 1-butyl-3-methylimidazolium chloride (BMIM Cl) in the presence of H_2, compared to the original used soda lignin. As a result, lignin antioxidant activity was enhanced [55].

In a one pot system, lignin can also be directly transferred to some aliphatic hydrocarbons in the so-called lignin to liquid process (LTL process). Kleinert et al. have introduced this method for fully liquefying lignin using formic acid, ethanol and i-propanol [52].

Hydrodeoxygenation of BCD intermediates: Catalytic hydrodeoxygenations are mostly appended to the BCD process. They are used to produce aromatic hydrocarbons and alky-cycloalkanes with low oxygen content and high stability. Within these kinds of processes, C—O bond-breaking occurs, and C—O, C—C and C—H bonds as well as the aromatic rings will be saturated. While oxygen is partially and selectively removed, acidity of the intermediates is lowered resulting also in an inhibition of repolymerization. Hereby, mixtures of aromatics and alkanes with narrower product distribution are generated, which contain oxygen-poor, low molecular weight molecules, giving access to biofuels and bulk chemicals [3].

Consequently, a significant part of the developments is found in patent applications. Processes combining several internal steps are also published. Mild BCD at up to 240°C is followed by selective hydrocracking with superacid catalyst. The resulting oxygen-containing depolymerized lignin of alkylphenols will further be catalytically etherficated with methanol or ethanol leading to reformatted partially oxygenated gasoline [20, 22]. Also the developments for liquid biofuel synthesis by subsequent BCD and HDO processes to C7-C11 alkylaromatic compounds and paraffins should be mentioned [21, 56]. Cobalt and molybdenum catalysts [21, 56] or iron oxides [57] in the presence of H_2 are utilized. Subsequent catalytic oxidation with peroxides of BCD intermediates is rising the production of useful platform and fine chemicals [18].

4.3. Utilizing black liquor and lignocellulose feedstock

Depolymerization of black liquor without isolation of lignin might be advantageous over solid lignin processing since the stage of precipitation could be bypassed, reducing the costs of the whole process chain. As kraft black liquor presents an alkali medium, BCD depolymerization of lignin can be carried out similarly to isolated lignin in order to obtain small phenolic compounds. A direct comparison for the conversion of solid soda lignin and soda black liquor revealed a higher conversion degree for black liquor. Yields of BCD-oil and BCD-oligomers were significantly higher, with less formation of char [36]. The application of catalytically active reagents, such as phenols or hydrogen peroxide, also markedly increases the amount of small phenolic compounds and prevents repolymerization to undesirable products (residual lignin and coke) [58].

Current developments also include the direct transformation of woody biomass and other lignocellulosic feedstock for the production of aromatic building blocks. An especially interesting concept is the process of transferring hydrogen equivalents from cellulose fraction into lignin. Copper-doped basic heterogeneous catalysts are well suited for the conversion of both lignin and cellulose-derived feedstocks into alcoholic species where humin formation is suppressed [13].

Furthermore and beyond the previous examples, we want to mention a remarkable development for reductive lignocellulosic fractionation at this point. The innovative catalytic lignin's first LF process aims at the direct and selective conversion of lignin directly from wood pellets, allowing a high conversion into monomeric and some dimeric alkyl phenols next to small oligomers. In the presence of a Ru on carbon catalyst (Ru/C) in methanol under H_2 atmosphere at elevated temperature, lignocellulose sawdust is efficiently delignified through simultaneous solvolysis and catalytic hydrogenolysis, resulting in a carbohydrate pulp and a lignin oil [59, 60].

5. Conclusion and closing remarks

Current research and development activities address the optimization of product specifications from BCD cleavage with regard to low molecular weight alkylphenols and alkylnaphthenes possessing narrow molecular weight distribution. Reactive additives, such as phenols or organic alcohols, are used to suppress repolymerization and condensation of product intermediates and, thus, to avoid char formation. However, complex mixtures consisting of molecules with relatively high oxygen content restricts its implementation in the usual process chains of the chemical and process industries. Therefore, concurrent or subsequent hydrodeoxygenation or hydrolysis of BCD-oils or BCD-oligomers is applied for further refinement and gives rise to the use as bulk chemicals or fuels. Significant effort has been made in recent years to advance developments in this area. But yet, industrial implementation has not taken place. Therefore, one has to ask at this point whether the utilization of lignin as low-priced products in the form of platform chemicals appears to be economical. Material use is usually in competition with its energetic use in the form of black liquor or other biorefinery lignin feedstocks. The production of fine chemicals and specificities by means of the BCD process might offer the possibility to provide even high-priced products made of lignin. In particular, functional monomeric derivatives, such as vanillin, syringaldehyde or apocynin, are interesting candidates. Even a cascade of uses, for example, low molecular weight aromatics in the field of fine and platform chemicals and medium and high molecular weight fractions as polymer additives, is worth mentioning here.

In order to increase the technology readiness level of the BCD process and, thus, to develop an industry-relevant process, it is necessary to establish an overall approach regarding material and energy efficiency as well as to examine its technical and economic feasibility. Still one of the biggest limitations for its commercial application is the obtainment of complicated product mixtures. The low selectivity to products and subsequent required separation and purification are the main hurdles for lignin cleavage processes. Therefore, new innovative approaches and technologies especially for downstream processing, like membrane techniques, have to be considered.

Author details

Björn Rößiger, Gerd Unkelbach and Daniela Pufky-Heinrich*

*Address all correspondence to: daniela.pufky-heinrich@igb.fraunhofer.de

Fraunhofer Center for Chemical-Biotechnological Processes (CBP), Leuna, Germany

References

[1] Kamm B, Schmidt M, Hirth T, Schulz M. Lignocellulose-based chemical products and product family trees. In: Biorefineries—Industrial Processes and Products: Status Quo and Future Directions. Weinheim: Wiley; 2006. pp. 67-84

[2] Zakzeski J, Bruijnincx PCA, Jongerius AL, Weckhuysen BM. The catalytic valorization of lignin for the production of renewable chemicals. Chemical Reviews. 2010;**110**(6):3552-3599. DOI: 10.1021/cr900354u

[3] Graglia M, Kanna N, Esposito D. Lignin refinery: Towards the preparation of renewable aromatic building blocks. ChemBioEng Reviews. 2015;**2**:377-392. DOI: 10.1002/cben.201500019

[4] Li C, Zhao X, Wang A, Huber GW, Zhang T. Catalytic transformation of lignin for the production of chemicals and fuels. Chemical Reviews. 2015;**115**(21):11559-11624. DOI: 10.1021/acs.chemrev.5b00155

[5] Xu C, Arancon RAD, Labidi J, Luque R. Lignin depolymerisation strategies: Towards valuable chemicals and fuels. Chemical Society Reviews. 2014;**43**:7485-7500. DOI: 10.1039/C4CS00235K

[6] Dos Santos PSB, Erdocia X, Gatto DA, Labidi J. Bio-oil from base-catalyzed depolymerization of organosolv lignin as an antifungal agent for wood. Wood Science and Technology. 2016;**50**(3):599-615

[7] Mahmood N, Yuan Z, Schmidt J, Xu C. Hydrolytic liquefaction of hydrolysis lignin for the preparation of bio-based rigid polyurethane foam. Green Chemistry. 2016;**18**:2385-2398. DOI: 10.1039/C5GC02876K

[8] Rößiger B, Röver R, Unkelbach G, Pufky-Heinrich D. Production of bio-phenols for industrial application: Scale-up of the base-catalyzed depolymerization of lignin. Green and Sustainable Chemistry. 2017;**7**(3):193-202. DOI: 10.4236/gsc.2017.73015

[9] Web of Knowledge [Internet]. 2017. Available from: https://www.wcs.webofknowledge.com/RA/analyze.do [Accessed: 2017-11-01]

[10] Abdullah B, Muhammad SAFAS, Mahmood NAN. Production of biofuel via hydrogenation of lignin from biomass. In: New Advances in Hydrogenation Processes—Fundamentals and Applications. Rijeka, Croatia: InTech; 2017. DOI: 10.5772/66108

[11] Azadi P, Inderwildi OR, Farnood R, King DA. Liquid fuels, hydrogen and chemicals from lignin: A critical review. Renewable and Sustainable Energy Reviews. 2013;**21**:506-523. DOI: 10.1016/j.rser.2012.12.022

[12] Joffres B, Laurenti D, Charon N, Daudin A, Quignard A, Geantet C. Thermochemical conversion of lignin for fuels and chemicals: A review. Oil & Gas Science and Technology. 2013;**68**(4):753-763. DOI: 10.2516/ogst/2013132

[13] Deuss PJ, Barta K. From models to lignin: Transition metal catalysis for selective bond cleavage reactions. Coordination Chemistry Review. 2016;**306**(2):510-532. DOI: 10.1016/j.ccr.2015.02.004

[14] Wang H, Tucker M, Ji Y. Recent development in chemical depolymerization of lignin: A review. Journal of Applied Chemistry. 2013;**2013**:1-9. DOI: 10.1155/2013/838645

[15] Gillet S, Aguedo M, Petitjean L, Morais ARC, da Costa Lopes AM, Łukasik RM, Anastas PT. Lignin transformations for high value applications: Towards targeted modifications using green chemistry. Green Chemistry. 2017;**19**:4200-4233. DOI: 10.1039/C7GC01479A

[16] Espacenet [Internet]. 2017. Available from: https://www.worldwide.espacenet.com/search; Results?ST=singleline&locale=en_EP&submitted=true&DB=&query=lignin+depolymerization&Submit=Search [Accessed: 2017-11-07]

[17] Wayman M, Lora JH. Process for depolymerization and extraction of lignin [Patent]. Canada. CA 1147105; 1983

[18] Joshi SS, Kulkarni BD, Tomer AS, Jadhav MR. A Continuous process for depolymerization of lignin to industrially useful chemicals [Patent]. Council of Scientific and Industrial Research. WO 2014/068590 A1; 2014

[19] Seayad J, Saravanan G, Ramalingam B. Process for the depolymerization of lignin under neutral oxidative conditions [Patent]. Singapore: Agency for Science, Technology and Research. WP 2017/171652 A1; 2017

[20] Shabtai JS, Zmierczak WW, Chornet E. Process for conversion of lignin to reformulated, partially oxygenated gasoline [Patent]. US 6172272; 2001

[21] Zmierczak WW, Miller JD. Processes for catalytic conversion of lignin to liquid bio-fuels and novel bio-fuels [Patent]. US 7964761 B2. University Of Utah Research Foundation; 2011

[22] Shabtai J, Zmierczak WW, Chornet E, Johnson D. Process for converting lignins into a high octane blending component [Patent]. US 20030115792 A1; 2003

[23] Chaudharyab R, Dhepe PL. Solid base catalyzed depolymerization of lignin into low molecular weight products. Green Chemistry. 2017;**19**:778-788. DOI: 10.1039/C6GC02701F

[24] Gasser CA, Hommes G, Schäffer A, Corvini PF. Multi-catalysis reactions: New prospects and challenges of biotechnology to valorize lignin. Applied Microbiology and Biotechnology. 2012;**95**(5):1115-1134. DOI: 10.1007/s00253-012-4178-x

[25] Crestini C, Crucianelli M, Orlandi M, Saladino R. Oxidative strategies in lignin chemistry: A new environmental friendly approach for the functionalisation of lignin and lignocellulosic fibers. Catalysis Today. 2010;**156**(1-2):8-22. DOI: 10.1016/j.cattod.2010.03.057

[26] Huber GW, Iborra S, Corma A. Synthesis of transportation fuels from biomass: Chemistry, catalysts, and engineering. Chemical Reviews. 2006;**106**(9):4044-4098. DOI: 10.1021/cr068360d

[27] Kang S, Li X, Fan J, Chang J. Hydrothermal conversion of lignin: A review. Renewable and Sustainable Energy Reviews. 2013;**27**:546-558. DOI: 10.1016/j.rser.2013.07.013

[28] Schmiedl D, Endisch S, Pindel E, Rückert D, Reinhardt S, Unkelbach G, Schweppe R. Base catalyzed degradation of lignin for the generation of oxy-aromatic compounds—Possibilities and challenges. Erdöl Erdgas Kohle. 2012;**128**(10):357-363

[29] Lavoie JM, Baré W, Bilodeau M. Depolymerization of steam-treated lignin for the production of green chemicals. Bioresource Technology. 2011;**102**(7):4917-4920. DOI: 10.1016/j.biortech.2011.01.010

[30] Jensen MM, Madsen RB, Becker J, Iversen BB, Glasius M. Products of hydrothermal treatment of lignin and the importance of ortho-directed repolymerization reactions. Journal of Analytical and Applied Pyrolysis. 2017;**126**:371-379. DOI: 10.1016/j.jaap.2017.05.009

[31] Katahira R, Mittal A, McKinney K, Chen X, Tucker MP, Johnson DK, Beckham GT. Base-catalyzed depolymerization of biorefinery Lignins. ACS Sustainable Chemistry & Engineering. 2016;**4**(3):1474-1486. DOI: 10.1021/acssuschemeng.5b01451

[32] Miller JE, Evans L, Littlewolf A, Trudell DE. Batch microreactor studies of lignin and lignin model compound depolymerization by bases in alcohol solvents. Fuel. 1999;**78**(11):1363-1366. DOI: 10.1016/S0016-2361(99)00072-1

[33] Roberts VM, Stein V, Reiner T, Lemonidou A, Li X, Lercher JA. Towards quantitative catalytic lignin depolymerization. Chemistry: A European Journal. 2011;**17**(21):5939-5948. DOI: 10.1002/chem.201002438

[34] Toledano A, Serrano L, Labidi J. Organosolv lignin depolymerization with different base catalysts. Journal of Chemical Technology and Biotechnology. 2012;**87**(11):1593-1599. DOI: 10.1002/jctb.3799

[35] Rodriguez A, Salvachúa D, Katahira R, Black BA, Cleveland NS, Reed M, Smith H, Baidoo EEK, Keasling JD, Simmons BA, Beckham GT, Gladden JM. Base-catalyzed depolymerization of solid lignin-rich streams enables microbial conversion. ACS Sustainable Chemistry & Engineering. 2017;**5**(9):8171-8180. DOI: 10.1021/acssuschemeng.7b01818

[36] Erdocia X, Prador R, Corcuera MA, Labidi J. Effect of different organosolv treatments on the structure and properties of olive tree pruning lignin. Journal of Industrial and Engineering Chemistry. 2014;**20**(3):1103-1108. DOI: 10.1016/j.jiec.2013.06.048

[37] Beauchet R, Monteil-Rivera F, Lavoie JM. Conversion of lignin to aromatic-based chemicals (L-chems) and biofuels (L-fuels). Bioresource Technology. 2012;**121**:328-334. DOI: 10.1016/j.biortech.2012.06.061

[38] Pufky-Heinrich D, Rößiger B, Röver R, Unkelbach G. Process scale-up for the production of bioaromatic compounds from lignin. Chemie Ingenieur Technik. 2016;**88**(9):1229-1230. DOI: 10.1002/cite.201650156

[39] Unkelbach G, Schmiedl D, Graf J, Schweppe R. Untersuchung der hydrothermalen Spaltung von Lignin/Produktabtrennung. Chemie Ingenieur Technik. 2008;**80**(9):1269-1270. DOI: 10.1002/cite.200750476

[40] Rinaldi R, Jastrzebski R, Clough MT, Ralph J, Kennema M, Bruijnincx PCA, Weckhuysen BM. Paving the way for lignin valorisation: Recent advances in bioengineering, biorefining and catalysis. Angewandte Chemie. 2016;**55**(29):8164-8215. DOI: 10.1002/anie.201510351

[41] Kozliak EI, Kubátová A, Artemyeva AA, Nagel E, Zhang C, Rajappagowda RB, Smirnova AL. Thermal liquefaction of lignin to aromatics: Efficiency, selectivity, and product analysis. ACS Sustainable Chemistry & Engineering. 2016;**4**(10):5106-5122. DOI: 10.1021/acssuschemeng.6b01046

[42] Mattsson C, Andersson SI, Belkheiri T, Åmand LE, Olausson L, Vamling L, Theliander H. Using 2D NMR to characterize the structure of the low and high molecular weight fractions of bio-oil obtained from LignoBoost™ kraft lignin depolymerized in subcritical water. Biomass and Bioenergy. 2016;**95**:364-377. DOI: 10.1016/j.biombioe.2016.09.004

[43] Toledano A, Serrano L, Pineda A, Romero AA, Luque R, Labidi J. Microwave-assisted depolymerisation of organosolv lignin via mild hydrogen-free hydrogenolysis: Catalyst screening. Applied Catalysis B: Environmental. 2014;**145**:43-55. DOI: 10.1016/j.apcatb.2012.10.015

[44] Vigneault A, Johnson DK, Chornet E. Base-catalyzed depolymerization of lignin: Separation of monomers. Canadian Journal of Chemical Engineering. 2007;**85**(6):906-916. DOI: 10.1002/cjce.5450850612

[45] Greminger DC, Burns GP, Lynn S, Hanson DN, King CJ. Solvent extraction of phenols from water. Industrial & Engineering Chemistry Process Design and Development. 1982;**21**(1):51-54. DOI: 10.1021/i200016a010

[46] Long J, Zhang Q, Wang T, Zhang X, Xu Y, Ma L. An efficient and economical process for lignin depolymerization in biomass-derived solvent tetrahydrofuran. Bioresource Technology. 2014;**154**:7-10. DOI: 10.1016/j.biortech.2013.12.020

[47] Europa HORIZON 2020 [Internet]. 2017. Available from: https://ec.europa.eu/research/participants/data/ref/h2020/wp/2014_2015/annexes/h2020-wp1415-annex-g-trl_en.pdf. [Accessed: 2017-11-22]

[48] Nguyen TDH, Maschietti M, Belkheiri T, Åmand EL, Theliander H, Vamling L, Olausson L, Andersson SI. Catalytic depolymerisation and conversion of Kraft lignin into liquid products using near-critical water. Journal of Supercritical Fluids. 2014;**86**:67-75. DOI: 10.1016/j.supflu.2013.11.022

[49] Beckham GT, Biddy MJ, Chmely SC, Sturgeon M. Hydroxide catalysts for lignin depolymerization [Patent]. US 20140107381 A1; 2014

[50] Beckham GT, Biddy MJ, Chmely SC, Sturgeon M. Hydroxide Catalyst for lignin depolymerization [Patent]. US 9631146 B2; 2017

[51] Cheng F, Brewer CE. Producing jet fuel from biomass lignin: Potential pathways to alkylbenzenes and cycloalkanes. Renewable and Sustainable Energy Reviews. 2017;**72**:673-722. DOI: 10.1016/j.rser.2017.01.030

[52] Kleinert M, Barth T. Phenols from lignin. Chemical Engineering & Technology. 2008;**31**(5):736-745. DOI: 10.1002/ceat.200800073

[53] Rutten C, Ramirez A, Duque JP. Hydrotreating and hydrothermal treatment of alkaline lignin as technological valorization options for future biorefinery concepts: A review. Chemical Technology and Biotechnology. 2017;**92**(2):257-270. DOI: 10.1002/jctb.5103

[54] Long J, Xu Y, Wang T, Yuan Z, Shu R, Zhang Q, Ma L. Efficient base-catalyzed decomposition and in situ hydrogenolysis process for lignin depolymerization and char elimination. Applied Energy. 2015;**141**:70-79. DOI: 10.1016/j.apenergy.2014.12.025

[55] Zhang S, Liu L, Fang G, Yan N, Ren S, Ma Y. Hydrogenolysis and activation of soda lignin using [BMIM]Cl as a catalyst and solvent. Polymer. 2017;**9**(7):279. DOI: 10.3390/polym9070279

[56] Zmierczak WW, Miller JD. Processes for catalytic conversion of lignin to liquid bio-fuels and novel bio-fuels [Patent]. US 8299310 B2; 2012

[57] Liu CF, Chen CC, Chen C, YU PJ, Chang YH, Wan HP, Lee HT. Method for preparing phenolic compounds [Patent]. US 8648218; 2014

[58] Fernández-Rodríguez J, Erdocia X, De Hoyos-Martinez PL, Alriols M, Labidi J. Small phenolic compounds production from Kraft black liquor by lignin depolymerization with different catalytic agents. Chemical Engineering Transactions. 2017;**57**:133-138. DOI: 10.3303/CET1757023

[59] Van den Bosch S, Schutyser W, Vanholme R, Driessen T, Koelewijn SF, Renders T, De Meester B, Huijgen WJJ, Dehaen W, Courtin CM, Lagrain B, Boerjan W, Sels BF. Reductive lignocellulose fractionation into soluble lignin-derived phenolic monomers and dimers and processable carbohydrate pulps. Energy & Environmental Science. 2015;**8**(6):1748-1763. DOI: 10.1039/C5EE00204D

[60] Renders T, Van den Bosch S, Koelewijn SF, Schutyser W, Sels BF. Lignin-first biomass fractionation: The advent of active stabilisation strategies. Energy & Environmental Science. 2017;**10**(7):1551-1557. DOI: 10.1039/C7EE01298E

Chapter 5

Computational Modeling Methods for Understanding the Interaction of Lignin and Its Derivatives with Oxidoreductases as Biocatalysts

Jans Alzate-Morales, Rodrigo Recabarren,
Isabel Fuenzalida-Valdivia, Jefferson Romero and
Angel T. Martinez

Additional information is available at the end of the chapter

http://dx.doi.org/10.5772/intechopen.70554

Abstract

This chapter will be presented as follow. First, a brief introduction to structure and characterization of lignin and its derivatives is presented, as well as their importance as chemical scaffolds for obtaining value-added products in chemical, food, pharmaceutical and agriculture industry. Second, an extensive review of different reports using computational modeling methods—like molecular dynamics simulations, quantum mechanics and hybrid calculation methods, among others—in the understanding of enzyme-substrate interaction and biocatalysis will be presented. Third, and as last part of chapter, some hand picked examples from literature will be chosen as successful cases where the interplay between experiment and computation has given as a result protein engineered oxidoreductases with improved catalytic capabilities.

Keywords: lignin, oxidoreductases, QM-MM, docking, molecular dynamics, biocatalysis

1. Introduction

Lignin, one of the more abundant and recalcitrant substrates available in biosphere, is a complex aromatic heteropolymer formed by radical polymerization of guaiacyl, syringyl, and *p*-hydroxyphenyl units linked by β-aryl ether linkages, biphenyl bonds and heterocyclic linkages, among others [1, 2]. This natural substrate and its derivatives are of upmost interest in biotechnological and chemical industry because the high-value products that can be obtained from their chemical or biological transformation. The principal strategies for depolymerizing lignin to produce valuable chemicals are catalytic reduction, cracking or hydrolysis reactions, catalytic oxidation and enzymatic reactions [3–6]. The latter approach has proven to be more efficient compared to existing physical and chemical pretreatments of biomass that involve the use of costly equipment and reagents, while biocatalytic processes may represent an alternative environmentally friendly approach with lower energy requirements and cost [7]. In this regard, the isolation of lignin-degrading microbial strains may lead to the discovery of novel biocatalysts, like peroxidases and laccases among other enzymes that could be potentially useful for lignin valorization [8, 9]. Obtaining compounds derived from biotransformation processes, with interesting and new applications in chemical, cosmetic, pharmaceutical and food industries, has been the primary interest of several research groups across the chemistry and biology research community. Microorganisms, or their enzymatic systems, catalyze biotransformation processes and these are usually carried out by growing cultures, purified enzymes or immobilized cells, among others [10]. *Those biological chemical reactions can be used in order to introduce chiral centers, to resolve racemates, to convert a particular functional group among several groups with similar reactivities, to functionalize a nonactivated carbon regioselectively and to convert labile molecules due to the mild reaction conditions under which they take place* [10]. A very good example are the ligninolytic enzymes from white-rot fungus *Phanerochaete chrysosporium*, that have been used in several biotransformation and bioremediation processes like the oxidation, by the extracellular enzyme lignin peroxidase (LiP) of diphenyl compounds biphenyl, biphenylene, dibenzofuran, dibenzo-*p*-dioxin and diphenyl ether, which constitute an important class of environmentally persistent pollutants [11]. More recently, the transformation of three anti-inflammatory drugs such as diclofenac, ibuprofen and naproxen were carried out by pellets of *P. chrysosporium* in fed-batch bioreactors operating under continuous air supply or periodic pulsation of oxygen [12]. The authors found that, under operating conditions of the reactor, high elimination percentages for the pharmaceutical compounds (80–99%) were achieved. On the other hand, high-value products derived from the natural substrate of ligninolytic enzymes, lignin, have also been reported. *First-generation fine chemicals can be obtained from lignosulfonate by oxidation, vanillin being the most important fine chemical directly produced from the hydrolysis-oxidation* [13]. *Vanillin is the main component of the natural vanilla extract. It is used as a flavoring agent in foods, beverages, pharmaceuticals and in the fragrance industry* [14]. Besides its value as flavoring agent, vanillin is the starting reagent in the synthesis of several *second-generation* and important fine chemicals. It can be used in synthetic processes leading to several pharmaceutical chemicals like cyclovalone (digestant or choleretic) [15], etamivan (analeptic and a central nervous system and respiratory system stimulant) [16] and levodopa (an antiparkinson agent) [17]. According to the authors, *even though only a very few of the first-generation fine chemicals from oxidation of lignosulfonates have been isolated, a huge variety of organic processes to pharmaceutical chemicals are based on them. Future discovery and optimization of more selective depolymerization processes of lignosulfonates may extend the employment of such fine chemicals as building blocks* [13]. However, the chemical processes for obtaining vanillin (and other important and related chemical building blocks) from lignosulfonates have been very questioned due to the huge amounts of waste effluents produced that, combined with the growing public awareness on environmental issues, are leading to unsustainable effluent-treatment costs [18–20]. A clean and friendly environmental alternative, for the synthesis of such kind of valuable chemical building blocks, is the biotransformation catalyzed by ligninolytic enzymes from *P. chrysosporium* or other microorganisms. Of special interest are oxidoreductases that take advantage from the incorporation of different cofactors—such as heme, flavin and metal ions—to catalyze redox reactions. As recently reported, classical fungal oxidoreductases comprise basidiomycete ligninolytic peroxidases, and ascomycete and basidiomycete multicopper oxidases (MCO, mainly laccases) with different redox potentials and abilities to act on lignin-derived products [21]. Among those, ligninolytic peroxidases have been known for some 40 years, and representatives of the three main types—lignin peroxidase (LiP, EC 1.11.1.14), manganese peroxidase (MnP, EC 1.11.1.13) and versatile peroxidase (VP, EC 1.11.1.16)—have been extensively studied due to their biotechnological potential for the chemical modification and degradation of lignin and other recalcitrant compounds [22]. The molecular biology and structure-function of these lignin-degrading heme peroxidases have been already carefully described [23]. Other interesting oxidoreductases are the so-called dye-decolorizing peroxidases (DyPs, EC 1.11.1.19) that have been recently described and structurally characterized in basidiomycetes [24]. Its structural convergence with ligninolytic peroxidases, besides a long-range electron transfer (LRET) mechanism for oxidation of bulky lignin-derived and dye substrates, make them worth for study as another example of evolutionary convergence between phylogenetically unrelated enzymes oxidizing recalcitrant structures including lignin [21]. Finally, different copper-containing oxidoreductases have been related to lignocellulose degradation, being one of the most studied the laccases (EC 1.10.3.2) that, together with peroxidases, are the most thoroughly studied oxidoreductases in wood-rot fungi and, by far, the largest number of biotechnological applications have been reported for [25, 26]. The present chapter summarizes the most relevant experimental/computational evidence that has been published recently about these key oxidoreductases as biocatalysts, that are helping in the biotransformation and valorization of lignin, and its derivatives, as substrates into high-value products. Moreover, it will be emphasized the use of computational modeling methods in the understanding of substrate-enzyme interactions and their helping role in protein engineering as a key tool for tune the efficiency and specificity of this important kind of enzymes.

2. Computational modeling methods in enzyme-substrate studies

Different computational methodologies could be used in order to study and to understand the underlying mechanisms that make biocatalysts so efficiently degrading pollutants or biotransforming substrates into high-value products. Therefore different computational methods currently available could be of key importance in understanding those sophisticated processes taking place at the molecular level. All of them differ both in computational cost as well as in accuracy. For instance, methods like molecular docking attempt to predict the structure (or structures) of the intermolecular complex formed between two or more molecules. Docking is widely used to suggest the binding modes of protein inhibitors. Most of the docking algorithms are able to generate a large number of possible structures, so they also require a means to score each structure to identify those of most interest. The 'docking problem' is thus concerned with the generation and evaluation of plausible structures of intermolecular complexes [27, 28]. The main disadvantage of docking methods is their scoring function, because it may become insufficient to adequately represent all binding forces participating in the protein-substrate binding. However those methods offer us an important structural starting point for the application of more accurate methods that can take into account the dynamical behavior, energetics profile and electronic characteristics of the system under study. Once one has the structural starting point for protein-substrate obtained from molecular docking, the next computational method that can be applied is molecular dynamics (MD) simulations. This approach is very powerful in the description of very large systems like DNA-drug complexes [29, 30], polymer structures [31, 32], membrane proteins [33, 34], protein-ligand complexes [35, 36], etc. The simplicity of the potential energy function of this method makes to run very long simulations properly, in the nano and micro second time scale, and to extract valuable structural information and energy data from trajectories obtained. Other more refined molecular dynamics approaches like MD-FEP allow to calculate absolute enzyme-substrate binding free energies [37]. As a matter of fact, in all these methods it is mandatory to perform a parameterization of the ligand prior to run the molecular dynamics simulations. This is a severe step back in the simulation process and the successful and accurate representation of enzyme-substrate binding process, because the quality of the simulation is strongly dependent on the quality of the ligand parameterization. This problem is totally alleviated using quantum mechanics (QM) or quantum mechanics/molecular mechanics (QM/MM) methods, which lie at the other end of the computational methods spectrum as the more computational expensive approaches like *ab initio* Car-Parrinello molecular dynamics, QM and derived electronic structure methods in general. These methods are very accurate in the prediction of the reactivity, structural and thermodynamic properties of small molecular systems, but their application to the understanding of enzyme-substrate interaction and to the enzyme catalysis is very scarce yet. This can be explained in part by the high computational cost they demand and the limitations to include explicit solvation effects and conformational searching in a fast and cost affective way. On the other hand, in the last two decades, the hybrid QM/MM calculations have became the more popular methods within computational chemistry community. The broad range of applications goes from calculation of spectroscopic and excited-states properties, studies of enzymatic reactions, surface and chemical catalysis and, in a minor extend, protein-ligand interactions (drug design). The successfully application of the QM/MM method in those areas relies on the combination of accurate QM methods and fast MM computational algorithms. For a complete review of the method and a survey of biological applications, see Refs. [38–40] and references therein. In the next sections, the role and support of/from those computational methodologies in the experimental study and understanding of main oxidoreductases as biocatalysts degrading lignin and its derivatives will be reviewed.

2.1. Laccase

As pointed out before, laccases are the most studied member of oxidoreductases because their broad ranges of chemical and biotechnological applications [41, 42]. From the computational point of view, there is also a growing amount of research that have attempted to understand their catalytic mechanism and also to tuned their substrate affinity with biotechnological, environmental or chemical application purposes. Regarding laccase catalytic mechanism, it has been studied for decades but is still not completely elucidated, especially in terms of the reduction of dioxygen to water. Its key structural features have been also under investigation by several groups using techniques such as X-ray diffraction, electron paramagnetic resonance (EPR) spectroscopy and site-directed mutagenesis. For a more complete view of this topic see Ref. [43]. In 2013, the available literature about theoretical studies on the active-site structure, spectroscopic and thermodynamic properties, and reaction mechanism of multicopper oxidases (MCO) were collected by Rulíšek *et al.* [44]. They emphasized that complicated electronic structure of the trinuclear copper cluster (TNC) make it hard to obtain quantitatively correct spectroscopic data from the density functional theory (DFT) methods. Therefore, multireference wave function methods (e.g. CASPT2 and MRCI-S) would be necessary to obtain quantitative splitting of the lowest doublet and quartet states, reliable excitation energies of ligand-field states, and qualitatively correct predictions of the *g* tensors for these spin-frustrated systems. More recently, this same group reported a work that combined quantum mechanical and molecular mechanical free-energy perturbation (QM/MM-FEP) methods in combination with explicit solvent simulations to study the reaction mechanism of the multicopper oxidases. They studied redox potentials (RP), acidity constants, isomerization reactions, as well as water and O_2 binding reactions. A full reaction mechanism of the multicopper oxidases with atomic detail was proposed. They also showed that the two copper sites in the protein communicate so that redox potentials and acidity constants of one site are affected by up to 0.2 V or 3 pK_a units by a change in the oxidation state of the other site [45]. On the other hand, Hong *et al.* [46] studied the redox potentials and reorganization energies of the type 1 (T1) Cu site in four multicopper oxidases namely laccase from *Trametes versicolor*, a laccase-like enzyme isolated from *Bacillus subtilis*, CueO required for copper homeostasis in *Escherichia coli* and the small laccase (SLAC) from *Streptomyces coelicolor*. They used first principles DFT and QM/MM simulations to calculate redox potentials and reorganization energies that were in good agreement with experiments. Moreover, they proposed a mutated MCO with improved performance taking into account that an estimate of the tunability of the T1 Cu potential, and as a relatively high cathodic redox potential can be achieved with Cu in a well-defined protein environment. In a similar way, geometrical distortions (distances and angles) in the copper coordination sphere were introduced in a model of the three-coordinated T1 Cu site of laccase from *T. versicolor* in order to evaluate the effect on redox potential. All calculations were done within a DFT framework, where the best approach to increase redox potential was by distortion of the dihedral angle ω (defined as $C_{methylthiolate}$ –S–Cu–N_{ImA}), rationalized as a decrease in the overlap of imidazole orbitals in the redox-active molecular orbital (β-LUMO) [47]. More recently, the pH dependence and mutants on the laccase redox potentials at T1 site were studied with QM/MM approaches. First, the protein optimized geometries were obtained with RI-BP86/def2-SVP/def2-TZVP(Cu):CHARMM QM/MM model and then the RPs were obtained at the M06/6-311++G**/SDD(Cu) level of theory. The authors found that the oxidation state of the TNC affected the T1 site RP by about 0.2–0.3 V, depending on the protein protonation state. Moreover, they also calculated the RP of a F463M laccase mutant protein to probe if their adopted methodology could describe the change upon mutation of this residue. Notably, they were able to reproduce the experimental change in the RP of a related enzyme (0.09–0.1 V) very closely (0.08 V) [48]. Other relevant research works have been published with the aim to understand the differential affinity of several substrates against laccase T1 site and to explain the electron transfer pathways and the enzyme residues involved in that process. For instance, Bello *et al.* [49] performed MD simulations of three substrates (2,6-dimethoxyphenol (DMP), syringaldazine (SGZ) and 2,2-azino-bis(3-ethylbenzothiazoline-6-sulfonic acid) (ABTS)) bounded to T1 site of MCOs from *Thermus thermophilus, B. subtilis* and *E. coli.* The authors found that binding modes of the electron-donor molecules to the electron transfer binding site were primarily attributed to hydrophobic contacts that are supposed to play an important role in substrate specificity. Moreover, some MCO-substrate complexes showed an electron-donor molecule conformation in which an electron could be directly transferred to the histidines coordinating T1 Cu, although for others additional electron transference pathways could be also feasible through the participation of charged residues during electron transfer. More recently, five lignin-related compounds, namely 2,6-dimethoxyphenol, ferulic acid, guaiacol, sinapic acid and vanillyl alcohol, were selected to demonstrate the key binding mechanisms between *T. versicolor* laccase and lignin. The authors performed molecular docking and molecular dynamics simulations to study and compare the interaction between various lignin model compounds and laccase with the aim to proposed the molecular basis for their binding, which might be helpful in the understanding of lignin-degrading reaction [50]. Based on molecular docking binding energies, of enzyme-substrate complexes, and their stability through 10 ns MD simulations, the authors concluded that hydrophobic interactions seemed to be necessary to the interaction of lignin/lignin model compounds with laccase, while hydrogen bonds (H-bonds) were alternative. In a similar research work, Awasthi *et al.* [51] also performed molecular docking, MD simulations and, additionally, molecular mechanics-Poisson Boltzmann surface area (MM-PBSA) analyses of several lignin model compounds namely sinapyl alcohol (monomer), guaiacyl 4-O-5 guaiacyl (dimer), syringyl β-O-4 syringyl β-O-4 sinapyl alcohol (trimer) and guaiacyl β-O-4 syringyl β-β syringyl β-O-4 guaiacyl (tetramer) bounded to plant and fungal laccases from *Populus trichocarpa* and *T. versicolor,* respectively. The authors concluded, from structural changes at the C-terminal region, differences in the interacting residues of the binding site, different binding modes of tetramer at the active site, differences in the root mean square deviation (RMSD), interaction energies, hydrogen-bond interactions and binding free-energy analyses during MD simulations; that their computational results supported a situation favorable for degradation of lignin polymer by fungal laccase meanwhile their synthesis is by plant laccase. The inhibition of laccases binding site has also been studied. For instance, the inhibition of laccase enzymatic catalytic activity by formetanate hydrochloride (FMT), a carbamate pesticide, was investigated by cyclic voltammetry and quantum-chemical calculations based on DFT with a protein fragmentation approach. The authors aimed to develop enzymatic electrochemical biosensors for inexpensive analytical detection of pesticides [52]. In other similar work, the authors performed molecular docking calculations to understand the inhibition mechanism of medicarpin on the lignin degradation enzyme laccase from *T. versicolor.* Those preliminary results would be useful in developing wood-preserving formulations based on medicarpin, according to authors [53]. Finally, Christensen *et al.* applied multiple-seed MD simulations to a *T. versicolor* laccase in response to variable ionic strengths, temperatures and glycosylation status. *The persistence of backbone hydrogen bonds was identified as a key descriptor of structural response to environment, whereas solvent-accessibility, radius of gyration and fluctuations were only locally relevant* [54]. When moving to study of laccases from rational protein design through directed evolution and computational approaches, the work done by the groups of Guallar and Camarero is a reference point. For instance, in 2015 Monza *et al.* [55] reported a novel computational approach that combined efficient conformational sampling and quick reactivity scoring in order to understand how substrate oxidation was improved during a directed evolution experiment of a fungal laccase from *Pycnoporus cinnabarinus*. Two substrates were used to screen the laccase activity, namely ABTS and DMP. First, the enzyme-substrate conformational space nearby the T1 pocket was sampled with Protein Energy Landscape Exploration (PELE) [56], a Monte Carlo algorithm that combines protein structure prediction algorithms with ligand and protein structural perturbations [57]. Afterwards, the lowest energy conformations were randomly selected and their reactivity was scored by estimating the amount of spin density localized on the substrate with QM/MM calculations. Authors drew three main conclusions from computational calculations: (1) the mutations accumulated during directed evolution increased laccases enzymatic activity by affecting substrate binding rather than the metal redox potential; (2) large conformational sampling revealed significant changes in the protein-ligand energy landscape upon mutation and (3) the oxidation rate of a target substrate can be improved by fine-tuning the binding event. Moreover, the oxidation (substrate's spin density) increase as a result of an enhanced electrostatic stabilization of the radical species could be confirmed by quantum-chemical calculations. In a very recent work, researchers from same group evaluated the robustness of abovementioned computational approach to estimate activity, emphasizing the importance of the binding event in laccase reactivity [58]. To do so, they used SGZ and four phenols derivatives, with 4OH, 4MeO, 4Me and 4Cl group substituents, as substrates bounded into the T1 Cu site of *Myceliophthora thermophila* laccase (MtL) and *P. cinnabarinus* laccase (PcL). Authors concluded that redesigning substrate binding at the T1 pocket, guided by *in silico* methodologies, is a more consistent option that concentrated efforts on increasing the redox potential of the enzyme. The strengths and weaknesses of the protocol were also discussed as well as the significance of the methodology in protein engineering. In Section 3, two additional successful examples of protein engineering aided by computer simulations, and reported by same group, will be analyzed and discussed with more detail.

2.2. Versatile peroxidase

Computational work in VPs has been mostly focused on identifying the type of amino acid radicals that these enzymes use to degrade substrates. QM calculations at a DFT level (mainly B3LYP functional) have been used together with experimental data from multifrequency electronic paramagnetic resonance (EPR) and pulse electron nuclear double resonance (ENDOR) experiments to study the nature of the tryptophan radical (Trp164) of VP from *Pleurotus eryngii* [59]. Here, Pogni *et al.* through the calculations of g-tensors, hyperfine (hf) couplings and spin densities identified that the radical form of Trp164 most likely corresponds to a neutral radical and not a cationic radical as it was thought to be for Trp171 from LiP [60]. However, it is very likely that the tryptophan cationic radical is indeed an intermediate which then becomes deprotonated at the indole nitrogen by Glu243, which is forming a hydrogen bond at this position, to form the more stable neutral radical [59]. This study was followed by a more robust QM/MM investigation on the same system and also considering the mutant W164Y [61], where EPR parameters, structural information and spin densities were compared with experimental data to confirm the nature of the neutral tryptophan radical. The calculations using the B3LYP/CHARMM level of theory showed that deprotonation of the cation radical (though not detected experimentally) occurs through an exothermic reaction with a barrier height of approximately 2–4 kcal/mol. QM/MM MD simulations were also performed to assess the stability of the tyrosyl radical. It was observed a spontaneous proton transfer of the phenoxyl hydrogen atom to Glu243, however, the hydrogen bond formed between the phenoxyl oxygen in the W164Y variant and the newly protonated Glu243 was found to be not stable featured by the formation of new hydrogen bonds with different water molecules. The more rigid hydrogen-bond network found for the tryptophan neutral radical and the more flexible vicinity around the tyrosyl radical can be correlated with the fact that this last mutant exhibits a slower self-reduction process and oxidation of substrates, which could be attributed to a slower re-protonation of the phenoxyl oxygen to recover the reduced Tyr residue. The analysis of the electrostatic potential at this site revealed the importance of specific residues that influence the EPR parameters, which in the case of $Trp164^{\bullet}$ were found to be Glu243 and Lys253, while for $Tyr^{\bullet}$, Glu243, Arg257 and Lys253 were identified. Following this study, in a more recent investigation [62], the flexibility and the formation of new hydrogen-bond networks in three different double variants of *P. eryngii* VP was studied in more detail by QM/MM MD simulations. The calculations showed that the incorporation of specific amino acids at the vicinity of the tyrosyl radical such as Leu and Glu (W164Y/R257L and W164Y/R257E mutants) helped to rigidize the hydrogen bond between the $Tyr^{\bullet}$ species with Glu243, but the variant W164Y/R257A produced the same effect that the single mutant W164Y where the hydrogen bond with Glu243 is easily broken. This can be attributed to the fact that in the first two double mutants neither Leu257 nor Glu257 compete to form hydrogen bonds with the phenoxyl oxygen of $Tyr^{\bullet}$, as it is in the case of the single mutant through the amino acid Arg257. Besides, the inclusion of the less bulky Ala residue instead of Arg would also promote the entrance of more water molecules that facilitate the breaking of the hydrogen bond with Glu243. Thus, and in the light of these results, QM/MM methods were crucial to model these types of amino acid radicals where effects of the protein environment such as electrostatic interactions and steric effects are highly important for the stabilization of the radical species. Other recent computational studies [63] have explored the binding of veratryl alcohol (endogenous substrate of ligninolytic enzymes) at the surface of *P. eryngii* VP, aiming at modeling the recognition event of this molecule and then the LRET pathway to the heme group. For this, exploration of the ligand binding process through the PELE [56] method mentioned above, allowed the observation of protein-ligand complexes to initiate the study of the LRET. This latter aspect was investigated at a QM/MM level using the e-pathway approach [64], in which different residues are ranked according to their electron affinity, and in this way, the most favorable amino acids for hosting the radical can be identified. The results from this study revealed that Trp164, Trp244 and Phe198 are directly involved in the LRET, fact that was confirmed by site-directed mutagenesis experiments. More recently, Sáez-Jiménez *et al.* [65] reported an experimental and computational study of an evolved VP from *P. eryngii* that was obtained by directed evolution. The authors could provide a structural-functional basis for the improved alkaline stability of the 2-1B variant (containing E37K/H39R/V160A/T184 M/Q202L/D213A/G330R mutations). Using 20 ns MD simulations of native VP and 2-1B in solution and pH = 8.0, and starting from solved crystal structures, they could explain the structural enzyme integrity of the variant (with penta-coordinated heme iron) under alkaline condition. As alkaline inactivation has been related to the formation of a hexa-coordinated iron complex with the distal histidine residue, they monitored the evolution of that distance looking at possible correlations with the stability of the enzyme. The authors concluded that introduction of three basic residues in VP (Lys37, Arg39 and Arg330) led to new connections between heme and helix B (where the distal histidine residue is located), and formation of new electrostatic interactions, that avoided the hexa-coordination of the heme iron and thereby the alkaline inactivation of the variant.

2.3. Lignin peroxidase (LiP)

Similar to versatile peroxidase, QM/MM calculations at the level B3LYP/CHARMM have been carried out to elucidate the nature of the tryptophan radical (Trp171) present in LiP from *P. chrysosporium* [60]. These calculations were based on the experimental work performed by Smith *et al.* [66], where catalytic activity toward veratryl alcohol (VA) was engineered in *Coprinus cinereus* peroxidase (CiP). This was achieved by introducing the catalytic tryptophan residue (Trp178 in CiP) and some specific acid residues that form part of the microenvironment around it, mimicking the surroundings of Trp171 in *P. chrysosporium* LiP. Smith *et al.* also studied different LiP variants (E168Q, E250Q and E250Q/E168Q) to compare the results obtained with the new-engineered CiP. Thus, in the theoretical work of Bernini *et al.* [60] three systems were studied namely pristine LiP, the LiP variant E250Q/E168Q and the CiP variant D178W/R257E/R271D to allow direct comparison with the experimental data available for the last two systems. The values for g-tensors and hf coupling constants were calculated for the corresponding neutral and cationic radicals. This allowed confirming that neutral radicals ($Trp171^{\bullet}$ and $Trp168^{\bullet}$) were present in both LiP and CiP variants, respectively. In the former case, the mutations provided a rather neutral electrostatic potential around Trp171 stabilizing the neutral radical, however, in the case of the CiP variant, since the electrostatic potential is more similar to pristine LiP, the stabilization of the radical comes from the presence of K260, residue that is absent in pristine LiP. This later residue would facilitate the entrance of more water molecules around the radical for its proper stabilization through hydrogen-bonding interactions. No experimental data are available in the case of pristine LiP to establish the nature of the Trp171 radical, though QM/MM calculations showed that deprotonation of Trp171 would be energetically unfavorable and that $Trp171^{\bullet +}$ might exist in pristine LiP due to the highly negative electrostatic potential around it [60]. QM and QM/MM calculations have also been carried out to understand the binding and the electron transfer mechanisms of VA at Tyr181 (redox-active amino acid) in *Trametopsis cervina* LiP [67]. It was observed that VA is stabilized by sandwich π stacking interactions with Phe89 and Tyr181 showing a spin density distribution largely maintained on the VA ligand. QM/MM calculations including in the QM region the heme group and its axial ligands, VA, Tyr181 and Phe89, showed that two unpaired electrons are present in the iron-oxo moiety while another unpaired electron is shared among the porphyrin ring, Tyr181 and VA, observing, therefore, electron transfer from the Tyr181-VA complex to the porphyrin radical, giving rise to compound II. An interesting characteristic of *T. cervina* LiP is the formation of adduct between Tyr181 and VA through covalent bonding. Besides experimental findings, its existence was confirmed by EPR-based quantum calculations and it was further characterized by QM calculations to study the most favorable binding position of VA at the ring of Tyr181. On the other hand, in a similar research work [68], the binding of VA at Trp171 in *P. chrysosporium* LiP was studied with different computational methods and QM/MM calculations were used to compute spin densities for different VA-Trp171$^{\bullet+}$ complexes at different separation distances. The goal was to obtain a cut-off distance for the occurrence of the electron transfer between these two fragments. This information was then used to analyze MD simulations of close-distance complexes that allowed identifying crucial interactions for the binding of VA at Trp171. Other recent QM/MM studies [69] have been focused on calculating the redox potential of compound I in LiP, looking for identifying specific amino acids that influence its oxidative power, and in this way propose new mutants with higher oxidative abilities, or with the capacity of functioning at softer pH conditions. Castro *et al.* [69] identified that acidic (mainly Asp and Glu) residues around the oxoferryl complex tend to lower the redox potential by electrostatic stabilization while positively charged residues destabilize the complex and therefore increase the redox potential. Two amino acids were *in silico* mutated (E40Q and D183N) which were chosen based on electrostatic and structural considerations and gave estimates of redox potential increases of 140 and 190 mV, respectively, and therefore represent novel mutations that can be tested experimentally. In an alternative approach, experimental work together with QM calculations have been used to engineer new mutants of *P. chrysosporium* LiP (isoenzyme H8) that increase the lignin-degrading efficiency by stabilizing radicals that relay the electron transfer to the heme [70]. Pham *et al.* found that Trp251 is a site where radical products (guaiacol radicals) from a non-phenolic lignin dimer are covalently bound, causing therefore suicidal inhibition of the degradation process. DFT-based QM calculations were used to calculate ΔG° barriers for the critical redox centers H176/Heme, W171 and W251, which confirmed W251 as a possible electron relay amino acid in the LRET process. Thus, mutants such as A242D and T208D that would tend to stabilize the W251 radical by electrostatic and hydrogen-bonding interactions were proposed and characterized, with the former one exhibiting until 21.1- and 4.9-fold higher increments in k_{cat} and k_{cat}/k_M values, respectively. In order to get more insights into the electron transfer mechanism operating in LiP, QM/MM calculations to analyze the most favorable electron transfer pathway have been performed using the e-pathway approach, as was done for versatile peroxidase [63]. It was confirmed the involvement of Trp251 in the electron transfer to the heme, proposing the final route Trp171-Phe205-Trp251 to be the one involved in the LRET. The computational results were supported by the experimental mutation W251A, which provided threefold lower catalytic efficiency toward the oxidation of VA. This finding allowed concluding that both VP and LiP, which come from different basidiomycetes, use an analog pathway in the LRET. Some research works have been done with regard to application of other computational methods to understand the stability of wild type LiP and mutants or binding of lignin compounds to its heme group or reactive surface Trp residue. For example, Gerini *et al.* investigated the dynamical and structural properties of LiP and its W171A mutant in aqueous solution using MD simulations. They found that in both cases, the enzyme retained its overall backbone structure and all its non-covalent interactions in the course of the MD simulations. *Moreover, these authors performed steered molecular dynamics docking simulations which have shown that LiP natural substrate (VA) could easily approach the heme edge through the access cannel* [71]. With a similar goal, some molecular dynamics calculations were performed on LiP and horseradish peroxidase (HRP) structures in order to evaluate the effect of calcium ions on protein structure [72]. Chen *et al.*

determined and verified, by MD simulations, the robustness and stability of the docking predicted 3D structures of laccase-lignin, LiP-lignin and MnP-lignin complexes [73]. From the point of view of electronic structure methods, Elder *et al.* used *ab initio* molecular orbital calculations for studying the LiP active site and VA at the STO-3G level using the unrestricted Hartree-Fock approximation for open-shell species [74]. In that book chapter, the author carefully studied the electronic and energetic properties of reactant species in LiP active site (Compounds I and II and $VA^{\bullet+}$ as well as their complexed forms). Due to the general biological significance of the metalloporphyrins and the enzyme systems in which they occur, including LiP, the author made a revision of all previous reports where these compounds have been studied using the methods of theoretical chemistry. Earlier the same author reported the oxidation of a lignin model compound via electron transfer. *They found that oxidation of VA appears to be controlled by relative endothermicity associated with a proton transfer step within each of the possible oxidative pathways. The transfer of an electron from the lignin model compound to the* $VA^{\bullet+}$ *was slightly endothermic, occurring by way of a non-bonded intermediate, with slightly lower energy than either products or reactants* [75]. Finally, ten Have *et al.* used calculated ionization potentials (IP) to determine the oxidation of vanillin precursors by LiP. *The conversion of a series of phenolic derivatives, O-methyl ethers, to the Cα-Cβ cleavage product 3,4-dimethoxybenzaldehyde by LiP allowed them to hypothesize that similar compounds with an IP < 9.0 eV would undergo the same Cα-Cβ cleavage reaction upon incubation with LiP* [76].

2.4. Manganese peroxidase

Computational research in manganese peroxidase has being centered on the study of the binding interactions of this enzyme with lignin model compounds as well as the role played by the C-terminal tail in the catalytic differences seen among MnPs subfamilies. For the former analysis, computational studies using molecular docking have been carried out to describe the binding modes of lignin (using a lignin model substrate) to LiP, laccase and MnP from *P. chrysosporium*, and MD simulations were performed to verify the stability of these systems. Chen *et al.* [73] determined for the lignin-MnP system that in the best docking pose the lignin model compound was positioned at the center of the binding pocket found by the cavity detection algorithm of Molegro Virtual Docker (MVD). The most important residues for the binding of lignin were Arg42, His173 and Arg177, all located next to the heme group. Hydrophobic contacts were the leading interactions. It is worth noting that according to the crystal structure of MnP [77], the residues Glu35, Glu39 and Asp179 are also implied in the binding of Mn^{2+} and the residue Arg42 could be involved in the stabilization of this ion through its interaction with Glu39. MD simulations of 3000 ps for this system were performed showing stability with respect to the starting conformation after 1000 ps, with a mean RMSD of 2.4 and 1.4 Å for the backbone of MnP and lignin, respectively. Also, the total energy during the molecular dynamics remained without major variations, confirming the stability of the system. The role of the C-terminal tail in the catalysis and stability of MnP from *Ceriporiopsis subvermispora* was studied through dynamic ligand diffusion and QM/MM techniques [78]. In a study of over 31 fungal genomes [79], 3 MnP subfamilies were defined depending on the length of the C-terminal tail: short, long and extralong MnPs. Experimental results showed that extralong tails of MnPs contribute to the Mn^{2+} oxidation and to a major acidic stability, but also showed that only short MnPs have the capacity of oxidizing ABTS independently (in the absence of Mn^{2+}) [78]. Fernández-Fueyo *et al.* using PELE [56] and electronic couplings calculations by the fragment charge difference (FCD) approach [80], provided computational information for the ABTS oxidation at the heme-propionate channel and the relation between the C-tail length and the MnPs catalytic properties. Since only small differences exist between long and extralong MnPs, there was no representative for the long MnP subfamily. Calculations were performed using the crystal structure of the extralong isoenzyme MnP6 [78] and its *in silico* shortened-tail form. Local exploration of the ABTS diffusion at the entrance of the heme-propionate channel using PELE showed that for the shortened-tail form of MnP6 the most favorable structures had ABTS closer to the active site. In contrast, for the extralong isoenzyme, the tool failed to provide a structure with good interaction energy, and ABTS was located at a longer distance from the active site. It was observed that the C-tail interferes with the entrance of the ligand into the heme-propionate channel and that longer Gluh35-Asph179 distances enable the approach of ABTS in the shortened-tail form. The C-tail in the extralong MnP seems to limit the mobility of the residues at the entrance of the channel, and also Asph179 and Gluh35 block the access of ABTS even in the most energetically favourable position. Moreover, QM/MM calculations (B3LYP/OPLS2005) using the FCD approach were performed in the best positions predicted for both systems, showing electronic coupling values of one order of magnitude higher for the shortened-tail MnP (4.315E−5 eV) than for the extralong MnP (4.65E-6 eV), demonstrating the higher capacity of the shortened-tail MnP to oxidize ABTS. Rational enzyme engineering of MnP6 from *C. subvermispora* has also been carried out with the help of computational methods [81]. Here, this highly stable peroxidase at low pH was rationally mutated and activated for the oxidation of ABTS based on the analysis of MnP4, a less stable but ABTS-oxidizing MnP from *Pleurotus ostreatus* [82]. Acebes *et al.* started the engineering research using PELE for the active site inspection on both systems, ABTS-MnP6 and ABTS-MnP4. Results of these explorations showed that in the energetically minimum structure of MnP4 located at the main heme channel, two histidines, H220 and H142, interacted forming hydrogen bonds with ABTS's negatively charged sulfonates. These two histidines were not found in MnP6, but instead G139 and N218 were found in their respective positions. Thus, *in silico* G139H and N218H mutations were built in MnP6. Further PELE exploration of the ligand diffusion of ABTS on MnP4 and MnP6 double mutant showed similar energy values, distance and binding modes between these two systems. These results were also supported by electronic coupling calculations using the FCD method, where this double mutant presented an increase of the electronic coupling value of almost three orders of magnitude compared to wild type MnP6 and about ~four times lower than in MnP4. Experimental results confirmed the activation of MnP6 toward the oxidation of ABTS.

2.5. Dye-decoloring peroxidase

DyP-type peroxidases were described 10 years ago like a new family of the heme-peroxidases [83]. The structural difference of them with the other class of fungal, plant and animal peroxidases is that they do not have well conserved amino acids around the heme moiety. Initial studies in DyPs exhaustively characterized this region and after being reported the cloning and expression of *Auricularia auricula-judae* DyP (*Aau*DyP) in *E. coli* [84], studies of the substrate oxidation sites in *Aau*DyP were reported. In this context, computational studies have helped to obtain a better characterization of *Aau*DyP and to give molecular details in the different applications that have been explored in this family of peroxidases. In a first approximation, simulations using PELE [56] were performed to study the binding of the substrate RB19 to the structure of *Aau*DyP [85]. The main interactions between RB19 and DyP were found to be with residues Trp105, Tyr147/337, Trp207, Tyr285, Trp377 and the heme entrance channel. The investigation aimed at elucidating which of these residues were involved in the oxidation of the substrate. For this, QM/MM pairwise comparison between Trp377 and the others residues was carried out where spin densities were calculated after subtracting one electron from the systems. Other QM/MM calculations including the heme group in its compound I state and RB19 in the QM region showed that Trp377 is the preferential oxidation site due to a higher concentration of the spin density on this residue. However, analysis of the spin density also showed that other surface amino acids such as Tyr337 could also function as potential oxidizing sites but with lower activities [85]. This was confirmed by site-directed mutagenesis experiments and enzymatic kinetic characterization of the mutants W377S, Y147S, Y337S, Y147F/Y337F and G169L, which revealed the existence of two oxidation sites, a high turnover site (residue Trp377) and a low turnover site (probably the heme-access channel). The data from EPR experiments proved the existence of the radical species Tyr337$^{\bullet}$ and Trp377$^{\bullet}$ and encouraged to propose that the oxidation in DyP-type peroxidases of bulky substrates, such as RB19, is ruled by a long-range electron transfer (LRET) mechanism. Finally, by means of the QM/MM e-pathway approach the residues Trp377, Pro310, Arg309, Arg306, Ile305 and His304, were identified as the residues involved in the LRET pathway in *A. aricula-judae* DyP [85]. Other studies of the mutants W377S, Y147S, Y337S and Y147F/Y337F in *A. aricula-judae* DyP evaluated by Baratto *et al.* [86] through EPR spectroscopy and QM/MM calculations have been performed. The calculation of *g*-tensor values via single point calculations on the B3LYP/AMBER-optimized geometries of the Tyr337$^{\bullet}$ and Tyr147$^{\bullet}$ radicals provided new insights into the molecular interactions with their surroundings. They found that Tyr337 presents a strong hydrogen-bonding interaction between the phenoxyl oxygen and the phenoxyl proton transferred to the carboxylic acid oxygen of the nearby Glu354. On the other hand, Tyr147 is involved in weaker hydrogen-bonding interactions with solvent molecules and the Trp377 radical is present in its neutral form, similar to Trp164 in *P. eryngii* VP. Additional evidence from EPR spectroscopy experiments helped to conclude that Tyr337 is the secondary radical species formed in *A. aricula-judae* DyP at pH 3, which is simultaneously created with the Trp377 radical, which is the main radical site. Thus, if Trp377 is replaced, Tyr337 constitutes the main redox site, and once Tyr337 is removed, a potential LRET pathway originated from Tyr147 is activated. This fact evidenced the activation and deactivation of different LRET pathways in *A. aricula-judae* DyP [86]. Other more recent studies [87], have also attempted to characterize the oxidative sites present in *Aau*DyP by means of molecular dynamics simulations, where energy differences in the charge transfer processes among the oxidative centers have been identified. Trp377 is recognized as the most kinetically favorable amino acid for hosting the electron hole, confirming the previously mentioned findings, otherwise the residue Tyr229, or the cluster formed by Trp105-Tyr147-Tyr337, are also thermodynamically favored. Based on the structural information provided by the crystal structure of *A. aricula-judae* DyP, it was proposed as a protein engineering strategy, the enlargement of the heme pocket by the single mutations F359G and L357G, and the evaluation of the sulfoxidizing reactivity of mutants toward methyl-phenyl sulfide (MPS) and methyl-*p*-tolyl sulfide (MTS). It is worth mentioning that the tuning of the stereoselectivity in sulfoxidation reactions has many applications in organic synthesis and in enzymatic reactions. The mutant F359G was observed to be highly stereoselective and reactive producing the *S* sulfoxide of MPS and MTS, while native DyP had no sulfoxidation activity, and the L357G variant produced racemic mixtures of sulfoxides *R* and *S* enantiomers but with lower catalytic efficiencies. PELE simulations showed that the sulfoxidation ability present in the mutants was due to a better access of the substrates to the reactive heme cofactor. In addition to this, QM/MM calculations performed at the DFT M06-L(lacvp*)/OPLS level with QSite [88] elucidated that the high stereoselectivity and reactivity observed in the mutant F359G could be explained by the substantial difference between the spin density populations on the substrate molecules for the two enantiomers, (0.8% pro R-enantiomer, 21.8% pro S-enantiomer) in MPS and (12.5% pro R-enantiomer, 25.8% pro S-enantiomer) in MTS [89]. Other study including computational methodologies was reported by Strittmatter *et al.* [24]. Docking experiments were performed in *Aau*DyPI using N-heterocyclic derivatives like substrates to explore its binding interaction in the heme pocket. The observation of competitive inhibition by imidazole with regard to H_2O_2 conversion and changes in the Soret band of *Aau*DyPI in the presence of imidazole derivatives was correlated with the computational data and permitted to predict which compound was a good candidate for coordination or not with the heme iron [24].

3. Rational protein design successful cases

In this section, we aim to briefly present two relevant and successful cases of rational protein design, where computational tools played a key role in improvement of catalytic activity of laccase as a key representative of oxidoreductase family. The first case was reported in 2016 by Santiago *et al.* [90]. That research work described the computer-aided engineering of a double mutant laccase that showed improved kinetic parameters (k_{cat} and k_M), when compared to parent enzyme, for oxidation of several arylamines. The authors studied the biological oxidation of four substrates (aniline (ANL), *p*-phenylenediamine (PPD), ABTS and *N,N*-dimethyl-p-phenylenediamine (DMPD)) that are of industrial interest for synthesis of conductive polyaniline (PANI), in case of ANL, or precursors of high-value reagents like methylene blue as potential antimalarial agent, in case of DMPD. The computational protocol used to design the double mutant of laccase was as follows: (1) determining of the binding event of substrates nearby T1 site by means of Monte Carlo-based molecular simulations using software PELE [56]; (2) estimating the spin density fraction localized on the substrate by means of QM/MM calculations on randomly selected structures; and (3) designing the laccase-binding site in order to improve substrate oxidation rate. This computational protocol was validated experimentally by estimating the kinetic parameters of substrates and comparing them with parent laccase. Notably, in case of ANL a 2-fold k_{cat} increase was achieved by introducing two mutations at T1 site namely N207S/N263D that stabilized the oxidized form of ANL thereby boosting electron transfer to the Cu T1 site. Moreover, the new mutant was tested at working conditions for the enzymatic synthesis of PANI from ANL, which showed an increment of absorbance of 36% compared to the parent when production of PANI was followed at 800 nm. Finally, an *in silico* cross validation was performed for aniline derivative DMPD. After applying the same computational protocol and measuring the kinetic parameters for biological oxidation of this substrate, by mutant and parent laccases, the authors found that substrate spin density increased from 39 to 68% by the presence of N207S/N263D mutations in the engineered laccase which also was consistent with the observed increase in k_{cat} (from 459 ± 18 to 741 ± 48 s^{-1}). The authors concluded that applied computational approach provided a reliable *in silico* screening, reducing experimental validation to precisely design laccase mutants.

In a second case of rational protein engineering, the substrate binding pocket of a chimeric laccase (3A4) was redesigned in order to improve the oxidation of synapic acid (SA), a lignin-related phenol of industrial interest [91]. This time the authors started using an experimentally protocol, iterative saturation mutagenesis [92], for screening more than 15,000 clones, from which two variants (C14F12 and CA32F1) and parent 3A4 were purified and characterized. Then, they rationalized the effect of mutations using computational methods. To do so, the 3D structures for 3A4 (mutation V162A), C14F12 (mutations V162A and F392N) and CA32F1 (mutations V162R, T164E and F392N) were built from coordinates of PM1L (PDB code: 5ANH), which shares 98% sequence identity with the chimeric variant 3A4. The 3D models were energy minimized with a force field and, in case of triple mutant variant, a 20 ns MD simulation was necessary to assure the structural stability of the model. With respect to SA, it was modeled in its protonated and deprotonated form, SAH and SA^-, respectively, due to its $pK_a = 4.9$ and MD simulations and experiments were performed at pH = 5.0 and 3.0. Then the binding modes of SA, and other substrates, on the T1 site of chimeric proteins were sampled with PELE [56]. Finally, and after filtering structures based on their energy from PELE simulations, the atomic Mulliken spin density of the substrate was computed on randomly selected structures that were minimized with QM/MM methodology including in the QM part the substrate, the T1 copper atom, and residues H394, C450, H455, I452 and F460; meanwhile the rest of structure was modeled classically with OPLS force field. The main results from computational simulation, and their relation with experimental findings, in this research work can be resumed as follow: (1) In 3A4, SAH binds with high affinity inside T1 site interacting with residues P393 and N207 through its carboxylic acid moiety and with residues A162, P163 and the side chain of S264 through its phenolic group. Its average spin density was about 88%. On the other hand, SA^- showed fewer structures populating the T1 site and instead a second site was explored. In that site substrate bound in a different orientation and further away from the metal center due to the favorable interaction between the negatively charged carboxylate group, the backbone of G391 and F392 and the hydroxyl group of S387. The SA^- averaged spin densities in those sites were 87 and 61%, respectively. That roughly means that 3A4 laccase oxidized, preferentially, SAH instead of SA^-. (2) In C14F12, SA^- binds with more affinity to T1 site than SAH showing the same interactions of its carboxylic group with residues G391, N392 and S387's side chain. Notably, the mutation F392N created a favorable binding site for the negatively charged substrate with the N392 side chain; therefore substrate's phenolic group showed two main orientations (one inside T1 site and other pointing to the solvent) with improved spin densities of 93 and 48%, respectively. This, in turn, was in agreement with improvement of catalytic constants for C14F12 (k_{cat} = 251 s^{-1}) relative to 3A4 (k_{cat} = 156 s^{-1}) that could be due to improved oxidation of SA^- (that is abundant at this pH than at pH = 3). (3) In CA32F1, the SAH and SA^- species were also bound with high affinity to T1 site, showing spin densities of 99 and 86%, respectively. However the relative orientation of the side chains from R162, E164 and D205, and specially the new acid–basic pair formed by R162 and E164, affected the substrate oxidation. Overall, the authors concluded that QM and MD calculations were performed to rationalize the effect of the selected mutations, revealing the critical role of the residues of the enzyme pocket in order to provide the precise binding of the substrate that enabled an efficient electron transfer to the T1 Cu. Additionally, they stressed the usefulness of atomic simulations for unveiling the molecular determinants for the efficient oxidation of a target molecule by these kind of enzymes.

4. Perspectives and conclusions

This chapter aimed to collect and highlight the increasing evidence of computational modeling methods as supporting and validation tools in the study of oxidoreductases as biocatalysts and the characterization of their biotransformation products. We decided to focus on five enzymes that have been investigated intensively during last years due to their broad application as biocatalysts in chemical, agriculture, environmental and biotechnological applications. Those enzymes are LiP, MnP and VP, which are heme peroxidases in the peroxidase-catalase superfamily, DyP as a recently discovered heme peroxidase in the CDE superfamily (including chlorite dismutases, DyPs and the EfeB proteins of *E. coli*), and laccase as the most studied member of copper-containing oxidoreductases. The relevant literature was intensively searched for reports were computational methods—mainly MD simulations, QM and QM/MM—were presented as auxiliary tools to explain the experimental evidence in aspects like substrate binding and oxidation of lignin or derivatives against different binding sites reported for these enzymes, improvement of catalytic activity of a given oxidoreductase for a specific substrate, rational protein design and more recently computed-aided protein engineering. However, we believe that, despite the recent advances in hardware, software and algorithms, the use of computational modeling methods in the understanding of substrate binding and enzymatic catalytic processes between oxidoreductases and their substrates is yet scarce. Nevertheless, the two recently successful and relevant applications, that were discussed in last section, about the interplay between experimental and computational protocols applied to directed evolution of improved laccases as biocatalysts are very encouraging. This accumulated knowledge in rational protein design from presented case studies on laccase paves the way to other studies of the same kind in enzymes from oxidoreductases.

Undoubtedly, during next years the research in this area will be benefited from advance in computational software and hardware as well as from more accurate and predictive computational tools. Additionally, it is expected that more efficient biocatalysts are successfully and quickly generated from conjunction between these experimental/computational techniques presented here with the aim to improve catalytic processes that are urgently demanded in environmental, food, chemical, pharmaceutical and energy industries.

Acknowledgements

J. A-M. and I. F-V. thank financial support from project FONDECYT No. 1140618. J.R. and R.R. acknowledge support from CONICYT to perform their doctoral studies through scholarships No. 21160905 and No. 21130949, respectively.

Author details

Jans Alzate-Morales[1]*, Rodrigo Recabarren[1], Isabel Fuenzalida-Valdivia[1], Jefferson Romero[1] and Angel T. Martinez[2]

*Address all correspondence to: jalzate@utalca.cl

1 Centro de Bioinformática y Simulación Molecular (CBSM), University of Talca, Talca, Chile

2 Biological Research Centre (CIB), CSIC, Madrid, Spain

References

[1] Ralph J, Lundquist K, Brunow G, Lu F, Kim H, Schatz PF, et al. Lignins: Natural polymers from oxidative coupling of 4-hydroxyphenyl-propanoids. Phytochemistry Reviews. 2004 Jan 1;**3**(1-2):29-60

[2] Ragauskas AJ, Beckham GT, Biddy MJ, Chandra R, Chen F, Davis MF, et al. Lignin valorization: Improving lignin processing in the biorefinery. Science. 16 May 2014;**344**(6185):1246843

[3] Pandey MP, Kim CS. Lignin depolymerization and conversion: A review of thermochemical methods. Chemical Engineering and Technology. 2011 Jan 1;**34**(1):29-41

[4] Zhou C-H, Xia X, Lin C-X, Tong D-S, Beltramini J. Catalytic conversion of lignocellulosic biomass to fine chemicals and fuels. Chemical Society Reviews. 2011 Oct 17;**40**(11): 5588-5617

[5] Nanayakkara S, Patti AF, Saito K. Chemical depolymerization of lignin involving the redistribution mechanism with phenols and repolymerization of depolymerized products. Green Chemistry. 2014 Mar 24;**16**(4):1897-1903

[6] Dashtban M, Schraft H, Syed TA, Qin W. Fungal biodegradation and enzymatic modification of lignin. International Journal of Biochemistry and Molecular Biology. 23 May 2010;**1**(1):36-50

[7] Harmsen PFH, Huijgen W, Bermudez L, Bakker R. Literature Review of Physical and Chemical Pretreatment Processes for Lignocellulosic Biomass [Internet]. Wageningen: Wageningen UR—Food & Biobased Research; 2010 [cited 2017 Jul 25] p. Report No.: 1184. Available from: http://library.wur.nl/WebQuery/wurpubs/396201

[8] Picart P, Wiermans L, Pérez-Sánchez M, Grande PM, Schallmey A, Domínguez de María P. Assessing lignin types to screen novel biomass-degrading microbial strains: Synthetic lignin as useful carbon source. ACS Sustainable Chemistry & Engineering. 2016 Mar 7; **4**(3):651-655

[9] Tian J-H, Pourcher A-M, Peu P. Isolation of bacterial strains able to metabolize lignin and lignin-related compounds. Letters in Applied Microbiology. 2016 Jul 1;**63**(1):30-37

[10] Leuenberger HGW. Interrelations of chemistry and biotechnology—I. Biotransformation—A useful tool in organic chemistry. Pure and Applied Chemistry. 1990;**62**(4):753-768

[11] Hiratsuka N, Oyadomari M, Shinohara H, Tanaka H, Wariishi H. Metabolic mechanisms involved in hydroxylation reactions of diphenyl compounds by the lignin-degrading basidiomycete *Phanerochaete chrysosporium*. Biochemical Engineering Journal. 2005 May 1; **23**(3):241-246

[12] Rodarte-Morales AI, Feijoo G, Moreira MT, Lema JM. Biotransformation of three pharmaceutical active compounds by the fungus *Phanerochaete chrysosporium* in a fed batch stirred reactor under air and oxygen supply. Biodegradation. 2012 Feb 1;**23**(1):145-156

[13] Bjørsvik H-R, Liguori L. Organic processes to pharmaceutical chemicals based on fine chemicals from lignosulfonates. Organic Process Research and Development. 2002 May 1; **6**(3):279-290

[14] Calvo-Flores FG, Dobado JA. Lignin as renewable raw material. ChemSusChem. 2010;**3**(11):1227-1235

[15] Rumpel W. (A. V. Waldheim Chemisch-Pharmazeutische Fabrik). AT 180 258. 1954

[16] Kratzl K, Kvasnicka E. (Österreichische Stickstoffwerke AG). U.S. Patent 2,641, 282. 1953

[17] Britton EC, White HC. (Dow Chemical Company). U.S. Patent 2,- 605,282. 1952

[18] Berger RG. Flavours and Fragrances—Chemistry, Bioprocessing and Sustainability. Berlin: Springer; 2007. p. 294-295

[19] da Silva EAB, Zabkova M, Araújo JD, Cateto CA, Barreiro MF, Belgacem MN, et al. An integrated process to produce vanillin and lignin-based polyurethanes from Kraft lignin. Chemical Engineering Research and Design. 2009 Sep;**87**(9):1276-1292

[20] Hocking MB. Vanillin: Synthetic flavoring from spent sulfite liquor. Journal of Chemical Education. 1997 Sep 1;**74**(9):1055

[21] Martínez AT, Ruiz-Dueñas FJ, Camarero S, Serrano A, Linde D, Lund H, et al. Oxidoreductases on their way to industrial biotransformations. Biotechnology Advances [Internet]. 2017 Jun 15. Available from: http://www.sciencedirect.com/science/article/pii/S0734975017300629

[22] Martínez AT, Camarero S, Ruiz-Dueñas FJ, Martínez MJ. Biological lignin degradation. In: Beckham GT, editor. Lignin Valorization: Emerging Approaches. Chapter 8. Royal Society of Chemistry; United Kingdom: 2017. ISBN 9781782625544

[23] Martínez AT. Molecular biology and structure-function of lignin-degrading heme peroxidases. Enzyme and Microbial Technology. 2002 Apr 16;**30**(4):425-444

[24] Strittmatter E, Serrer K, Liers C, Ullrich R, Hofrichter M, Piontek K, et al. The toolbox of *Auricularia auricula-judae* dye-decolorizing peroxidase—Identification of three new potential substrate-interaction sites. Archives of Biochemistry and Biophysics. 2015 May 15;**574**:75-85

[25] Mate DM, Alcalde M. Laccase: A multi-purpose biocatalyst at the forefront of biotechnology. Microbial Biotechnology. 2016 Sep 1;**0**(0):000-000. DOI:10.1111/1751-7915.12422

[26] Pezzella C, Guarino L, Piscitelli A. How to enjoy laccases. Cellular and Molecular Life Sciences. 2015 Mar 1;**72**(5):923-940

[27] Meng EC, Gschwend DA, Blaney JM, Kuntz ID. Orientational sampling and rigid-body minimization in molecular docking. Proteins: Structure, Function, and Genetics. 1993 Nov;**17**(3):266-278

[28] Blaney JM, Dixon JS. A good ligand is hard to find: Automated docking methods. Perspect Drug Discov Des. 1993 Dec;**1**(2):301-319

[29] Bocian W, Kawęcki R, Bednarek E, Sitkowski J, Williamson MP, Hansen PE, et al. Binding of topotecan to a nicked DNA oligomer in solution. Chemistry—A European Journal. 2008 Mar 17;**14**(9):2788-2794

[30] Siu F-M, Che C-M. Persistence of camptothecin analog–topoisomerase I–DNA ternary complexes: A molecular dynamics study. Journal of the American Chemical Society. 2008 Dec 31;**130**(52):17928-17937

[31] Tallury SS, Pasquinelli MA. Molecular dynamics simulations of flexible polymer chains wrapping single-walled carbon nanotubes. The Journal of Physical Chemistry. B. 2010 Apr;**114**(12):4122-4129

[32] Luo Y-D, Chen J-H, Huang C-I, Chiu W-Y. Molecular dynamics study of TiO_2/poly(acrylic acid-co-methyl methacrylate) and Fe_3O_4/polystyrene composite latex particles prepared by heterocoagulation. Journal of Applied Polymer Science. 15 May 2010;**116**(4):2275-2284

[33] Chakrapani S, Sompornpisut P, Intharathep P, Roux B, Perozo E. The activated state of a sodium channel voltage sensor in a membrane environment. Proceedings of the National Academy of Sciences. 2010 Mar 5;**107**(12):5435-5440

[34] Lähdesmäki K, Ollila OHS, Koivuniemi A, Kovanen PT, Hyvönen MT. Membrane simulations mimicking acidic pH reveal increased thickness and negative curvature in a bilayer consisting of lysophosphatidylcholines and free fatty acids. Biochimica et Biophysica Acta (BBA)—Biomembranes. 2010 May;**1798**(5):938-946

[35] Amorim HLN, Netz PA, Guimarães JA. Thrombin allosteric modulation revisited: A molecular dynamics study. Journal of Molecular Modeling. 2010;**16**(4):725-735

[36] Li J, Zhu X, Yang C, Shi R. Characterization of the binding of angiotensin II receptor blockers to human serum albumin using docking and molecular dynamics simulation. Journal of Molecular Modeling. 2010 Apr 1;**16**(4):789-798

[37] Best SA, Merz KM Jr, Reynolds CH. Free energy perturbation study of octanol/water partition coefficients: Comparison with continuum GB/SA calculations. The Journal of Physical Chemistry. B. 1999;**103**(4):714-726

[38] Groenhof G. Introduction to QM/MM simulations. In: Monticelli L, Salonen E, editors. Biomolecular Simulations: Methods and Protocols [Internet]. Totowa, NJ: Humana Press; 2013. p. 43-66. Available from: https://doi.org/10.1007/978-1-62703-017-5_3

[39] Senn HM, Thiel W. QM/MM methods for biomolecular systems. Angewandte Chemie, International Edition. 2009;**48**(7):1198-1229

[40] Friesner RA. Combined quantum and molecular mechanics (QM/MM). Drug Discovery Today: Technologies. 2004 Dec;**1**(3):253-260

[41] Hamid M, Khalil-ur-Rehman. Potential applications of peroxidases. Food Chemistry. 2009 Aug 15;**115**(4):1177-1186

[42] Pollegioni L, Tonin F, Rosini E. Lignin-degrading enzymes. The FEBS Journal. 2015 Apr 1; **282**(7):1190-1213

[43] Mot AC, Silaghi-Dumitrescu R. Laccases: Complex architectures for one-electron oxidations. Biochemistry (Moscow). 2012 Dec 1;**77**(12):1395-1407

[44] Rulíšek L, Ryde U. Theoretical studies of the active-site structure, spectroscopic and thermodynamic properties, and reaction mechanism of multicopper oxidases. Coordination Chemistry Reviews. 2013 Jan 15;**257**(2):445-458

[45] Li J, Farrokhnia M, Rulíšek L, Ryde U. Catalytic cycle of multicopper oxidases studied by combined quantum- and molecular-mechanical free-energy perturbation methods. The Journal of Physical Chemistry. B. 2015 Jul 2;**119**(26):8268-8284

[46] Hong G, Ivnitski DM, Johnson GR, Atanassov P, Pachter R. Design parameters for tuning the Type 1 Cu multicopper oxidase redox potential: Insight from a combination of first principles and empirical molecular dynamics simulations. Journal of the American Chemical Society. 2011 Apr 6;**133**(13):4802-4809

[47] Vázquez-Lima H, Guadarrama P, Martínez-Anaya C. Geometric distortions on a three-coordinated T1 Cu site model as a potential strategy to modulate redox potential. A theoretical study. Journal of Molecular Modeling. 2012 Feb 1;**18**(2):455-466

[48] Götze JP, Bühl M. Laccase redox potentials: pH dependence and mutants, a QM/MM study. The Journal of Physical Chemistry. B. 2016 Sep 8;**120**(35):9265-9276

[49] Bello M, Correa-Basurto J, Rudiño-Piñera E. Simulation of the cavity-binding site of three bacterial multicopper oxidases upon complex stabilization: Interactional profile and electron transference pathways. Journal of Biomolecular Structure & Dynamics. 2014 Aug 3;**32**(8):1303-1317

[50] Chen M, Zeng G, Lai C, Li J, Xu P, Wu H. Molecular basis of laccase bound to lignin: Insight from comparative studies on the interaction of *Trametes versicolor* laccase with various lignin model compounds. RSC Advances. 2015 Jun 12;**5**(65):52307-52313

[51] Awasthi M, Jaiswal N, Singh S, Pandey VP, Dwivedi UN. Molecular docking and dynamics simulation analyses unraveling the differential enzymatic catalysis by plant and fungal laccases with respect to lignin biosynthesis and degradation. Journal of Biomolecular Structure & Dynamics. 2015 Sep 2;**33**(9):1835-1849

[52] Martins ACV, Ribeiro FWP, Zanatta G, Freire VN, Morais S, de Lima-Neto P, et al. Modeling of laccase inhibition by formetanate pesticide using theoretical approaches. Bioelectrochemistry. 2016 Apr 1;**108**:46-53

[53] Martínez-Sotres C, Rutiaga-Quiñones JG, Herrera-Bucio R, Gallo M, López-Albarrán P. Molecular docking insights into the inhibition of laccase activity by medicarpin. Wood Science and Technology. 2015 Jul 1;**49**(4):857-868

[54] Christensen NJ, Kepp KP. Stability mechanisms of a thermophilic laccase probed by molecular dynamics. PLoS One. 2013 Apr 29;**8**(4):e61985

[55] Monza E, Lucas MF, Camarero S, Alejaldre LC, Martínez AT, Guallar V. Insights into laccase engineering from molecular simulations: Toward a binding-focused strategy. Journal of Physical Chemistry Letters. 2015 Apr 16;**6**(8):1447-1453

[56] Borrelli KW, Vitalis A, Alcantara R, Guallar VPELE. Protein energy landscape exploration. A novel Monte Carlo based technique. Journal of Chemical Theory and Computation. 2005 Nov 1;**1**(6):1304-1311

[57] Cossins BP, Hosseini A, Guallar V. Exploration of protein conformational change with PELE and meta-dynamics. Journal of Chemical Theory and Computation. 2012 Mar 13; **8**(3):959-965

[58] Lucas MF, Monza E, Jørgensen LJ, Ernst HA, Piontek K, Bjerrum MJ, et al. Simulating substrate recognition and oxidation in laccases: From description to design. Journal of Chemical Theory and Computation. 2017 Mar 14;**13**(3):1462-1467

[59] Pogni R, Baratto MC, Teutloff C, Giansanti S, Ruiz-Dueñas FJ, Choinowski T, et al. A tryptophan neutral radical in the oxidized state of versatile peroxidase from *Pleurotus eryngii*. A combined multifrequency EPR and density functional theory study. Journal of Biological Chemistry. 2006 Apr 7;**281**(14):9517-9526

[60] Bernini C, Pogni R, Basosi R, Sinicropi A. The nature of tryptophan radicals involved in the long-range electron transfer of lignin peroxidase and lignin peroxidase-like systems: Insights from quantum mechanical/molecular mechanics simulations. Proteins: Structure, Function, and Bioinformatics. 2012 Mar;**80**(5):1476-1483

[61] Bernini C, Pogni R, Ruiz-Dueñas FJ, Martínez AT, Basosi R, Sinicropi A. EPR parameters of amino acid radicals in *P. eryngii* versatile peroxidase and its W164Y variant computed at the QM/MM level. Physical Chemistry Chemical Physics. 2011 Mar 2;**13**(11):5078-5098

[62] Bernini C, Pogni R, Basosi R, Sinicropi A. Prediction of hydrogen-bonding networks around tyrosyl radical in *P. eryngii* versatile peroxidase W164Y variants: A QM/MM MD study. Molecular Simulation. 3 May 2014;**40**(6):485-490

[63] Acebes S, Ruiz-Dueñas FJ, Toubes M, Sáez-Jiménez V, Pérez-Boada M, Lucas MF, et al. Mapping the long-range electron transfer route in ligninolytic peroxidases. The Journal of Physical Chemistry. B. 2017 Apr 27;**121**(16):3946-3954

[64] Guallar V, Wallrapp F. Mapping protein electron transfer pathways with QM/MM methods. Journal of the Royal Society Interface. 2008 Dec 6;**5**(Suppl 3):233-239

[65] Sáez-Jiménez V, Acebes S, Garcia-Ruiz E, Romero A, Guallar V, Alcalde M, et al. Unveiling the basis of alkaline stability of an evolved versatile peroxidase. The Biochemical Journal. 2016 Jul 1;**473**(13):1917-1928

[66] Smith AT, Doyle WA, Dorlet P, Ivancich A. Spectroscopic evidence for an engineered, catalytically active Trp radical that creates the unique reactivity of lignin peroxidase. Proceedings of the National Academy of Sciences. 2009 Sep 22;**106**(38):16084-16089

[67] Miki Y, Pogni R, Acebes S, Lucas F, Fernández-Fueyo E, Baratto MC, et al. Formation of a tyrosine adduct involved in lignin degradation by *Trametopsis cervina* lignin peroxidase: A novel peroxidase activation mechanism. Biochemical Journal [Internet]. 2013 Apr 3 [cited 2013 Jun 14];Available from: http://www.biochemj.org/bj/452/bj4520575.htm

[68] Recabarren R, Fuenzalida-Valdivia I, Alzate-Morales J. Studying the binding mechanisms of veratryl alcohol to *P. chrysosporium* lignin peroxidase: Insights from theoretical approaches. Theoretical Chemistry Accounts. 2016 Mar 1;**135**(3):71

[69] Castro L, Crawford L, Mutengwa A, Götze JP, Bühl M. Insights into structure and redox potential of lignin peroxidase from QM/MM calculations. Organic & Biomolecular Chemistry. 2016 Feb 16;**14**(8):2385-2389

[70] Pham LTM, Kim SJ, Kim YH. Improvement of catalytic performance of lignin peroxidase for the enhanced degradation of lignocellulose biomass based on the imbedded electron-relay in long-range electron transfer route. Biotechnology for Biofuels. 2016 Nov 15;**9**:247

[71] Francesca Gerini M, Roccatano D, Baciocchi E, Nola AD. Molecular dynamics simulations of lignin peroxidase in solution. Biophysical Journal. 2003 Jun;**84**(6):3883-3893

[72] Banci L, Carloni P, Diaz A, Savellini GG. Molecular dynamics calculations on peroxidases: The effect of calcium ions on protein structure. JBIC. Journal of Biological Inorganic Chemistry. 1996 Jun 1;**1**(3):264-272

[73] Chen M, Zeng G, Tan Z, Jiang M, Li H, Liu L, et al. Understanding lignin-degrading reactions of ligninolytic enzymes: Binding affinity and interactional profile. PLoS One. 2011 Sep 29;**6**(9):e25647

[74] Elder T, Young DC. Molecular orbital calculations on the interaction of veratryl alcohol with the lignin peroxidase active site. In: Lignin: Historical, Biological, and Materials Perspectives [Internet], ACS Symposium Series. Vol. 742. American Chemical Society; 1999. p. 186-204. Available from: http://dx.doi.org/10.1021/bk-2000-0742.ch007

[75] Elder T. Oxidation of a lignin model compound by the veratryl alcohol cation radical. Results from molecular orbital calculations. Holzforschung—International Journal of the Biology, Chemistry, Physics and Technology of Wood. 1997;**51**(1):47-56

[76] ten Have R, Rietjens IMC, Hartmans S, Swarts HJ, Field JA. Calculated ionisation potentials determine the oxidation of vanillin precursors by lignin peroxidase. FEBS Letters. 1998 Jul 3;**430**(3):390-392

[77] Sundaramoorthy M, Gold MH, Poulos TL. Ultrahigh (0.93Å) resolution structure of manganese peroxidase from *Phanerochaete chrysosporium*: Implications for the catalytic mechanism. Journal of Inorganic Biochemistry. 2010 Jun 1;**104**(6):683-690

[78] Fernández-Fueyo E, Acebes S, Ruiz-Dueñas FJ, Martínez MJ, Romero A, Medrano FJ, et al. Structural implications of the C-terminal tail in the catalytic and stability properties of manganese peroxidases from ligninolytic fungi. Acta Crystallographica. Section D, Biological Crystallography. 2014 Nov 22;**70**(Pt 12):3253-3265

[79] Floudas D, Binder M, Riley R, Barry K, Blanchette RA, Henrissat B, et al. The paleozoic origin of enzymatic lignin decomposition reconstructed from 31 fungal genomes. Science. 2012 Jun 29;**336**(6089):1715-1719

[80] Voityuk AA. Electronic coupling for charge transfer in donor–bridge–acceptor systems. Performance of the two-state FCD model. Physical Chemistry Chemical Physics. 2012 Sep 26;**14**(40):13789-13793

[81] Acebes S, Fernandez-Fueyo E, Monza E, Lucas MF, Almendral D, Ruiz-Dueñas FJ, et al. Rational enzyme engineering through biophysical and biochemical modeling. ACS Catalysis. 2016 Mar 4;**6**(3):1624-1629

[82] Fernández-Fueyo E, Ruiz-Dueñas FJ, Martínez MJ, Romero A, Hammel KE, Medrano FJ, et al. Ligninolytic peroxidase genes in the oyster mushroom genome: Heterologous expression, molecular structure, catalytic and stability properties, and lignin-degrading ability. Biotechnology for Biofuels. 2014 Jan 3;**7**:2

[83] Sugano Y. DyP-type peroxidases comprise a novel heme peroxidase family. Cellular and Molecular Life Sciences. 2009 Apr 1;**66**(8):1387-1403

[84] Linde D, Coscolín C, Liers C, Hofrichter M, Martínez AT, Ruiz-Dueñas FJ. Heterologous expression and physicochemical characterization of a fungal dye-decolorizing peroxidase from *Auricularia auricula-judae*. Protein Expression and Purification. 2014 Nov 1; **103**:28-37

[85] Linde D, Pogni R, Marina C, Lucas F, Guallar V, Baratto MC, et al. Catalytic surface radical in dye-decolorizing peroxidase: A computational, spectroscopic and site-directed mutagenesis study. The Biochemical Journal. 2015;**466**:253-262

[86] Baratto MC, Sinicropi A, Linde D, Sáez-Jiménez V, Sorace L, Ruiz-Duenas FJ, et al. Redox-active sites in *Auricularia auricula-judae* dye-decolorizing peroxidase and several directed variants: A multifrequency EPR study. The Journal of Physical Chemistry. B. 2015 Oct 29;**119**(43):13583-13592

[87] Bauß A, Langenmaier M, Strittmatter E, Plattner DA, Koslowski T. Thermodynamic integration networks and their application to charge transfer reactions within the AauDyPI fungal peroxidase. The Journal of Physical Chemistry. B. 2016 Jun 9;**120**(22):4937-4944

[88] QSite, version 5.8, Schrödinger. New York, NY: LLC; 2012

[89] Linde D, Cañellas M, Coscolín C, Davó-Siguero I, Romero A, Lucas F, et al. Asymmetric sulfoxidation by engineering the heme pocket of a dye-decolorizing peroxidase. Catalysis Science & Technology. 2016 Aug 8;**6**(16):6277-6285

[90] Santiago G, de Salas F, Lucas MF, Monza E, Acebes S, Martinez ÁT, et al. Computer-aided laccase engineering: Toward biological oxidation of arylamines. ACS Catalysis. 2016 Aug 5;**6**(8):5415-5423

[91] Pardo I, Santiago G, Gentili P, Lucas F, Monza E, Medrano FJ, et al. Re-designing the substrate binding pocket of laccase for enhanced oxidation of sinapic acid. Catalysis Science & Technology. 31 May 2016;**6**(11):3900-3910

[92] Reetz MT, Carballeira JD. Iterative saturation mutagenesis (ISM) for rapid directed evolution of functional enzymes. Nature Protocols. 2007;**2**(4):891-903

Chapter 6

Conversion of Lignin to Heat and Power, Chemicals or Fuels into the Transition Energy Strategy

Myriam Adela Amezcua-Allieri and Jorge Aburto

Additional information is available at the end of the chapter

http://dx.doi.org/10.5772/intechopen.71211

Abstract

Energy transition toward low carbon, high sustainable and efficient generation and distribution systems will change the supply matrix of the world and create new opportunities but challenges still remain. Energy generation from biomass, or bioenergy, is one of such renewable sources and its use might be generalized in the following years. Bioenergy is a very promising strategy to provide energy not only for mobility but also for onsite places for heat and power generation. Besides, bioenergy differentiates from other renewable energies that biomass may be the source of a myriad of molecules enabling the bio-based economy and allowing the replacement in an extent of solvents, petrochemicals, and polymers produced by the petroleum industry. Biomass is generally composed of some large polymers found in nature such as cellulose, hemicellulose, proteins, starch, chitin, and lignin. The latter is a complex phenylpropanoid biopolymer conferring mechanical strength to plant cell walls and one of major spread in nature along with cellulose and chitin. Lignin has a plenty of potential uses in modern bio-based economy, from conventional paper industry uses to more challenging conversion to useful chemicals, materials, and clean biofuels. This chapter undertakes a rapid overview on lignin applications in order to describe the basis of a lignin-based economy.

Keywords: lignocellulosic materials, lignin, biorefinery, bioeconomy, heat and thermal power, bio-based chemicals, biofuels

1. Introduction

Although the world energy balance is dominated by fossil fuels, one major issue raised is how to change the future energy matrix toward more sustainable and renewable sources of energy.

The inclusion of renewables—wind, solar, geothermal, water and biomass—in the energy matrix has been marginal because of high costs and underdeveloped technologies. However, recognition of the damaging environmental impact from excessive dependence on fossil fuels, along with the growing concerns about the supply of some fossil fuels to meet rising global demand for energy, has brought into focus, the need for a cleaner and more diversified energy mix. Considering the challenges of the twenty-first century, including energy transition toward low carbon, bioenergy plays a key role in the diversification of sources of energy supply. In addition, the use of alternative fuels and bio-based chemicals will be increased to complete the energy matrix.

Biomass can be used directly (e.g., fuel wood for heating and cooking) or indirectly by converting it into a liquid or gaseous fuel (e.g., alcohol from sugar crops or biogas from animal waste). The main biomass processes utilized in the future are expected to be direct combustion of residues and wastes for electricity and heat generation, bioethanol and biodiesel as liquid fuels, and combined heat and power production from energy crops. The future of biomass electricity generation lies in biomass-integrated gasification/gas turbine technology, which offers high-energy conversion efficiencies. Biomass will compete favorably with fossil mass for niches in the chemical feedstock industry [9]. Hence, bioenergy is recognized as the energy derived from the conversion of biomass and where the biomass may be used directly as fuel, or processed into solids, liquids, and gases [16].

Biomass is mainly composed of three macromolecular biopolymers known as cellulose, hemicellulose, and lignin. Cellulose is composed of β-D-glucopyranose units linked by (1-4) glucosidic bonds and different polymorphs has been described from high to amorphous crystallinities [22]. On the other hand, hemicellulose is mostly amorphous and generally classified into four groups: (1) xyloglycans (xylans), (2) mannoglycans (mannans), (3) xyloglucans and (4) mixed-linked β-glucans [12]. They are considered the most abundant biopolymers on superior plants on earth just followed by lignin.

In this chapter, we present the actual and potential contribution of lignin as a source of solid, gas or liquid biofuels but also as a platform for bio-based chemicals and materials.

2. Lignin

Lignin is an amorphous and highly branched polyphenolic polymer with phenylpropane units, which may be present in varying amounts in biomass, comprising around 8–30% of biomass weight (**Table 1**). The complex structure of lignin contains numerous ether linkages, hydroxyl and methoxy groups and, therefore, a high-oxygen content. This cross-linked macromolecule is composed of three types of monolignols or lignin precursors, also known as phenolic monomers, including p-coumaryl, coniferyl and sinapyl alcohol (**Figure 1**). These precursors correspond to the p-hydroxyphenyl (H), guaiacyl (G) and syringyl (S) units in lignin, respectively [7]. Lignin is considered the clue agent that gives robustness and strength to the cellulose microfibril systems surrounded by hemicellulose [10, 11].

Biomass	Cellulose (%)	Hemicellulose (%)	Lignin (%)	Calorific value (GJ/ton)	Density of compressed material (kg/m³)
Wood (*Pinus* spp.)	43.3	25.9	29.6	18.3	906.8
Agave bagasse (*Agave atrovirens*)	45.9	26.6	8.7	15.5	868
Sugarcane bagasse (*Saccharum officinarum*)	32.3	35.2	22.4	19.1	721.7
Corn stubble (*Zea mays*)	39.1	30.9	18.6	15.6	928.2
Wheat Straw (*Triticum* spp.)	42.9	28.9	21.6	17.9	899

Table 1. Biomass composition, calorific values and densities of compressed materials [18].

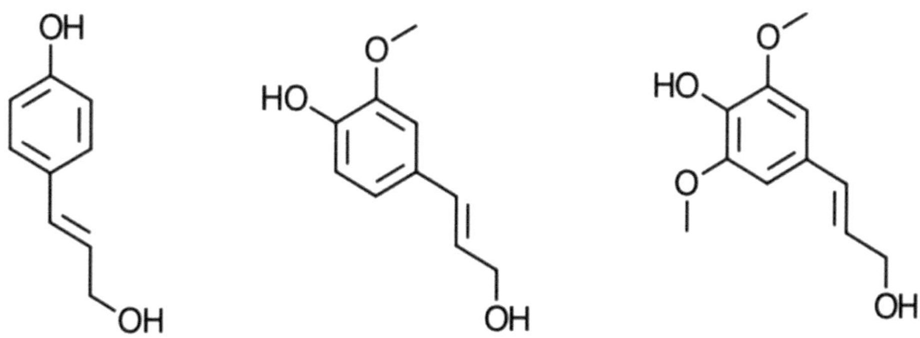

Figure 1. Lignin precursors from left to right: p-coumaryl, coniferyl and sinapyl alcohols.

The main classification of wood includes softwood (gymnosperm as pine or spruce) and hardwood (angiosperm as oak and walnut); the former is usually cheaper than the latter because about 80% of all timber comes from softwood. Softwoods have a wide range of applications and are found in building components (e.g., windows, doors), furniture, medium-density fiberboard (MDF), paper, Christmas trees and so on. Hardwood is usually used in high-quality furniture, decks, flooring and construction that need to have a long life. Lignin present in softwood usually has more coniferyl alcohol (75%) and less p-coumaryl (5%) and sinapyl alcohol content (20%) than that present in hardwood (50, 10 and 40%, respectively; [24]).

Lignin in biomass may be valued through the generation of thermal and electric power as well as the production of materials and chemicals (**Figure 2**). Here, we focus on industrial uses of biomass-containing lignin and processed or extracted lignin. Therefore, we focus on sawmill, agro-industrial and agricultural activities. Concerning forestry activities in sawmills, we refer to FAO definitions. Please keep in mind that there is not sufficient data and mass balance may be not complete. Hence, roundwood refers to all quantities of woods removed from trees inside the forests and from trees outside the forests. It includes all woods removed with or without bark, including woods removed in its round form or split form and considered fuel wood, sawlogs and veneer

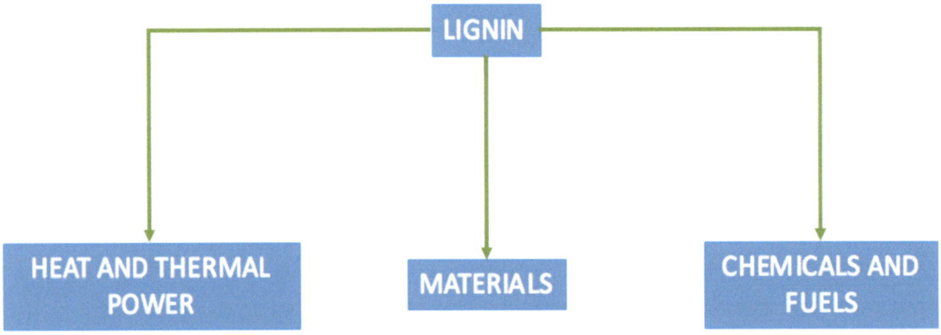

Figure 2. Scheme of lignin uses.

logs, wood pulp, round and split and other industrial roundwood. All fuel wood is roundwood that will be used for cooking, heating or power production, while wood pulp is the fibrous material prepared from pulpwood, wood chips, particles, residues or recovered paper by mechanical and/or chemical processes for the manufacturing of paper and other cellulose products [13].

According to FAO, the world production of industrial roundwood, wood fuel and wood pulp is shown in **Table 2**. This forestry biomass might be considered as the available raw material for processing into power, materials, chemicals and fuels. If we look at specific countries, USA and China lead the potential production of roundwood in coming years (**Figure 3**). Nevertheless, European countries such as France and UK as well as Latin American countries such as Chile and Mexico together represent the potential production of China but with a smaller population.

Regarding regional plant production, *Agave* is a succulent genus belongs to the monocot family Agavaceae [29] and the most important plants in Mexico are *A. tequilana, A. angustifolia,* and *A. atrovirens,* which are used for the production of spirit and fermented non-distilled beverages such as Tequila, Mezcal and Pulque, respectively. Agave plants have a lower content of lignin ranging from 7 to 16% [18, 29], and are considered easier to process when compared to wood species (lignin content around 30%) and Gramineae such as cane sugar, corn, wheat (lignin content around 18–22%).

As we mentioned earlier, lignin is the third most abundant biopolymer on earth, when we assess the occurrence on superior plants and it plays an important role in the transition energy global agenda (**Figure 4**). OECD countries, in **Figure 4**, possess an interesting potential of roundwood-derived lignin production that can help to achieve the Paris 21st COP's agreement.

Wood product	**World production (thousand m^3)**	**World production (metric tons)**
Roundwood	1.836,541	1,240,906
Wood fuel	1.863.828	1,259,343
Wood pulp	255,930	172,926

Table 2. World production of roundwood, fuel wood and wood pulp [13].

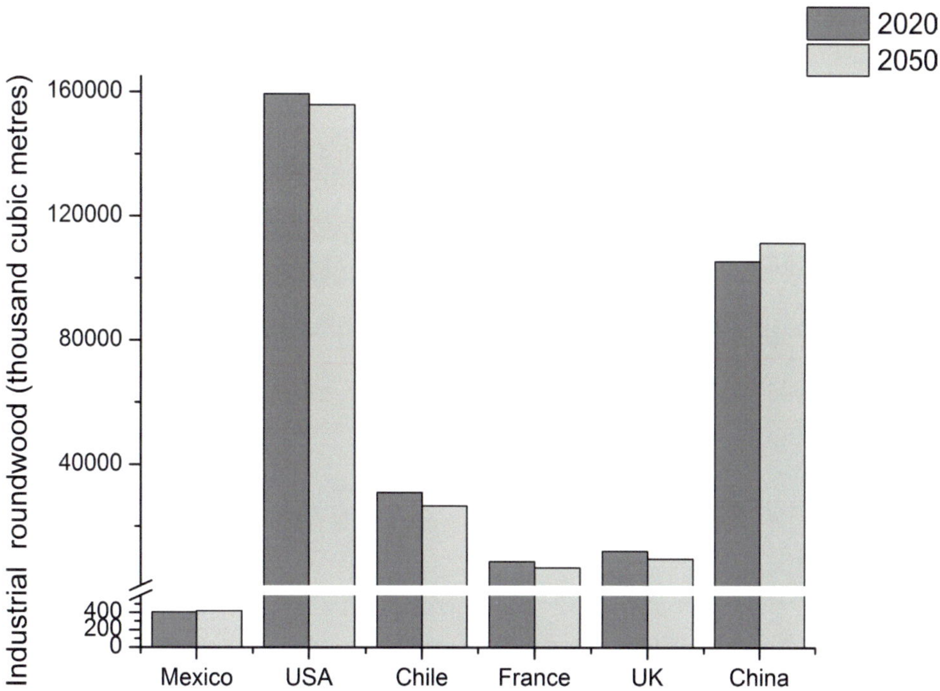

Figure 3. Industrial roundwood potential production in selected countries between 2020 and 2050.

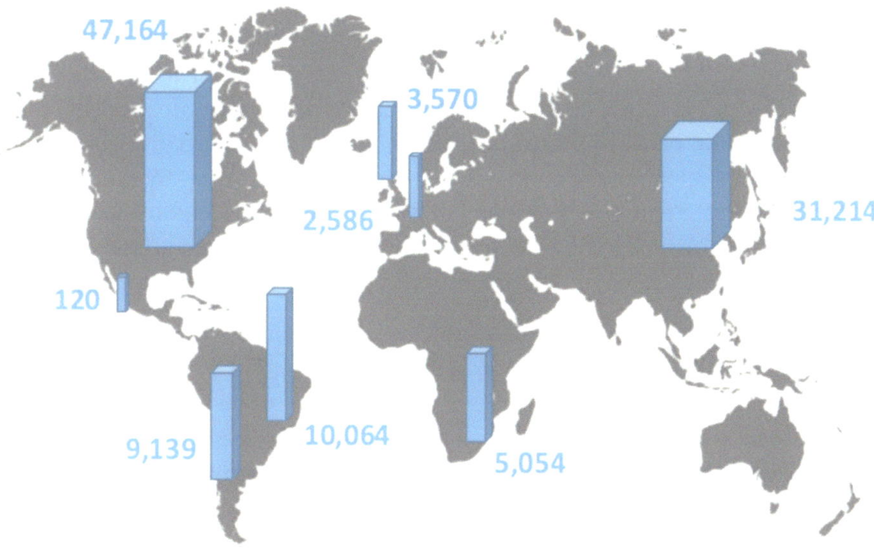

Figure 4. Lignin potential production from roundwood products in selected countries (1000 cubic meters; [13]).

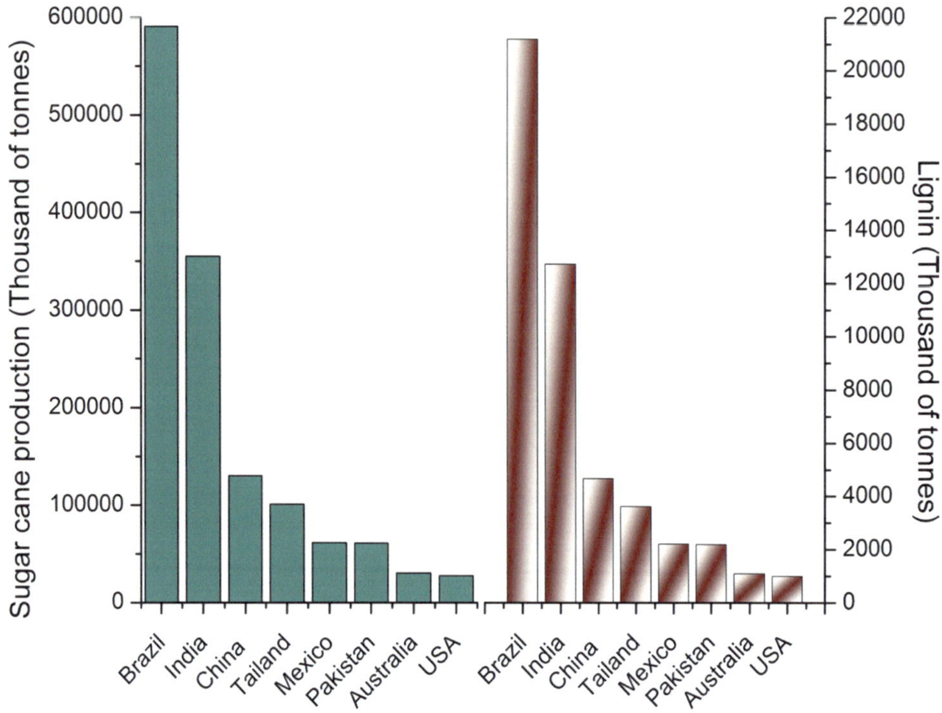

Figure 5. Sugarcane production in selected countries in 2013 and estimation of lignin content based on own data [30].

Sugarcane is another interesting crop since its industrial processing is already done in an integrated biorefinery, where all plant is valued into sugar, molasses, heat and electric power, and in some cases, into ethanol as oxygenated agent in gasoline (**Figure 5**). Indeed, since the earlier 1970s, countries like Brazil has undergone an intensive program of sugarcane production because there is an enormous need of alternative fuel for gasoline, considering Brazil's dependency of oil and gas importations during that time. Other countries like India, Mexico and, in a lesser extent, the USA produce sugarcane and sugarcane-related products but they also have an important potential of value-added lignin from the bagasse.

Traditional crops such as maize and wheat may be considered as a source of biomass and lignin for energy and chemical production (**Figure 6**). Their production is spread all around the world but some countries like Mexico has forbidden the use of cereal grains, rich in starch and considered as a cultural value, for energy uses. Nevertheless, we must consider that such biomass already has conventional and important value-added uses in land fertility, erosion mitigation and livestock breeding. The partial use of such biomass may be accompanied with integral policies that protect and assure the quality of renewable soil as well as the mitigation of disturbance on conventional value chains and stakeholders.

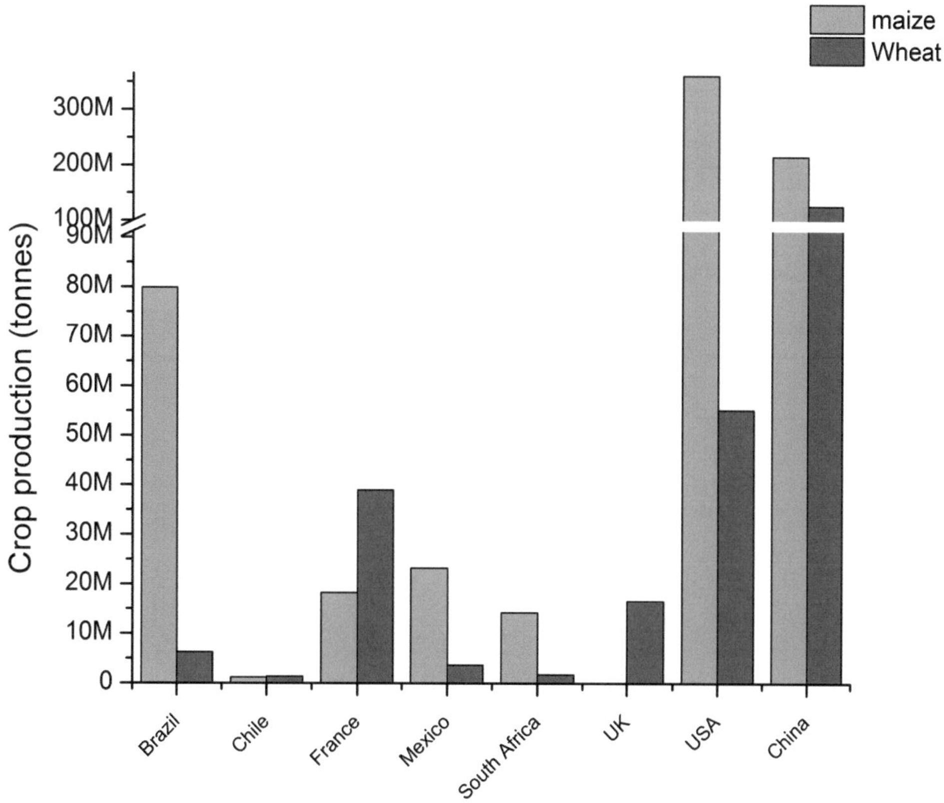

Figure 6. Maize and wheat production in selected countries in 2014 [13].

3. Generation of heat and power

According to the Chemical Sciences Roundtable of the National Research Council [21], burning biomass to produce energy and heat is nothing new, but doing so at a large scale still cannot economically compete with coal and natural gas. Local, small-scale biomass-to-power systems may prove to be the most efficient way of generating energy from biomass. Already, small-scale production of biogas from biomass and on-site co-generation of electricity and heat is widespread in Europe. Farm-sized units are in operation in the United States as well. One interesting example is sugarcane-producing mills which find its residue, sugarcane bagasse, a reliable and affordable source of thermal and power generation for auto-consumption needs with an installed capacity between 10 and 50 MW. Sugarcane has 16% of fiber or bagasse (dry basis) with almost 9% of lignin ([1, 6, 18]; see **Figure 5**). Then, we estimated the potential energy supply considering a caloric value of 19.1 GJ/ton for sugarcane bagasse (see **Table 1** and **Figure 7**). We obtained for Mexico, a 188 PJ/year potential energy from sugarcane bagasse, which is lower but within the range reported by IRENA and REMBIO (Bioenergy Mexican Network in Spanish; [15].

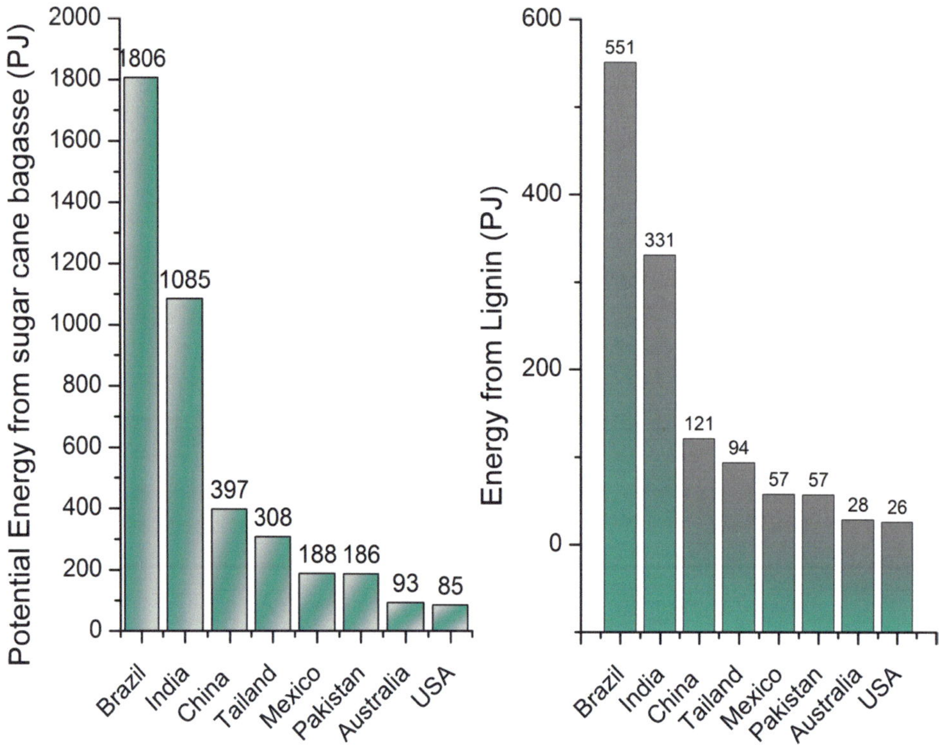

Figure 7. Potential energy from sugarcane bagasse and its isolated lignin in main producing countries.

In developing countries such as Mexico, Chile and even China, the predominant fuel in rural areas is still wood fuel from forest clearing and recollection (**Figure 8**), but traditional generation of heat for cooking through open fires or stoves generates a lot of smoke, carbon black, has low conversion efficiencies with 10–20%, and biomass is often sourced unsustainably. On the other hand, industrialized countries have developed more sophisticated forms as pellets and bricks produced from wood processing industries [20].

Nowadays, the biomass feedstock can be quiet different depending on regional or international distribution for specific needs and consumptions like residential (<0.5 MW) to medium (0.5–10 MW) and large plant power capacities (>10 MW; **Table 3**). According to International Energy Agency (IEA) [16], residential thermal generation as well as medium and large power generation with biomass is already economic and environmental competitive with other fuels such as coal and natural gas.

Indeed, renewable energies for heat and power generation are increasingly competing with fossil fuels since the reduction of capex and opex costs altogether with their better performance, even without government subsidies but carbon tax implementation might accelerate their adoption. Many factors affect the capital cost for renewable project including market, government policy, technology availability and maturity and capacity factors such as scale size. Solar photovoltaic and wind are considered as raw feedstock, free in contrast with the heat and power generation from biomass where its availability, logistics, conditioning, quality/volume, regulatory issues

http://dx.doi.org/10.5772/intechopen.71211

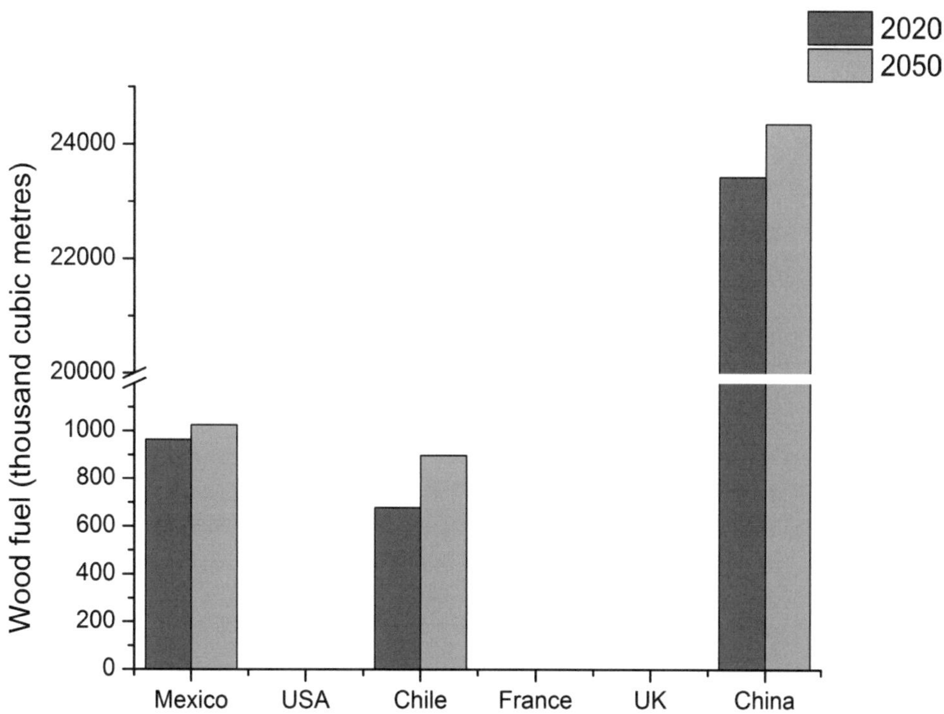

Figure 8. Wood fuel potential production in selected countries between 2020 and 2050.

	Biomass			
	Wastes	**Processing residues**	**Locally collected feedstocks**	**Internationally traded feedstocks**
	Organic waste, sewage sludge, manure	**Timber residues, black liquor, bagasse, rice husk**	**Agricultural residues, roundwood, energy crops**	**Roundwood, wood chips, biomass pellets, biomethane**
Feedstock costs (USD/GJ)	<0	0–4	4–8	8–12
Plant capacity	0.5–50	0.5–50	10–50	50
Typical power generation efficiency (%)	14–18	14–18	18–33	28–40
Capex (USD/kW)	6000–9800	6000–9800	3900–5800	2400–4200

Table 3. Some biomass feedstocks to produce power and heat, and related costs [16].

among others affect its cost as energy source. These costs may hugely vary depending also on scale and country, but some weighted costs show that all three technologies may compete on the field but government policy must assure equal conditions based on competitiveness (**Figure 9**). We observed that biomass is one of the more competitive alternatives for power generation in all world regions in contrast with fossil fuels and other renewable energies, but usually the lack of

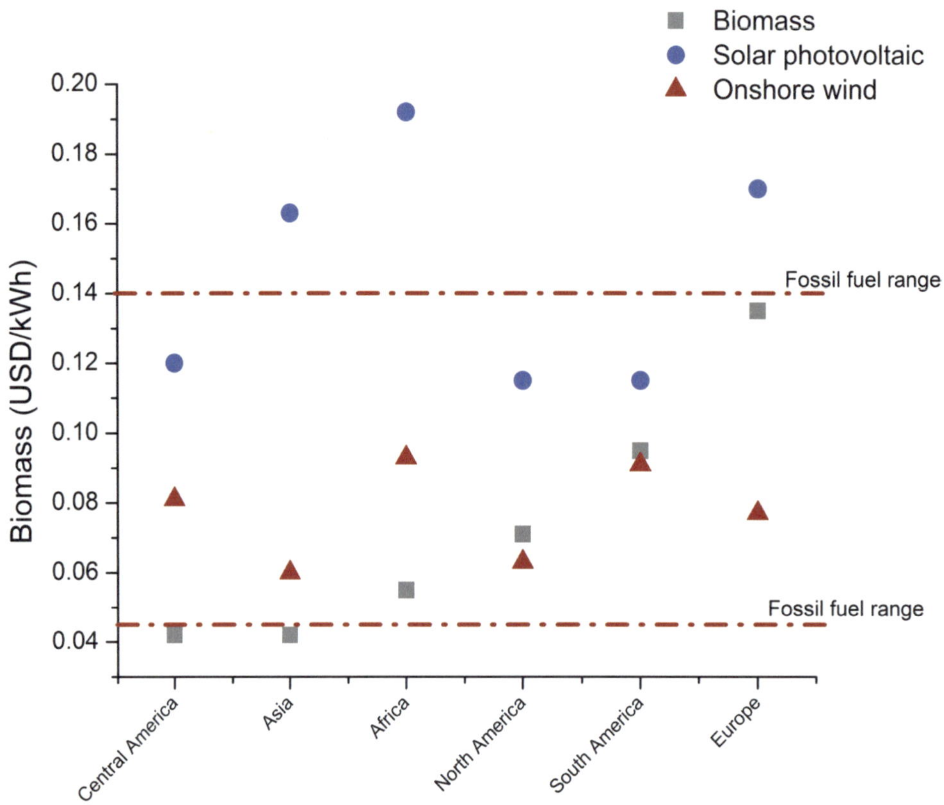

Figure 9. Weighted average cost of electricity (USD/kWh) by world regions and compared with fossil fuel power generation costs [15, 16].

an affordable and reliable source of biomass, related with social issues, may increase considerable associated risk, which makes the credits and the biomass costs more expensive. Wood- and lignin-based biomass might play an important role in the generation of sustainable heat and power, but they are needed policies that help to transit on this sinuous road.

4. Chemicals and fuels

Lignin as a polymeric material and pulping wood derivative has followed the development of paper industry all around the world since the end of nineteenth century. In such industry, lignin must be separated in order to get access to the more valuable cellulose. Hence, the alkali and acidic procedures are the more expanded method in industry and they allow obtaining of lignin derivatives such as alkali, kraft, lignosulfonate and organosolv lignin, respectively (**Figure 10**). An effective method to separate lignin is the organosolv process, which uses alcohol mixtures with water at high temperature near 200°C, sometimes with the aid of an alkali or acid in order to improve the lignin removal and wood pulp quality. Today, by far Kraft lignin dominates the market over lignosulfonate and organosolv lignin with a 90% share.

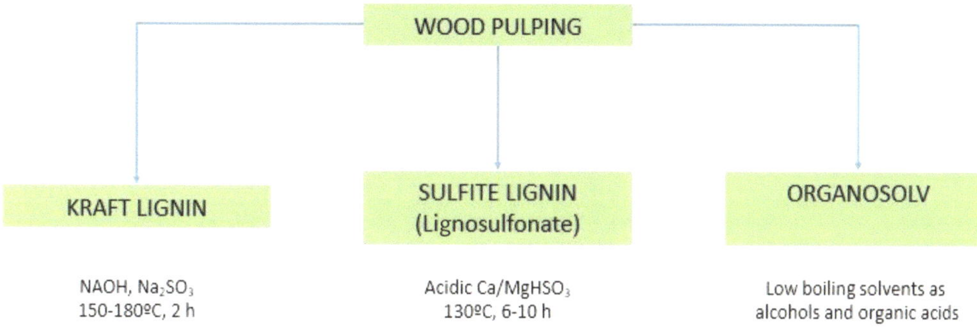

Figure 10. Scheme of conventional production of lignin derivatives in paper industry.

Krfat lignin is used in oil refineries as carbon crackers, cement additive and production of biofuels, BTX and other specialty chemicals such as vanillin and phenols. Lignosulfonate chemical is used in textile industries, thanks to its water solubility and functional properties as binders, dispersants, emulsifying and complexing agents. Nevertheless, the need for a sulfur-free agent is reconsidering lignosulfonate market hegemony for cleaner and sustainable sources of lignin [5].

Lignin may also be recovered by solubilizing the recalcitrant crystalline cellulose and amorphous hemicellulose using chemical, thermochemical, biological and/or enzymatic approaches [3, 7, 19, 23, 25]. We have recently developed a mixed chemical-enzymatic pretreatment of lignocellulosics to solubilize almost all cellulose and hemicellulose, except a lignin-rich polymer as seen by FTIR analysis (**Figure 11**). Indeed, we can see the empty cells where cellulose used to be leaving an interesting hole material with infrared characteristic signals at 4000–2500 cm^{-1} νO–H,1700–1740 cm^{-1} νC=O carbonyl and carboxylic groups, 1590–1610 cm^{-1} νC=C aromatic skeletal and 1126 cm^{-1} νC–O–C ether groups.

Here, lignin might be the source of materials, energy, biofuels or bio-based chemicals but we may consider two routes to achieve this: (1) its conditioning to obtain materials or bioenergy and (2) its depolymerization followed by conversion to polymers, fuels or bulk or fine chemicals (**Figure 12**). The simplest approach considers a biorefinery where lignin may serve as an energy source through direct combustion, gasification or pyrolysis, as discussed later in the section, but also for the production of materials such as activated carbon and/or biochar [8, 14].

Indeed, biochar production during pyrolysis of biomass or the fraction of lignin must be considered in an integrated biorefinery, where it helps for the economy and environmental impact of the whole process. Additional studies are needed in order to quantify the carbon capture by soil and microbes, reduction of CO_2 emissions due to bio-oil displacement of fossil fuels and decreased use of fertilizers, as well as reduction of NO_x emission from soils by better soil aeration [5, 17].

The present and future transition into a less depending fossil fuels economy needs not only renewable energy but also renewable raw materials like biomass and lignin to satisfy the need for materials and chemicals used in our everyday life. Biomass is the only renewable source known today that may support not only the generation of heat and electric power but also the processing and manufacturing of fuels and goods. Solar energy also has indeed the potential to achieve this; it does every day in leaves, algae and microorganisms such as microalgae and cyanobacteria [2]. Nevertheless, its development is still incipient but full of encouraging challenges.

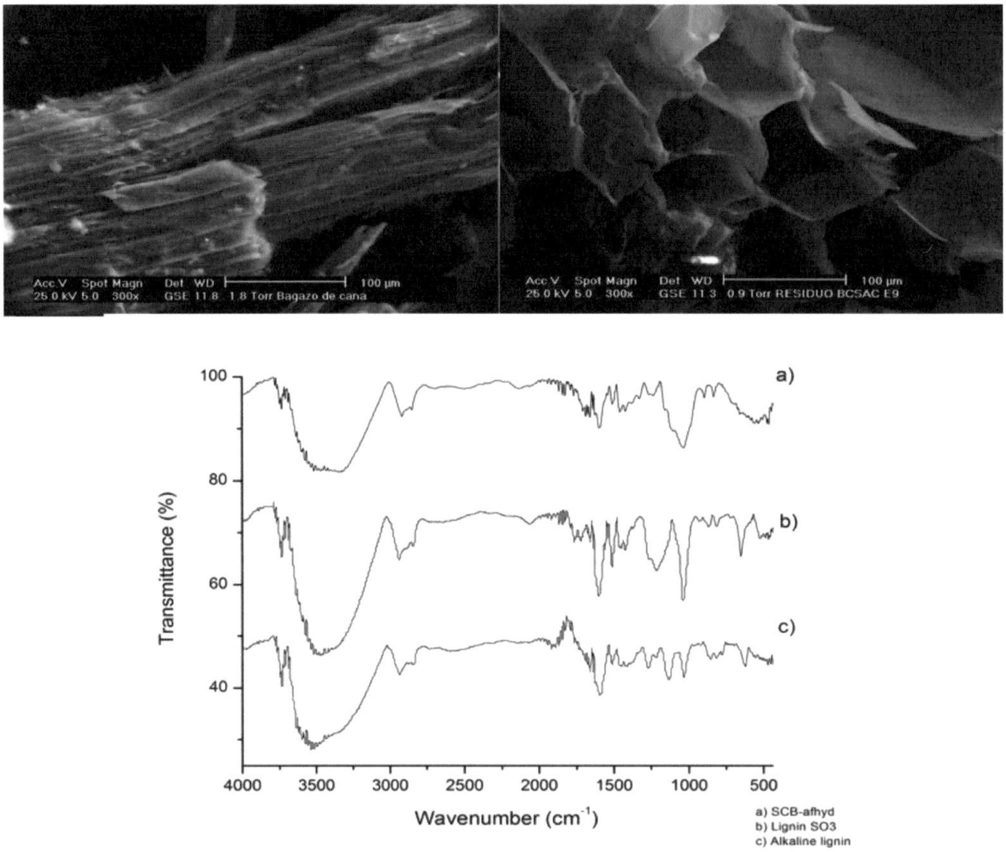

Figure 11. Scanning electronic micrographs of sugarcane bagasse before (top left) and after (top right) chemical and enzymatic pretreatment. FTIR spectra of pretreated (a) sugarcane bagasse, (b) lignosulfonate lignin and (c) alkali lignin.

A modern and integrated biorefinery will process low-value biomass into energy for self-consumption and export excess to the grid. More valuable biomass may be used for the production of edible or nonedible oils, protein-rich flour for fish/shrimp farms and used to convert the lignocellulosics into fuels and chemicals [4]. We have identified four additional market niches for lignin derivatives: (1) conventional paper-related derivatives; (2) specialized chemicals; (3) commodity chemicals; and (4) biofuels (**Table 4**).The first two have already established markets since low value-added generic industrials to high value-added flavor and fragrance additives. The latter use has a high demanding worldwide market for aromatic and phenol-based compounds that may be obtained from lignin. This approach must be considered in a biorefinery scheme since their high price may account for the financial sustainability and competiveness of the business. Indeed, the flavor and fragrance industry may reconsider inverting the balance between the use of aromatic compounds from fossil source against biomass or lignin raw materials, that is, compounds like phenylacetic acid, benzoates, phenylacetates, cinnamates, phenols, phenyl esters, alcohols and ethers, among many others [27].

In present and near future terms, biorefineries must be additionally considered to balance the production of biofuels and chemicals that can be used without any or less change in current

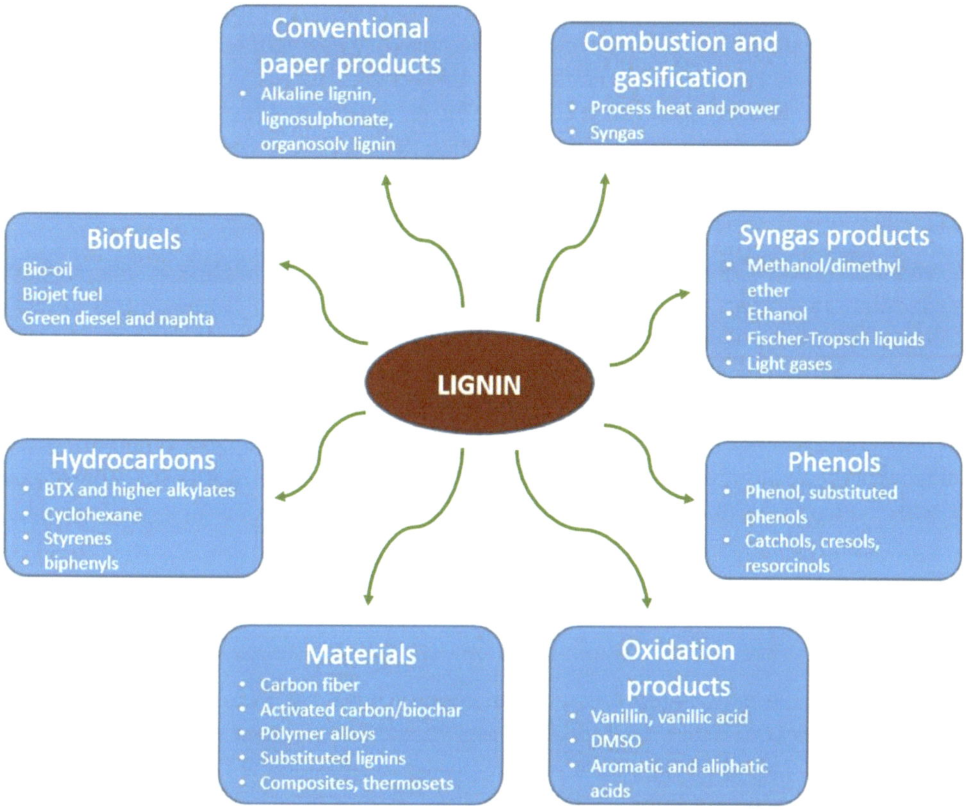

Figure 12. Scheme of conventional and potential products from lignin. BTX: benzene, toluene and xylene isomers and DMSO: dimethyl sulfoxide.

Market niches	Lignin derivative	Market volume (Mtons/year)	Price (USD/ton)
Conventional paper related derivatives	Lignosulfonates	1,000,000	350
	Kraft lignin	100,000	364
	Organosolv lignin	1000	919
Specialized chemicals	Lignin-based vanillin	3200	1200
	Crude oil-based vanillin	12,800	1200
	Natural vanillin	60	600,000
Commodity chemicals	BTX	102	1200
	Phenol	8	1500
Biofuels and fuels	Bio-jet fuel-based on lignin/ biomass	1–50% in mix with fossil jet fuel	1300–6400
	Fossil oil-based jet fuel	252,456	400

Table 4. Market volumes (Mtons) and price (USD/tons) of some conventional and emerging chemicals and fuels [26].

established technologies, known as drop-in fuels and chemicals. In this sense, lignin as a source of chemical compounds such as benzene, toluene, and xylenes might be relevant today to some petrochemical platforms but the demand is insignificant related to other higher volume markets as biofuels or specialized chemicals. Additionally, environmental and health issues will certainly become more stringent in near future, limiting the use of such compounds. Nevertheless, the society will need more advanced chemicals, building units, and materials, and lignin may be a reliable and sustainable source of new upcoming technology platforms.

Lignin-based biofuels may be obtained basically from three routes: (1) hydrothermal depolymerization and excess oxygen elimination; (2) slow or fast pyrolysis to depolymerize the lignin into a bio-oil that must be processed to eliminate excess of oxygen, and (3) gasification followed by biomass to liquids (BTL) process or Fischer-Tropsch process for the obtention of different fuel fractions or even alcohols [3, 28]. Even if these technologies represent the today and near future options to satisfy the need for clean, sustainable and neutral carbon fuels; the real issue is the deployment of such technologies and their spread adoption in order to comply with the world weather change commitments.

5. Final remarks

Lignin is a versatile polyphenol raw material comporting a significant presence in biomass. Due to its chemical nature, it is difficult to isolate and process into chemical commodities, specialized chemicals, thermal and/or electric power and also advanced biofuels. The actual energy transition requires a significant change of paradigm between the past unsustainable fossil energy source to more advanced environmental, health and socially committed sources of reliable, renewable and sustainable forms of energy; among those, bioenergy from biomass and lignin will play an important role. As a part of biomass, lignin has plenty of potential uses in a modern bio-based economy, from conventional paper products to more challenging processes such as its conversion to useful chemicals, materials, and clean biofuels in a biorefinery scheme.

Author details

Myriam Adela Amezcua-Allieri and Jorge Aburto*

*Address all correspondence to: jaburto@imp.mx

Division of Biomass Conversion, Instituto Mexicano del Petróleo, Mexico City, Mexico

References

[1] Aburto J, Martínez T, Murrieta F. Technical and economical evaluation of bioethanol production from lignocellulosic residues. Tecnología, Ciencia, Educación. 2008;**23**(1):23-30

[2] Alstrum-Acevedo JH, Brennaman MK, Meyer TJ. Chemical approaches to artificial photosynthesis. 2. Inorg. Chem. 2005;**44**(20):6802-6827

[3] Amezcua-Allieri M, Gutiérrez-Villegas JC, Aburto J. Combustibles avanzados de aviación. Instituto Mexicano del Petróleo, Copyrights 03-2017-022410275400-01; 2017. Available online on: http://rtbioenergia.org.mx/wp-content/uploads/2016/12/Divulgacion_Biojetfuel-booklet-vf.pdf [consulted on May 2017]

[4] Amezcua-Allieri M, Sánchez-Durán T, Aburto J. Study of chemical and enzymatic hydrolysis of cellulosic material to obtain fermentable sugars. Journal of Chemistry. 2017;**9**. Article ID 5680105. DOI: 10.1155/2017/5680105

[5] Azadi P, Inderwildi OR, Farnood R, King DA. Liquid fuels, hydrogen and chemicals from lignin: A critical review. Renewable and Sustainable Energy Reviews. 2013;**21**:506-523

[6] Barrera I, Amezcua-Allieri MA, Estupiñan L, Martínez T, Aburto J. Technical and economical evaluation of bioethanol production from lignocellulosic residues in Mexico: Case of sugar cane and blue agave bagasses. Chemical Engineering Research and Design. 2016;**107**:91-101

[7] Barrera I, Guzmán N, Peña E, Vázquez T, Cerón-Camacho R, Folch J, Honorato J, Aburto J. Ozonolysis of alkaline lignin and sugarcane bagasse: Structural changes and their effect on saccharification. Biomass and Bioenergy. 2017;**94**:167-172

[8] Carrot SPJM, Ribeiro-Carrott MML. Lignin—From natural adsorbent to activated carbon: A review. Bioresource Technology. 2007;**98**:2301-2312

[9] Demirbas A. Energy Sources Part A: Recovery Utilization and Environmental Effects. 2017;**39**(8):754-760

[10] Chen Z, Wan C. Biological valorization for converting lignin into fuels and chemicals. Renewable and Sustainable Energy Reviews. 2017;**73**:610-621

[11] Bilal M, Asgher M, Iqbal HMN, Hu H, Zhang H. Biotransformation of lignocellulosic materials into value-added products: A review. International Journal of Biological Macromolecules. 2017;**98**:447-458

[12] Ebringerová A. Structural diversity and application potential of hemicelluloses. Macromolecular Symposia. 2006;**232**:1-12

[13] Food and Agriculture Organization of the United Nations (FAO). Forest products. In: FAO Statistics Series Co. 205. 2014. Rome, 2016

[14] Holladay JE, Bozell JJ, White JF, Johnson D. Top value-added chemicals from biomass. In: Volume II. Results of screenings for potential candidates from biorefinery lignin. Pacific Northwest National Laboratory. United States of America: PNNL-16983; 2007

[15] International Energy Agency (IEA). Renewable power generation costs in 2014. 2015

[16] International Energy Agency (IEA). Technology roadmap. Bioenergy for heat and power; 2012

[17] Laird DA, Brown RC, Amonette JE, Lehmann J. Review of the pyrolysis platform for coproducing bio-oil and biochar. Biofuels, Bioproducts and Biorefining. 2009;**3**(5):547-562

[18] Honorato A, Trejo S, Aburto J. Base de datos de materiales lignocelulósicos. Project 151370. Mexico: CONACyT; 2017

[19] Merino Pérez O, Almazán V, Martínez-Palou R, Aburto J. Screening of ionic liquids for pretreatment of Taiwan grass in Q-Tube minireactors for improving bioethanol production. Waste and Biomass Valorization. 2016;**8**(3):733-742

[20] Mustafa B, Kahraman B. Wood as an Energy Source: Potential Trends, Usage of Wood, and Energy Politics. Energy Sources, Part A: Recovery, Utilization, and Environmental Effects. 2006;**28**(9):837-844

[21] National Research Council (US). Chemical Sciences Roundtable. Opportunities and Obstacles in Large-Scale Biomass Utilization. The Role of the Chemical Sciences and Engineering Communities A Workshop Summary. Washington, DC: National Academies Press (US). ISBN-13: 978-0-309-27864-5 ISBN-10: 0-309-27864-3; 2012

[22] O'Sullivan AC. Cellulose: The structure slowly unravels. Cellulose. 1997;**4**:173-207

[23] Rabemanolontsoa H, Saka S. Various pretreatments of lignocellulosics. Bioresource Technology. 2016;**199**:83-91

[24] Saka S. Chapter 2. Chemical composition and distribution. In: NSH D, Nobuoi S, editors. Wood and Cellulosic Chemistry. 2nd ed. New York: Marcel Dekker, Inc.; 2000

[25] Sindhu R, Binod P, Pandey A. Biological pretreatment of lignocellulosic biomass – An overview. Bioresource Technology. 2016;**199**:76-82

[26] Smolarski N. High-Value Opportunities for Lignin: Unlocking its Potential. Frost & Sullivan; 2012. Available online on: https://www.greenmaterials.fr/wp-content/uploads/2013/01/high-value-opportunities-for-lignin-unlocking-its-potential-market-insights.pdf Consulted online May 2017

[27] Surburg H, Panten J. Common Fragrance and Flavor Materials. 5th ed. Weinheim: Wiley-VCH Verlag; 2006

[28] Wang W-C, Tao L, Markham J, Zhang Y, Tan E, Batan L, Warner E, Biddy M. Review of biojett fuel conversion technologies. In: Technical Report NREL/TP-5100-66291. National Renewable Energy Laboratory (NREL), United States of America; 2016. Available online on: www.nrel.gov/publications Consulted online on June 2017

[29] Escamilla-Treviño LL. Potential of plants from the Genus Agave as bioenergy crops. Bioenergy Research. United States of America. 2012;**5**(1):1-9

[30] Unión Nacional de Cañeros. http://caneros.org.mx/ Consuted online on July 2017

Chapter 7

Biomass Gasification: A Review of Its Technology, Gas Cleaning Applications, and Total System Life Cycle Analysis

Kenji Koido and Takahiro Iwasaki

Additional information is available at the end of the chapter

http://dx.doi.org/10.5772/intechopen.70727

Abstract

Gasification technology presents one option for energy-conversion technique from woody biomass contaminated by radionuclides released in March 2011 during the Fukushima Dai-ichi Nuclear Power Plant accident. The gasification process converts carbonaceous materials into combustible gases, carbon dioxide, and residues. Owing to their small-scale distributed configuration, woody biomass gasification plants are suitable for gasifying Japan's biomass and have been installed increasingly in Japan recently. This chapter reviews current trends of gasification and life cycle assessment (LCA) of total systems, including gas cleaning.

Keywords: biomass gasification, gas cleaning, desulphurisation, radioactive materials, life cycle assessment

1. Introduction

Lignin is a crosslinked macromolecular material based on a phenylpropanoid monomer structure. Vascular plant species such as fern-related plants, gymnosperms, and angiosperms have lignin structures. Woody plants such as softwoods and hardwoods are composed, respectively, of 25–35% and 20–25%. Recently, small-scale woody biomass power generation plants have been installed increasingly in Japan. Such plants are suitable for gasifying Japan's biomass because of their small-scale, distributed characteristics.

To utilise small-scale biomass more effectively, biomass plants must be installed near energy source and demand sites in order to shorten transport distances. This is especially effective for small-scale woody biomass plants. Here small scale is defined as a plant scale of at most

2 MWe. Power generation in such small-scale plants requires biomass gasification technology in order to obtain higher thermal efficiency. Gasification is a process that converts carbonaceous materials into carbon monoxide, hydrogen, carbon dioxide, and gaseous hydrocarbons (producer gas). Producer gas can be supplied as fuel to internal combustion engines and power generators. To maximise the efficiency of woody biomass conversion, producer gas should be utilised not only for power generation but also for thermal production from the producer gas' sensible heat. Cogeneration system for heat and power production is called combined heat and power (CHP).

Recently, many kinds of biomass gasification combined heat and power (BGCHP) systems have been developed (mainly in Europe). These micro/small-scale CHP systems can be connected and integrated to achieve an appropriate plant scale according to the biomass supply and heat and power demands. Further description of BGCHP is presented in section 2.4.

Woody biomass in Fukushima was contaminated by radionuclides released between 12 and 31 March 2011 due to the Fukushima Dai-ichi Nuclear Power Plant accident. To safely utilise the contaminated woody biomass in Fukushima as biomass gasification (BG) fuel, the radioactivity of products and by-products such as offgas and ashes in ash bins and on filters must be investigated.

This chapter reviews current trends of gasification and life cycle assessment (LCA) of total systems, including gas cleaning. Previously, there have been no reviews of small-scale gasification process for hydrogen production and CHP and no critical review of LCA of BG processes. Gas cleaning is one of the most important processes in BG systems for controlling contaminants in producer gases and preserving the catalysts of fuel cells and gas engines.

2. Gasification technology

A biomass gasifier is comprised of four reaction zones, i.e. drying, pyrolysis, combustion, and reduction. The produced gas (syngas) contains impurities to be cleaned utilising a bag filter, activated carbon, scrubber, etc. This chapter describes the fates of two contaminants, radionuclides and sulphur, during the gasification process. The radioactivity of caesium-137 (^{137}Cs) and caesium-134 (^{134}Cs) in fly ash over a bag filter was observed using a germanium semiconductor detector. The fate of sulphur was also reviewed because sulphur often triggers fuel cell catalyst poisoning and gas engine erosion. The LCA of a total system based on the energy profit ratio and environmental impact is then reviewed.

2.1. Biomass feedstock

Biomass feedstocks are classified based on several factors: moisture content, material, and form, as presented in **Table** 1. Biomass is broadly divided into three groups: dry, wet, and other. Dry biomass is classified as woody or herbaceous and wet biomass as sludge/excreta, common food, or other. Each classification has three sub-categories: waste, unutilised, and produced.

http://dx.doi.org/10.5772/intechopen.70727

Moisture	Classification	Wastes	Unutilised	Produced
Dry	Woody	Construction wastes, Timber offcuts	Forest thinnings, Remaining forest timbers, Damaged trees	Short-rotation woody crops (eucalyptus, willow, etc.)
	Herbaceous		Crop residues (rice / wheat straw, rice husk)	Grasses (Napier grass, sorghum, Miscanthus etc.)
Wet	Sludge /Excreta	Sewage sludge, Livestock excreta		
	Common food	Food-processing wastes, Kitchen wastes		
Other	Other	Molasses, Waste food oil	Landfill gas	Cultivated maize, Cultivated sugar cane

Table 1. Classification of biomass feedstocks.

Among these, dry woody/herbaceous biomasses are used as feedstocks for BG plants in Japan. For instance, woody biomasses contain waste woods (construction wastes and timber offcuts) and unutilised woods (forest thinnings, remaining timbers, and damaged trees) composed of cedar, cypress, pine, etc. Short-rotation woody crops (eucalyptus, willow, etc.) are categorised as produced woods. Energy crops such as willow are expected to be cultivated in land fallow and used as biomass because non-food-producing farmlands have been recently abandoned in Japan [1].

For unutilised herbaceous biomass, crop residues such as rice/wheat straw and rice husks [2] are available, while for produced herbaceous biomass, grasses such as Napier grass, sorghum, and Miscanthus are usable. Above all, rice husks contain abundant silica. The ash by-product from gasifiers has potential use in nanomaterials [3].

2.2. Gasifier

There are various kinds of gasifiers (e.g., fixed-bed, fluidised, pressurised, etc.). This chapter describes the simplest fixed-bed gasifiers. Fixed-bed gasifiers have a long history and have established confidence through experience with small-scale biomass gasification reactors [4]. **Figure 1** schematically illustrates fixed-bed updraft and downdraft gasifiers. The flows of biomass and producer gases are counter-current for updraft and co-current for downdraft. The gasification reaction is comprised of four main reaction zones. Heat released during the combustion process is used for drying, pyrolysis, and reduction processes in the gasifier.

For both gasifier types, biomass fuels are supplied into each gasifier followed by a drying process. The dried biomass is then pyrolysed to release volatiles and gases. These products are combusted partially or completely using air fed with heat released from the combustion process. The gases, tars, and chars resulting after combustion are reduced in the reduction zone. The positional relationship of these processes is shown schematically in **Figure 2**.

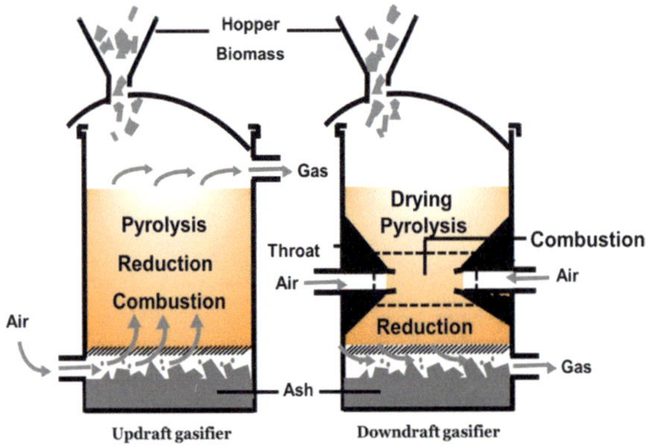

Figure 1. Schematic of updraft and downdraft fixed-bed gasifiers [5].

2.3. Gasification mechanisms

Gasification proceeds through many kinds of simultaneous or consecutive complex reactions. Reactions corresponding to each reaction zone are shown in this section. **Figure 3** represents several main reaction pathways in biomass gasification reactions. The reactions for each pathway (or reaction zone) are presented in **Table 2** based on the literature [7, 8].

2.3.1. Drying

Biomass moisture at ordinary temperatures becomes water vapour during the drying process at about 100–250 °C. The conversion occurs owing to heat transfer between hot gases from the oxidation and biomass in the drying zone [4]. First, the moisture on the biomass surface evaporates followed by inherent moisture evaporation. The vapour produced is used for reduction reactions in the reduction zone, including the water-gas reaction.

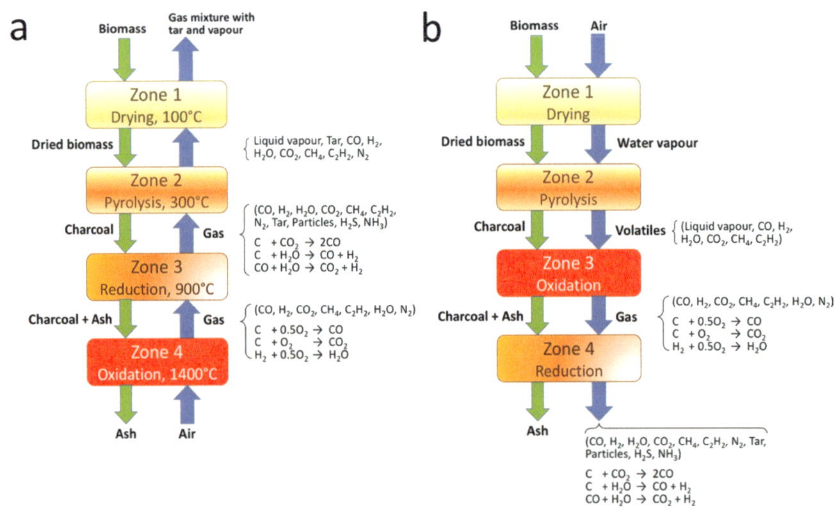

Figure 2. Conceptual diagram of multiple steps in fixed-bed (a) updraft and (b) downdraft gasifiers [6].

http://dx.doi.org/10.5772/intechopen.70727

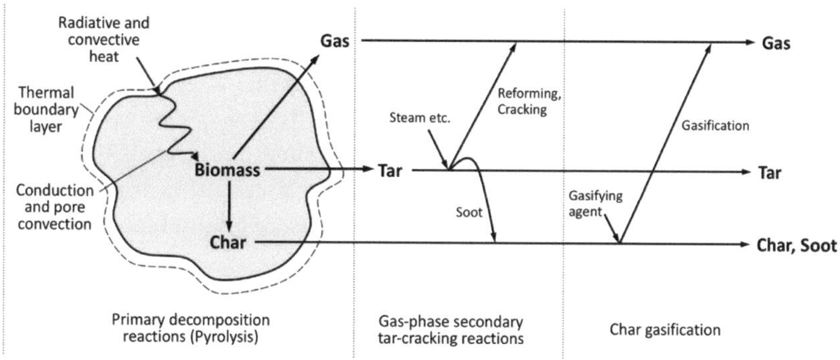

Figure 3. Biomass gasification pathway proposed based on Ref. [10, 11].

2.3.2. Pyrolysis

In the absence of oxygen, volatiles with weaker molecular bonds begin to be thermally decomposed at 200–240 °C; this continues up to 400 °C. The volatiles are vaporised to produce gases, tars, and chars. The chars are also pyrolysed with gas production and weight loss of about 30% [9]. Pyrolysis takes place owing to heat transfer from radiation, convection, and conduction to the biomass as shown in **Figure 3a**. The pyrolysis reactions are represented as Reaction (R1) in **Table 2**.

Name of reaction	Chemical reaction	$\Delta H^0_{r(298)}$ (kJ/mol)	$\Delta G^0_{r(298)}$ (kJ/mol)	No.
Pyrolysis	$C_xH_yO_z \rightarrow aCO_2 + bH_2O + cCH_4 + dCO + eH_2 + fC_{2+} + \text{char} + \text{tar}$			(R1)
Partial oxidation	$C + 0.5\,O_2 \rightarrow CO$	−111		(R2)
Complete oxidation	$C + O_2 \rightarrow CO_2$	−394		(R3)
Steam-tar reforming	$C_nH_m + 2nH_2O \rightarrow (2n + m/2)\,H_2 + nCO_2$			(R4)
Hydrogenating gasification	$C + 2H_2 \leftrightarrow CH_4$	123.7	168.6	(R5)
Boudouard equilibrium	$C + CO_2 \leftrightarrow 2CO$	205.3	140.1	(R6)
Water-gas shift (WGS)	$CO + H_2O \leftrightarrow CO_2 + H_2$	−41.47	−28.5	(R7)
Heterogeneous WGS	$C + H_2O \leftrightarrow CO + H_2$	130.4	89.8	(R8)
Steam reforming of methane	$CH_4 + H_2O \leftrightarrow CO + 3H_2$	172.6	118.4	(R9)
Dry reforming of methane	$CH_4 + CO_2 \leftrightarrow 2CO + 2H_2$	−74.9	−50.3	(R10)
Ethylene	$2CO + 4H_2 \leftrightarrow C_2H_4 + 2H_2O$	−104.3	−111.6	(R11)
Ethane	$2CO + 5H_2 \leftrightarrow C_2H_6 + 2H_2O$	−172.7	−212.7	(R12)
Propane	$3CO + 7H_2 \leftrightarrow C_3H_8 + 3H_2O$	−165.1	−293.2	(R13)
Butane	$4CO + 9H_2 \leftrightarrow C_4H_{10} + 4H_2O$	−161.9	−376.7	(R14)
H_2S formation	$S + H_2 \rightleftarrows H_2S$			(R15)
H_2S-COS equilibrium	$H_2S + CO \rightleftarrows COS + H_2$			(R16)

Table 2. Chemical reactions occurring in biomass gasification (gasifying agent: steam).

2.3.3. *Oxidation*

Heat is released during the oxidation of gases and of gaseous volatiles and chars produced in pyrolysis under supplied air. The heat is used for drying, pyrolysis, and other endothermic reactions [4]. Partial oxidation (R2) and complete oxidation (R3) occur. Reaction temperatures are around 600–900 °C for partial oxidation and around 800–1400 °C for complete oxidation. Partial oxidation releases 111 kJ/mol of heat while complete oxidation releases 394 kJ/mol.

2.3.4. *Reduction*

During the reduction process, the char and tar produced from oxidation release gases at around 600–950 °C via several reactions. The typical reactions include the heterogeneous water-gas shift (HWGS: $C + H_2O \rightarrow H_2 + CO$) and water-gas shift reaction (WGS: $CO + H_2O \leftrightarrows CO_2 + H_2$). Via the endothermic HWGS reaction, char or tar reacts with water vapour (derived from biomass and air) to produce CO and H_2 at temperatures greater than 750 °C. The WGS reaction is an exothermic reversible reaction and has an equilibrium point for CO, H_2O, CO_2, and H_2 concentrations. The higher the reaction temperature, the greater the amount of gaseous products produced on the equation's right-hand side. Reduction is a totally endothermic reaction because the water-gas reaction is dominant.

2.4. Combined heat and power gasification

Gasification power generation systems are different in terms of gasifier type, gas cleaner for dust/tar/hydrogen sulphide removal (as described in sections 3.2 and 3.3), heat exchanger type, and power generator type (gas turbine, internal combustion engine, fuel cell, etc.). A BGCHP requires seven elemental processes: (i) pretreatment, (ii) storage, (iii) gasifier, (v) gas cleaner, (iv) gas cooler, (vi) gas engine, and (vii) power generator, as shown in **Figure 4**. BGCHP is a process that cogenerates heat and power from syngas produced in the gasifier. **Figure 4** shows a simplified schematic of BGCHP. Generally, biomass power generation from direct combustion has an efficiency of ca. 30% at most. However, via BGCHP, a total efficiency of ca. 85% is attained with thermal-load-following operation.

There are many kinds of small-scale gasifiers manufactured by European companies such as in Germany (e.g., Burkhardt [12], Spanner [13], Entrade [14], etc.), Finland (e.g., Volter [15]), and Sweden (e.g., Cortus Energy AG [16]).

2.5. Biohydrogen production via gasification

Biohydrogen via thermochemical conversion is obtained by gasifying woody biomass followed by a gas cleaning process. Biomass gasification typically yields producer gases such as H_2 (14–25%), CO (15–24%), CO_2 (12–15%), CH_4 (2.0–2.5%), H_2S (<100 ppmv), and COS (50 ppmv) [6]. For H_2 production, gas cleaning is required for gases other than H_2. Generally, CO is converted into H_2 and CO_2 with steam by a shift converter packed with nickel or nickel oxide catalysts.

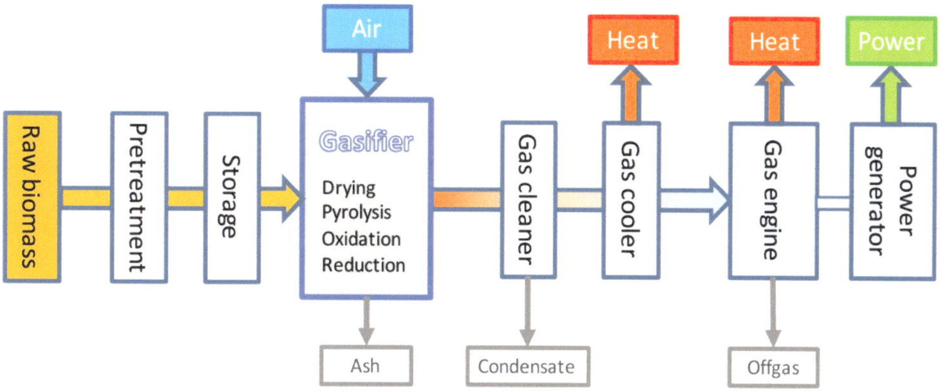

Figure 4. Flow diagram of power generation from biomass gasification.

3. Gas cleaning

3.1. Contaminants

Contaminants in syngas generally include particulate matter, condensable hydrocarbons (i.e. tars), sulphur compounds, nitrogen compounds, alkali metals (primarily potassium and sodium), hydrogen chloride (HCl), and radioactive nuclides. In this review, the sulphur compounds and radioactive nuclides in Fukushima, Japan are the focus.

Syngas ($CO + H_2$) has many uses ranging from heat and/or power applications (e.g., CHP) to many kinds of synthetic fuels and chemicals as shown in **Figure 5**. During production, each contaminant triggers process inefficiencies, including not only corrosion pipe blockages but also rapid and permanent deactivation of catalysts [17].

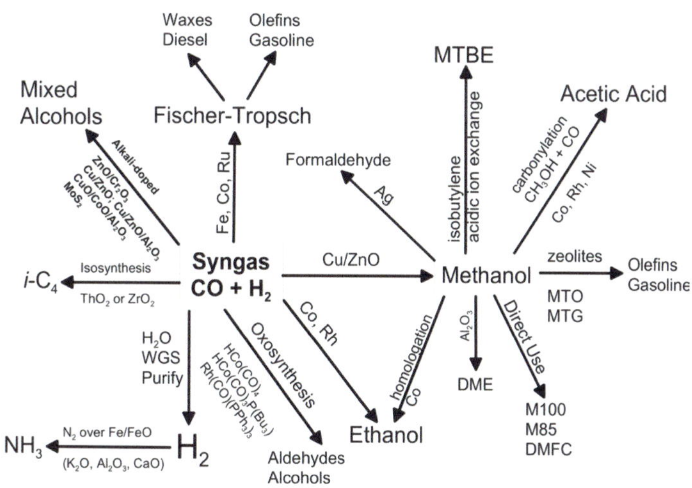

Figure 5. Different syngas transformation routes for synthesising fuels and other chemicals [18].

Contaminant levels depend on feedstock impurities and the syngas generation method used. The level of cleaning required may also be influenced substantially by end-use technology and/or emission standards [17]. **Table 3** shows the syngas cleaning requirements for some typical end-use applications.

3.2. Sulphur

There are several gas cleaning technologies: absorption, adsorption, conversion, and biological transformation. Wiheeb et al. has described the adsorption process of H_2S. **Table 4** presents a summary of the characteristics of sulphur removal technologies [19]. In gasification processes, absorption and adsorption are employed well as gas/bag filter and scrubber processes, respectively.

3.2.1. Fate of sulphur during gas cleaning over adsorbents

The fate of sulphur during the gasification of lignin slurry was investigated by Koido et al. [20]. In their study, hydrogen sulphide removal from bio-synthesis gas over a nickel oxide catalyst supported by calcium aluminate ($NiO/CaAl_2O_4$) was investigated at high temperatures. They investigated the sulphur balance of the process at different operating temperatures (T = 750–950 °C), moisture contents of the lignin slurry (MC = 73–90 wt%), and catalyst loadings (CL = 0.00–0.61 g-catalyst/g-feedstock). The sulphur balance was 0.79, 0.04, 0.003, and 0.378 mmol/g-lignin for the gas, char, water-soluble fraction, and $NiO/CaAl_2O_4$ catalyst surface, respectively.

3.3. Radionuclides

3.3.1. Introduction

Radionuclides including ^{134}Cs and ^{137}Cs were released into the environment after the Fukushima Dai-ichi Nuclear Power Plant accident in March 2011. After the accident, decontamination was

Contaminants	**Applications**			
	Methanol synthesis (mg m^{-3})	**FT synthesis (μL L^{-1})**	**Gas turbine (μL^{-1})**	**IC engine (mg m^{-3})**
Particulate (soot, dust, char, ash)	<0.02	Not-detectable	<0.03 (PM_5)	< 50 (PM_{10})
Tars (condensable)	–[a]	<0.01[b]	—	—
Tars (heteroatoms, BTX)	<0.1	<1	—	<100
Alkalis	—	<0.01	<0.024	—
Nitrogen (NH_3, HCN)	<0.01	<0.02	<50	—
Sulphur (H_2S,COS)	<1	<0.01	<20	—
Halides (primarily HCl)	<0.1	<0.01	1	—

[a]Data are not available in the original literature.
[b]All values are at STP unless explicitly specified.

Table 3. Syngas cleaning requirements for some typical end-use applications [17].

http://dx.doi.org/10.5772/intechopen.70727

Classification [19]	Technology [19]	Characteristics [19]	Reference
Absorption (wet desulphurisation)	Conventional absorption process	• Absorption has been used in petroleum and gas industries to remove H_2S and CO_2 from sour natural gas and refinery gas. • The removal is called gas sweetening, which involves transferring of H_2S from a gaseous phase (feed) into a liquid phase (solvent). • The conventional process has higher sulphur removal but requires strict pH control of chemical concentrations and wastewater treatment, which could cause corrosion problems.	Taheri et al., [21]
	Membrane reactor	• Membranes can be used to purify biogas. • Membranes are not usually used for selective removal of H_2S, but rather to upgrade biogas to natural gas standards.	Dolejš et al., [22]
Adsorption (dry desulphurisation)	Carbonaceous adsorbents	• Activated carbon has a high specific surface area of more than 1000 m^2/g. • The surface area, pore volume, and surface chemistry promote numerous catalytic reactions.	Kazmierczak-Razna et al., [23]
	Metal oxide adsorbents	• The removal of H_2S at high temperature has received much attention owing to its potential in reducing H_2S concentration to 10 ppm. • Metal oxide: FeO, Cu_2O, MnO, ZnO CoO, NiO and MoO, alkaline earths (CaO, SrO, and BaO), and alkalis (Li_2O, K_2O and Na_2O). • The process has higher thermal efficiency through sensible heat utilisation and easy treatment of wastewater but requires either catalyst exchange or regeneration.	Abdoulmoumine et al., [24]
Conversion	Claus process	• The Claus process is used in oil and natural gas refining facilities and removes H_2S by oxidising it to elemental sulphur. • Removal efficiency is about 95% using two reactors and 98% using four reactors.	Ibrahim et al., [25]
	Selective catalytic oxidation	• The selective catalytic oxidation of H_2S into elemental sulphur is one of the treatment methods employed for the removal of H_2S from Claus process tail gas. • The catalytic oxidation of H_2S can be performed above or below the sulphur dew point (180 °C).	Tasdemir et al., [26]
	Liquid redox sulphur recovery	• Liquid-phase oxidation systems convert H_2S into elemental sulphur through redox reactions by electron transfer from sources such as vanadium or iron reagents.	Kim et al., [27]
Biological transformation	Biological methods	• Microorganisms have been used for the removal of H_2S from biogas. • Ideal microorganisms would have the ability to transform H_2S to elemental sulphur.	Tóth et al., [28]

Table 4. Characteristics of each sulphur removal technology [20].

implemented in resident areas, and was completed over all residential area surfaces at the end of March 2017 [29, 30]. However, for utilisation of Fukushima's forest resources, which are contaminated by radioactive nuclides, utilisation of woody biomass as fuel for bioenergy production from gasification (such as heat, power, hydrogen, etc.) in the near future is a possible option. For this purpose, the mass balance of radionuclides must be revealed.

In particular, ^{137}Cs is almost distributed in argilliferous soils and fallen leaves in Fukushima. Of all the local ^{137}Cs in 2015, 87% was distributed in soil with 10% in fallen leaves in Japanese cedar forests, while 87% was distributed in soils with 11% in fallen leaves in *Quercus serrata* forests [31]. Moreover, 3% of ^{137}Cs was in Japanese cedar timbers (bark: 0.5%; boards: 0.4%; branches: 0.7%; and leaves: 0.6%), while 2% of ^{137}Cs was found in *Quercus serrata* timber (bark: 0.7%; boards: 0.2%; branches: 0.5%; and leaves: 0.1%) [31].

The radioactivity of gasification pellet fuel, products, and by-products was measured to clarify the fate of ^{134}Cs and ^{137}Cs from Fukushima's woody biomass during the biomass gasification process.

3.3.2. Material and methods

The measured samples were woody pellet fuel, gasification residue in ash bins (main ash), soot on ash filters (fly ash), and exhaust gas, which are produced from woody BGCHP systems (E3 unit, Entrade Energy) at the Spa Resort Abukuma in Nishigo village, Fukushima. The woody pellets are comprised of a mixture of Japanese cedar and *Quercus serrata* obtained from the Yamizo Mountains. Using these pellets, the radioactivity of products/by-products for each BGCHP process was observed. The exhaust gas was filtered by means of a high-volume air sampler (Shibata Scientific Technology Ltd., HV-500RD) for 30 min at suction flow rate of 500 L/min.

A germanium semiconductor detector (CANBERRA GC4020) was used to detect the radioactivity arising from radionuclides such as ^{134}Cs and ^{137}Cs. To minimise measurement errors, each measurement was taken over 3 h for the pellet fuels, the gasification residues, and the soot on filters, while 12 h for the filter of high-volume air sampler. Each radioactivity concentration was calculated by dividing the measured radioactivity by the sample mass/volume. The solid samples were placed into a vessel (100 mL) and measured.

3.3.3. Results and discussion

For ^{137}Cs, the radioactivity levels of solid samples were 20.6 Bq/kg (standard error, SE: 1.01 Bq/kg) for the woody pellets, 1333 Bq/kg (SE = 10.4 Bq/kg) for residue, and 5432 Bq/kg for soot from bag filters as presented in **Table 5**. The offgas radioactivity was not detectable when the limit of detection (LOD) was 0.002 Bq/m^3. All of the by-products were smaller than the criterion. For ^{134}Cs, the solid sample radioactivity levels were less than 4.10 Bq/kg for the woody pellets, 207 Bq/kg (SE = 7.58 Bq/kg) for residue, and 849 Bq/kg for soot from bag filters as presented in **Table 5**. The offgas radioactivity was not detectable. All of the by-products were smaller than the criterion. In this study, the biomass gasification plant was capable of keeping the radioactive nuclides (in the residue and the filters) within the plant. However, radionuclides should be monitored periodically.

Sample	^{137}Cs			^{134}Cs			Sample amount
	Mean	SE[a]	LOD[b]	Mean	SE[a]	LOD[b]	
Woody pellet (Bq/kg)	20.6	1.01	3.50	< 4.10	na	3.55	70.5 g
Residue (Bq/kg)	1333	10.4	15.0	207	7.58	15.4	39.6 g
Bag filter (Bq/kg)	5432	na	43.0	849	na	46.8	11.6 g
Offgas (Bq/m^3)	nd	na	0.002	nd	na	0.003	14.9 m^3

[a]SE = standard error.
[b]LOD = limit of detection.
na = not available.
nd = not detectable.

Table 5. Radioactivity levels in the E3 biomass gasification process.

4. Life cycle assessment of the biomass gasification process

Life cycle assessment (LCA) is a methodology that examines products and services "from cradle to grave" with a view to understanding system-wide environmental impacts. A cradle-to-grave LCA study of a product considers all life cycle stages from extraction or primary production of materials and fuels ('cradle') through production and use of the product to its final disposal ('grave'). The framework has been standardised by the International Organisation as ISO 14,044:2006 [32].

Recent studies concerning biomass gasification are summarised in **Table 6**. Recent LCA studies concerning biomass gasification are categorised into four groups: (i) biomass-based hydrogen (bio-H_2) production [33–38], (ii) biomass gasification combined heat and power (CHP) [39–44], (iii) other energy systems [45–48], and (iv) dynamic LCA [49]. This chapter covers the review of the LCA studies about biomass gasification.

4.1. Biomass gasification for hydrogen production

To evaluate the environmental performance of H_2 production via indirect gasification of short-rotation poplar, a LCA was implemented using process simulation for normal BG processes [33] and for BG with CO_2 capture by pressure swing adsorption [34]. From a life-cycle perspective, H_2 from poplar gasification generally arose as a good alternative to conventional, fossil-derived H_2 produced via steam methane reforming.

Moreno and Dufour [35] examined the environmental feasibility of four Spanish lignocellulosic wastes (vine and almond pruning, and forest wastes coming from pine and eucalyptus plantation) for the production of H_2 through gasification via LCA methodology using global warming potential, acidification, eutrophication and the gross energy necessary for the production of 1 Nm^3 of hydrogen as impact categories.

Kalinci et al. [36] performed LCA for stages from biomass production to the use of the produced hydrogen in proton exchange membrane fuel cell vehicles. Two different gasification

Study	Methodology				Economic impact	Environmental impact category												
	Location	FU	System boundary	Products		EC	AC	EP	GWP	HT	LU	OD	PO	RC	WC	ET	PM	IR
Susmozas et al. [33]		1 kg of H_2 produced from the plant with 99.9 vol% purity	Cradle-to-gate	Bio-H_2 from BG with CCS		X	X	X	X		X	X	X					
Susmozas et al. [34]	EU	1 kg of H_2 produced from the plant with 99.9 vol% purity	Cradle-to-gate	Bio-H_2 from indirect BG with CO_2 capture		X	X	X	X		X	X	X					
Moreno and Dufour [35]			Cradle-to-gate	H_2, CO_2, other emissions, wastes		X	X	X	X									
Kalinci et al. [36]			Cradle-to-gate	Bio-H_2	X				X									
Iribarren et al. [37]		1 m^3 stp of purified H_2	Cradle-to-gate	Bio-H_2		X	X	X	X		X	X	X					
El-Emam, et al. [38]		1 TJ of produced energy	Cradle-to-gate	Bio-H_2					X									
Elsner et al. [39]	Poland		Cradle-to-gate	CHP	X													
Adams and McManus [40]	UK	1 MJ (or kWh) of energy produced	Cradle-to-gate	Small-scale BG CHP		X			X	X				X	X	X	X	
Patuzzi et al. [41]	South-Tyrol, Italy	Thermal efficiencies	Cradle-to-gate	Small-scale BG CHP		X												
Oreggioni et al. [42]	EU		Cradle-to-gate	Small-scale BG CHP			X			X				X			X	
Klavina et al. [43]	Latvia	1 kg of untreated forest residue woodchips	Cradle-to-gate	Small-scale BG CHP		X	X	X	X		X	X				X		X
Kimming et al. [44]	EU	1 y supply of heat and power to a modern village of 150 households	Cradle-to-gate	Small-scale BG CHP		X	X		X		X							

Study	Methodology				Economic impact	Environmental impact category												
	Location	FU	System boundary	Products		EC	AC	EP	GWP	HT	LU	OD	PO	RC	WC	ET	PM	IR
Sreejith et al. [45]	State of Kerala in India	1 MJ energy content in the gaseous fuel	Cradle-to-gate	Producer gas generated from coconut shell gasification					X			X		X				
Parvez et al. [46]			Cradle-to-gate	Heat and syngas produced from gasification		X			X									
Kalina [47]	Poland	Cold/hot gas efficiency	Cradle-to-gate	Integrated BG dual fuel combined cycle power plant	X	X												
Wang et al. [48]	Harbin, China	1 y operation	Cradle-to-gate	Building cooling heating and power	X	X			X					X				
Yang and Chen [49]	China	1 MJ of primary energy produced in a BG plant (dynamic LCA)	Cradle-to-grave	Producer gases (CO_2, H_2, CO and CH_4) from BG	X				X									

Abbreviation: CCS: carbon dioxide capture; na: not available; CHP: combined heat and power; BG: biomass gasification; EC: energy consumption; AC: acidification; EP: eutrophication; GWP: global warming potential; HT: human toxicity; LU: land use; OD: ozone depletion; PO: photochemical oxidation; RC: resource consumption; WC: water consumption; ET: ecotoxicity; PM: particulate matter; IR: ionising radiation.

Table 6. Recent life cycle assessment studies on biomass gasification.

systems, a downdraft gasifier and a circulating fluidised bed gasifier (CFBG), are considered and analysed for H_2 production using actual data taken from the literature. Functional unit was 1 MJ/s H_2 production. Then, the costs of GHG emissions reduction are calculated.

Iribarren et al. [37] assessed environmental and thermodynamic performance of H_2 production via BG through a LCA and an exergetic analysis. The case study involves poplar gasification in a low-pressure char indirect gasifier, catalytic tar destruction, cold wet gas cleaning, syngas conversion and hydrogen purification. The system boundary covers from poplar cultivation to H_2 purification.

El-Emam et al. [38] focused on efficiency and environmental impact assessments of steam biomass gasification and gasification-solid oxide fuel cell (SOFC) integrated system for power and H_2 production. The environmental assessment is performed based on the carbon dioxide produced from the system with respect to the generated useful products.

4.2. Biomass gasification for CHP

Experimental and numerical analyses of a CHP installation (75 kWe of electrical power) was investigated by Elsner et al. [39], which is equipped with a biomass downdraft gasifier, gas purification system, and gas piston engine. The economic analysis was performed taking into account policies and regulations in the Polish energy market sector. They revealed that it is more profitable to consume the generated power and heat for its self-consumption rather than selling it on the market.

Using the techniques of LCA and net energy analysis, the study by Adams and McManus [40] quantified the energy, resource, and emission flows, to assess the net energy produced and potential environmental effects of BG using wood waste. The paper conducted a case study that uses waste wood from a factory for use in an entrained flow gasification CHP plant. Overall, small-scale biomass gasification is an attractive technology if the high capital costs and operational difficulties can be overcome, and a consistent feedstock source is available.

Patuzzi et al. [41] did an investigation resulting in an overview of the actual state-of-the-art of small-scale biomass gasification technology in Italy in terms of energy efficiency of the plants, effectiveness of the adopted solutions and characteristics of the products and by-products. In the study by Oreggioni et al. [42], a combined heat and power via BG, CHP with pre-combustion adsorptive carbon capture unit, and CHP with post combustion absorptive carbon capture unit were environmentally assessed.

Klavina, et al. [43] performed environmental performance analysis of biochar from woodchip pyrolysis, and woodchip CHP through midpoint category impact comparison using LCA. Kimming et al. [44] conducted a simplified LCA over four scenarios for supply of the entire demand of power and heat of a rural village. Three of the scenarios are based on utilisation of biomass in 100 kW_e CHP systems and the fourth is based on fossil fuel in a large-scale plant.

4.3. Biomass gasification for other energy systems

Sreejith et al. [45] investigated the suitability of coconut shell-derived producer gas as a substitute for coal gas from an environmental perspective using LCA. Thermochemical gasification in an air-fluidised bed with steam injection is the gaseous fuel production process. The study indicates that coconut shell-derived producer gas life cycle is capable of saving 18.3% of emissions for global warming potential, 64.1% for ozone depletion potential, and 71.5% for nonrenewable energy consumption. The analysis of energy and exergy consumptions is 62.9% for producer gas life cycle, while it is only 2.8% for coal gas life cycle.

In the study of Parvez et al. [46], air, steam, and CO_2-enhanced gasification of rice straw was simulated using Aspen Plus™ and compared in terms of their energy, exergy, and environmental impacts. The maximum exergy efficiency occurred in 800–900 °C. For CO_2-enhanced gasification, exergy efficiency was found to be more sensitive to temperature than CO_2/Biomass ratios. In addition, the preliminary environmental analysis showed that CO_2-enhanced gasification resulted in significant environmental benefits compared with steam gasification.

Kalina [47] presents theoretical study of the concept of a small-scale combined cycle system composed of natural gas fired micro turbine and Organic Rankine Cycle, integrated with thermal gasification of biomass. The main issues addressed in the paper are configuration of the ORC technology and allocation of generated electricity between natural gas and biomass. Energy and exergy allocation keys are demonstrated. An initial cash flow calculations are presented in order to assess financial performance of the plant.

Wang et al. [48] proposed a combined methodology of optimisation method and life cycle inventory for the biomass gasification based building cooling, heating, and power (BCHP) system. The system boundary of life cycle models includes biomass planting, biomass collection-storage-transportation, BCHP plant construction and operation, and BCHP plant demolition and recycle. Economic cost, energy consumption and CO_2 emission in the whole service-life were obtained. Then, the optimisation model for the biomass BCHP system including variables, objective function and solution method are presented.

Prior to large-scale crop-residue gasification application, the lifetime environmental performance should be investigated to plan sustainable strategies. As traditional static LCA does not include temporal information for dynamic processes, Yang and Chen [49] proposed a dynamic life cycle assessment approach, which improves the static LCA approach by considering time-varying factors, e.g., greenhouse gas characterization factors and energy intensity. Results show that the crop residue gasification project has high net global warming mitigation benefit and a short global warming impact mitigation period, indicating its prominent potential in alleviating global warming impact.

5. Conclusions

After the nuclear power plant accident in Fukushima, nearby forests were contaminated by the released radionuclides. Gasification technology can gasify the contaminated woody biomass and produce syngas and ash. Current trends in biomass gasification technologies and the subsequent gas cleaning process including desulphurisation and separation of radioactive substances were reviewed.

Life cycle assessment of the total gasification system is receiving increasing attention. It analyses the energy profit ratio and environmental impacts of a process of interest (e.g. greenhouse gas emission, acidification potential, photochemical oxidation, eutrophication potential, land competition, etc.).

The radioactivity of the syngas produced was quite low, and that of ash was high (within acceptable levels), implying that the gasification technology can be utilised as an option for energy conversion of contaminated woody biomass in Fukushima.

Acknowledgements

The author thanks all those involved in preparing this chapter, including Mr. Tachihara and Dr. Moritz for providing plant data. The authors acknowledge financial aid from Fukushima prefecture for overseas collaboration on renewable energy research and development.

Author details

Kenji Koido* and Takahiro Iwasaki

*Address all correspondence to: koido@sss.fukushima-u.ac.jp

Faculty of Symbiotic Systems Science, Fukushima University, Fukushima, Japan

References

[1] Morimoto H, Satoh S, Koido K. A simplified method of harvesting willow crops cultivated as an energy source in short rotation coppices in small-scale farmlands. Journal of the Japan Institute of Energy. 2016;**95**(1):152-155. DOI: 10.3775/jie.95.152

[2] Khonde R, Chaurasia A. Rice husk gasification in a two-stage fixed-bed gasifier: Production of hydrogen rich syngas and kinetics. International Journal of Hydrogen Energy. 2016;**41**(21):8793-8802. DOI: 10.1016/j.ijhydene.2016.03.138

[3] Shen Y. Rice husk silica derived nanomaterials for sustainable applications. Renewable and Sustainable Energy Reviews. 2017;**80**:453-466. DOI: 10.1016/j.rser.2017.05.115

[4] Susastriawan AAP, Saptoadi H, Purnomo. Small-scale downdraft gasifiers for biomass gasification: A review. Renewable and Sustainable Energy Reviews. 2017;**76**:989-1003. DOI: 10.1016/j.rser.2017.03.112

[5] Sikarwar VS, Zhao M, Fennell PS, Shah N, Anthony EJ. Progress in biofuel production from gasification. Progress in Energy and Combustion Science. 2017;**61**:189-248. DOI: 10.1016/j.pecs.2017.04.001

[6] Asadullah M. Barriers of commercial power generation using biomass gasification gas: A review. Renewable and Sustainable Energy Reviews. 2014;**29**:201-215. DOI: 10.1016/j.rser.2013.08.074

[7] de Lasa H, Salaices E, Mazumder J, Lucky R. Catalytic steam gasification of biomass: Catalysts, thermodynamics and kinetics. Chemical Reviews. 2011;**111**(9):5404-5433. DOI: 10.1021/cr200024w

[8] Moon J, Lee J, Lee U, Hwang J. Transient behavior of devolatilization and char reaction during steam gasification of biomass. Bioresource Technology. 2013;**133**:429-436. DOI: 10.1016/j.biortech.2013.01.148

[9] Kumazaki M, Nishiyama A, Kuki Y, Takebayashi M, Sawabe O. Woody Biomass Energy Power Generation. 1st ed. Nikkan Kogyo Shimbun: Tokyo; 2016

[10] Basu P. Biomass Gasification and Pyrolysis: Practical Design and Theory. 1st ed. Chichester: Elsevier; 2010

[11] Society of Chemical Engineers Japan (SCEJ), Japan Institute of Energy (JIE). Biomass Process Handbook. 1st ed. Tokyo: Ohmsha; 2012

[12] Burkhardt. Burkhardt Website. [Internet]. http://burkhardt-gruppe.de/hp2/Home.htm. Published 2017. [Accessed August 15, 2017]

[13] Spanner Re2. Spanner Website. [Internet]. http://www.holz-kraft.com/en/. Published 2017. [Accessed August 15, 2017]

[14] Entrade Energy. Entrade Brochure. [Internet]. http://www.entrade.co/assets/entrade_brochure_en_08_16_screen.pdf. Published 2016. [Accessed August 15, 2017]

[15] Volter. Volter Website. [Internet]. http://volter.fi/portfolio/volter-indoor-model/. Published 2017. [Accessed August 15, 2017]

[16] Cortus Energy. The Picture Illustrates the Basic Principles of the WoodRoll® process. [Internet]. http://www.cortus.se/woodroll_process.pdf. Published 2016. [Accessed August 15, 2017]

[17] Woolcock PJ, Brown RC. A review of cleaning technologies for biomass-derived syngas. Biomass and Bioenergy. 2013;**52**:54-84. DOI: 10.1016/j.biombioe.2013.02.036

[18] Spath PL, Dayton DC. Preliminary Screening — Technical and Economic Assessment of Synthesis Gas to Fuels and Chemicals with Emphasis on the Potential for Biomass-Derived Syngas. Golden, Colorado: National Renewable Energy Laboratory, NREL/TP-510-34929; 2003

[19] Wiheeb AD, Shamsudin IK, Ahmad MA, Murat MN, Kim J, Othman MR. Present technologies for hydrogen sulfide removal from gaseous mixtures. Reviews in Chemical Engineering. 2013;**29**(6):449-470. DOI: 10.1515/revce-2013-0017

[20] Koido K, Watanabe Y, Ishiyama T, Nunoura T, Dowaki K. Fate of sulphur during simultaneous gasification of lignin-slurry and removal of hydrogen sulphide over calcium aluminate supported nickel oxide catalyst. Journal of Cleaner Production. 2017;**141**:568-579. DOI: 10.1016/j.jclepro.2016.09.010

[21] Taheri M, Mohebbi A, Hashemipour H, Rashidi AM. Simultaneous absorption of carbon dioxide (CO_2) and hydrogen sulfide (H_2S) from CO_2–H_2S–CH_4 gas mixture using amine-based nanofluids in a wetted wall column. Journal of Natural Gas Science and Engineering. 2016;**28**:410-417. DOI: 10.1016/j.jngse.2015.12.014

[22] Dolejš P, Poštulka V, Sedláková Z, et al. Simultaneous hydrogen sulphide and carbon dioxide removal from biogas by water–swollen reverse osmosis membrane. Separation and Purification Technology. 2014;**131**:108-116. DOI: 10.1016/j.seppur.2014.04.041

[23] Kazmierczak-Razna J, Gralak-Podemska B, Nowicki P, Pietrzak R. The use of microwave radiation for obtaining activated carbons from sawdust and their potential application in removal of NO_2 and H_2S. Chemical Engineering Journal. 2015;**269**:352-358. DOI: 10.1016/j.cej.2015.01.057

[24] Abdoulmoumine N, Adhikari S, Kulkarni A, Chattanathan S. A review on biomass gasification syngas cleanup. Applied Energy. 2015;**155**:294-307. DOI: 10.1016/j.apenergy.2015.05.095

[25] Ibrahim S, Al Shoaibi A, Gupta AK. Effect of benzene on product evolution in a H2S/O2 flame under Claus condition. Applied Energy. 2015;**145**:21-26. DOI: 10.1016/j.apenergy.2015.01.094

[26] Tasdemir HM, Yasyerli S, Yasyerli N. Selective catalytic oxidation of H2S to elemental sulfur over titanium based Ti–Fe, Ti–Cr and Ti–Zr catalysts. International Journal of Hydrogen Energy. 2015;**40**(32):9989-10001. DOI: 10.1016/j.ijhydene.2015.06.056

[27] Kim K, Kim DY, Lee KR, Han J-I. Electricity generation from iron EDTA-based liquid redox sulfur recovery process with enhanced stability of EDTA. Energy Conversion and Management. 2013;**76**:342-346. DOI: 10.1016/j.enconman.2013.07.063

[28] Tóth G, Nemestóthy N, Bélafi-Bakó K, Vozik D, Bakonyi P. Degradation of hydrogen sulfide by immobilized *Thiobacillus thioparus* in continuous biotrickling reactor fed with synthetic gas mixture. International Biodeterioration & Biodegradation. 2015;**105**:185-191. DOI: 10.1016/j.ibiod.2015.09.006

[29] Hashimoto S, Ugawa S, Nanko K, Shichi K. The total amounts of radioactively contaminated materials in forests in Fukushima, Japan. Scientific Reports. 2012;**2**(416):1-5. DOI: 10.1038/srep00416

[30] Mallampati SR, Mitoma Y, Okuda T, Sakita S, Simion C. Preferential removal and immobilization of stable and radioactive cesium in contaminated fly ash with nanometallic Ca/CaO methanol suspension. Journal of Hazardous Materials. 2014;**279**:52-59. DOI: 10.1016/j.jhazmat.2014.06.044

[31] Forestry Agency. Investigative Report on Radioactive Nuclides Distribution Conditions over the Forests in 2016 FY. [Internet]. http://www.rinya.maff.go.jp/j/kaihatu/jyosen/attach/pdf/H28_jittaihaaku_kekka-1.pdf. Published 2016. [Accessed August 12, 2017]

[32] International Organization for Standardization. Environmental Management – Life Cycle Assessment–Requirements and Guidelines. ISO14044. 2006. p. 14044

[33] Susmozas A, Iribarren D, Dufour J. Life-cycle performance of indirect biomass gasification as a green alternative to steam methane reforming for hydrogen production. International Journal of Hydrogen Energy. 2013;**38**(24):9961-9972. DOI: 10.1016/j.ijhydene.2013.06.012

[34] Susmozas A, Iribarren D, Zapp P, Linβen J, Dufour J. Life-cycle performance of hydrogen production via indirect biomass gasification with CO_2 capture. International Journal of Hydrogen Energy. 2016;**41**(42):19484-19491. DOI: 10.1016/j.ijhydene.2016.02.053

[35] Moreno J, Dufour J. Life cycle assessment of hydrogen production from biomass gasification. Evaluation of different Spanish feedstocks. International Journal of Hydrogen Energy. 2013;**38**(18):7616-7622. DOI: 10.1016/j.ijhydene.2012.11.076

[36] Kalinci Y, Hepbasli A, Dincer I. Life cycle assessment of hydrogen production from biomass gasification systems. International Journal of Hydrogen Energy. 2012;**37**(19):14026-14039. DOI: 10.1016/j.ijhydene.2012.06.015

[37] Iribarren D, Susmozas A, Petrakopoulou F, Dufour J. Environmental and exergetic evaluation of hydrogen production via lignocellulosic biomass gasification. Journal of Cleaner Production. 2014;**69**:165-175. DOI: 10.1016/j.jclepro.2014.01.068

[38] El-Emam RS, Dincer I, El-Emam SH. Gasification of biomass for hydrogen and power production: Efficiency and environmental assessment. In: Dincer I, Colpan CO, Kizilkan O, Ezan MA, editors. Progress in Clean Energy, Novel Systems and Applications. Cham: Springer International Publishing; 2015;**2**:147-162. DOI: 10.1007/978-3-319-17031-2_12.

[39] Elsner W, Wysocki M, Niegodajew P, Borecki R. Experimental and economic study of small-scale CHP installation equipped with downdraft gasifier and internal combustion engine. Applied Energy. 2017;**202**:213-227. DOI: 10.1016/j.apenergy.2017.05.148

[40] Adams PWR, McManus MC. Small-scale biomass gasification CHP utilisation in industry: Energy and environmental evaluation. Sustain Energy Technol Assessments. 2014;**6**:129-140. DOI: 10.1016/j.seta.2014.02.002

[41] Patuzzi F, Prando D, Vakalis S, et al. Small-scale biomass gasification CHP systems: Comparative performance assessment and monitoring experiences in South Tyrol (Italy). Energy. 2016;**112**:285-293. DOI: 10.1016/j.energy.2016.06.077

[42] Oreggioni GD, Singh B, Cherubini F, et al. Environmental assessment of biomass gasification combined heat and power plants with absorptive and adsorptive carbon capture units in Norway. Int J Greenh Gas Control. 2017;**57**:162-172. DOI: 10.1016/j.ijggc.2016.11.025

[43] Klavina K, Romagnoli F, Blumberga D. Comparative life cycle assessment of woodchip uses in pyrolysis and combined heat and power production in Latvia. Energy Procedia. 2017;**113**:201-208. DOI: 10.1016/j.egypro.2017.04.055

[44] Kimming M, Sundberg C, Nordberg Å, et al. Biomass from agriculture in small-scale combined heat and power plants – A comparative life cycle assessment. Biomass and Bioenergy. 2011;**35**(4):1572-1581. DOI: 10.1016/j.biombioe.2010.12.027

[45] Sreejith CC, Muraleedharan C, Arun P. Life cycle assessment of producer gas derived from coconut shell and its comparison with coal gas: an Indian perspective. International Journal of Energy and Environmental Engineering. 2013;**4**(1):8. DOI: 10.1186/2251-6832-4-8

[46] Parvez AM, Mujtaba IM, Wu T. Energy, exergy and environmental analyses of conventional, steam and CO_2-enhanced rice straw gasification. Energy. 2016;**94**:579-588. DOI: 10.1016/j.energy.2015.11.022

[47] Kalina J. Techno-economic assessment of small-scale integrated biomass gasification dual fuel combined cycle power plant. Energy. 2017. DOI: 10.1016/j.energy.2017.05.009

[48] Wang J-J, Yang K, Z-L X, Fu C, Li L, Zhou Z-K. Combined methodology of optimization and life cycle inventory for a biomass gasification based BCHP system. Biomass and Bioenergy. 2014;**67**:32-45. DOI: 10.1016/j.biombioe.2014.03.026

[49] Yang J, Chen B. Global warming impact assessment of a crop residue gasification project—A dynamic LCA perspective. Applied Energy. 2014;**122**:269-279. DOI: 10.1016/j.apenergy.2014.02.034

Chapter 8

Lignin as Natural Antioxidant Capacity

Zahed Mahmood, Muhammad Yameen,
Muhammad Jahangeer, Muhammad Riaz,
Abdul Ghaffar and Irum Javid

Additional information is available at the end of the chapter

http://dx.doi.org/10.5772/intechopen.73284

Abstract

Cell wall of various plants contains an organic polymer lignin. Lignin makes significant structural constituents that support of vascular plants tissues, some types of algae and makes plants stiff and woody. Chemically lignin composed of cross-linked phenolic polymers and Kraft lignin dissolved in the unsaturated resin system. Lignin is a vastly available and well-studied antioxidant. They work as antioxidant against the biological, chemical and mechanical stress on plants. The substance that removes potentially damaged oxidizing agents in a living organism is regarded as antioxidant. Lignin contains a huge amount of phenolic constructions that enable it to act as an effectual antioxidant. Infrared spectroscopy (IR) observed decreases in some oxidative aging products and maintained the idea that lignin play role as an effective antioxidant. It was discovered that antioxidant action of lignin varies with the molecular characteristics, isolation method and genetic origin. Capacity of lignin to scavenge the radicals that are involved in the oxidation is affected by partial dispersion into polymer matrix. Antioxidant activities of various technical lignins are determined. Chemical changes in lignin could result in products with characteristics that can be used in special industrial divisions.

Keywords: kraft lignin, lignin as antioxidant, cytotoxicity, application of lignin, phenolic polymers, sources of lignin, and isolation of lignin

1. Introduction

The term lignin is derivative from the Latin word lignum, meaning wood. It was initially separated from wood, described in a scientific description made by the French scientist Payen 1838, and later specified its modern name in 1857 by Schulze. Lignin was previously labeled as an incrusting of cellulose and this opinion is insightful as lignification happens after the deposition of polysaccharides. Simply stated, it is similar to the medium for a fiber reinforced compound. Lignin plays several different roles with the cell wall like modifying the permeability and temperature stability, but its principal role is to serve as a structural material which enhances the strength and rigidity to plants. Lignin differentiates lignocellulosic material from other materials rich in polysaccharide by strengthening the polysaccharide framework of the cell wall. Its presentation is very beneficent that it permits trees to compete other plants for sunlight creating the major creatures on this planet.

Lignin contains phenolic hydroxyl group also that makes it capable to play role in the food industry as strong antioxidant, avoid removing colour of food, taste and active vitamins concentration and molecules stabilization that are involved in these features. Oxidized lignin comprises properties which are associated to lignosulfonates and can be used in different industrial applications, mostly as dispersants and fillers. Kraft lignin from wood sources in pulp industry was stated to be as effective as vitamin E to shelter the oxidation of corn oil. The monomers degradation of lignin and their derivatives afforded high-quality antioxidant and anti-inflammatory characteristics. Additionally, lignin itself has been demonstrated to carry antimicrobial and insecticidal characteristics. The lignins of different botanical origins like annual plants, coniferous trees and deciduous trees were separated and fractionated by special types of different techniques included delignification by alkali, Kraft process, fast pyrolysis and hydrolysis. For characterization of antioxidant action of naturally present Phenolic compounds, a method using diphenyl picrylhydrazyl (DPPH) as a responsive free radical is now familiar as one that offers an opportunity to find radical scavenging ability. The antioxidant activities were determined as the capability to scavenge the DPPH• and (2,2′-azino-bis (3-ethylbenzothiazoline-6-sulphonic acid)) $ABTS^{\bullet +}$ free radicals. The correlation between the structure of lignin and the antioxidant action was characterized by multivariate analysis, partial correlation, pair correlation, together with interrelated components regression.

Lignin constitutes 15–40% of dry weight of woody plants, it is highly present aromatic polymer on this planet and the second frequently present organic polymer after cellulose. Based on annual biomass growth rates, the overall production of lignin is 5.36×108 tons. That's why, lignin is a proved to be the fundamental source of aromatic compounds for the chemical industry, increasing after the conversion of modern age CO_2, and its effective operation resolves a possible puzzle in generating useful byproducts in a biorefinery organization. This reasoning is because if wood is converted to billion-ton scale to produce biofuels and biochemicals, then there will be 300 million tons upwards of lignin potentially accessible. To put this in viewpoint it is hardly the size of a global polymer market.

The monomeric units are built into the macromolecular lignin from the oxidative radical coupling of these substructures [1]. Lignification is originated when the phenolic hydroxyl hydrogen is inattentive by the peroxidase enzyme to form a free radical of phenoxy, typically referred to as dehydrogenative polymerization. This phenoxy free radical will then delocalize to both aromatic and side chain carbon atoms through the process of resonance stabilization.

Development regarding lignin description in the nineteenth century donated to new materials development created upon renewable resources of lignin. Enhanced skills linked with processing of wood, papermaking and cotton industry also take part in the developmental process of these polymers structural forms biomaterials [2] that might be combined in biorefineries [3] to develop a wide variety of production like resources, electricity, chemicals and fuel.

The improvement in lignin by main evolution of biorefinery productions and new developments to change this large biomolecule into different other valuable raw material [4]. Analysis of lignin impacts several changes to it from its native state making it difficult to have an exact structure of lignin within the cell wall. However, there are isolation methods that can be used to get a better idea of the characteristics of a less severely modified lignin. One such method involves the resizing of wood into a fine flour through ball milling, using cellulose degrading enzymes to remove the bulk of the polysaccharides, and then an acidosis reaction to break several of the lignin carbohydrate linkages. The separation of this enzymatic mild acidosis lignin (EMAL) conserves numerous features related to lignin applied as a standard to associate with different lignin separation methods. Analysis by means of quantitative nuclear magnetic resonance (NMR) spectroscopy, this exposes the quantity of functional groups related to the amount of lignin.

Many lignin's comprises 4 mmol aliphatic —OH groups associated to the side chain/gram of lignin and 0.2–1 mmol free phenolic groups/gram of lignin. Also, it is possible to analyse lignin inside the cell wall lacking the requirement of isolation. Method involves the crushing of wood into a fine dust and then consuming solvents for cell wall dissolution. Dissolved wood is examined using two-dimensional (2D) 13C-1H heteronuclear single quantum coherence (HSQC) NMR spectroscopy. This procedure can regulate the comparative concentrations of interunit lignin structures that helps in understanding the lignin structure.

There are two ways that happen through delignification. First one involves rupture of key lignin-carbohydrate associations. This breakage of bond will permit extractability of lignin as observed through mild acidosis of lignin where dilute HCl is utilized to break these associations. The second one includes breakage of some of the interunit bonds of lignin which may cause the reduction of molecular weight. Hence, delignification technologies dramatically influence lignin functionality and molecular weight. Consequently, the properties related to lignin solubility are altered. Numerous researches have shown that lignin can be fractionated by using several solvents with highly dissimilar solvent parameters. Minute quantity of lignin can be soluble like in toluene (non-polar), non-hydrogen bonding solvents like dichloromethane (CH_2Cl_2), whereas other may be soluble in extra polar solvents like aliphatic alcohols. The solubility difference of the fractions clarifies the heterogeneous nature of lignin even if it is from a softwood isolation that comprises 95% guaiacyl lignin. Delignification can alter the lignin by addition of the reactants such as Sulphur or alcohol to the lignin or enhances the molecular weight by reactions of lignin with itself through delignification. These separation processes result in lignin that comprise acidic groups, lose aliphatic OH groups, increase the free phenolic groups by breakage of the linkage, that comprises more C—C bonds within units.

2. History of Lignin

About 170 years ago, the French chemist, Anselme Payen (1795–1871) dried woods with nitric acid (HNO_3) and sodium hydroxide, recovering two dissimilar products (Marton et al., 1966).

The first one was denoted as "cellulose" by him and the other ingredients with high carbon concentration was slow as a coating material, in which the cellulose was implanted.

Discovery of lignin by Anselme Payen (1839) was initially defined as the encrusting material in wood. Scientists were confused by the nature of this widely present material [5]. Though lignin was observed to be rich in carbon concentration than carbohydrates, the chemistry remained unclear for a long time. Consequently, since the last century, the lignin structure was labeled as a complex polymer composed of asymmetrical branched units. Lignin is noticed in all vascular plants, commonly between the cells, within cells and cell walls. The polyphenolic lignin structure is well known for its role in woody material to give resistance to chemical and biological degradation. This is because of their hydrophobic nature and insolubility in aqueous systems stopping access of degrading chemicals and organisms [6].

Previously, the nature of the plentiful substantial was not known, and its chemical construction of its constituents persisted as unidentified for a period of time. The quality of lignin was explained by Bente, and as late as 1890, Benedikt and Bamberger observed that get together of cellulose, lignified important of lignin contains methoxyl groups. Lignin chemistry is in main slice because of the struggle of Peter Klason (1848–1937), who devoted high devotion to lignin and its properties containing its description, experimental bases show that but also with his awareness, he explains that lignin is prepared by coniferyl alcohol [7]. Several procedures discussed in a study were beneficial for research of lignin, and his procedure is highly in use [8]. Freudenberg also contributed to lignin study. He extracted lignin from wood by different method and described it by watchful diagnostic purpose. According to this study, he defined lignin as an amorphous, a type of structural order based upon similar blocks containing phenylpropane components. According to these results of oxidative dehydrogenation of lignin, he suggested that the chief linking systems would be C—C and ether linkages, latter otherwise alkyl-aryl ether linkages [9]. At the end, the structure of lignin formula has several studies for its clarification. Thus, Adler projected in 1961 the first formulae including 12 phenylpropanoic units connected by C—C and C—O bonds, while Freudenberg was familiar with more complex structure for lignin built on 18 units in 1965. NMR studies exposed lignin structure polymer has next been published by Ludwigand Nimz for softwood and beech lignin, individually. During last few years, increase in care on lignin is visible by the extensive number of reviews, books, publications and patents, containing a wide diversity of subjects and fields of application [10]. These are likely to clarify the structural procedures and reactivity of lignin. Some of them are the financial types for their use to produce polymers. There are some original contributions of Glasser and Sarkanen who had greatly shared on the studies of lignin's structure and have published a lot of researches publications on lignin and the linked financial parts.

3. Lignin functions

Lignin is an important constituent of the structural framework in plants forming part of the primary elements of the cell wall. By viewing the point of evolution, lignin has been attributed as the terrestrial variation that allows significant vertical growth. As an important part of the cell, lignin provides support to plant by communicating rigidity to the cell wall. The plant resists ecological stresses because of building blocks of cell wall [11]. Lignin provides rigidity to the plants but in aggregation with the hetero-polysaccharides, it enhances flexibility which is central for suitable response to dynamic loads from wind and snow. Additionally, lignin changes the polysaccharide network to make it resistant to foreign organisms. Lignin helps in the protection of woody tissues from microbial and fungal attack covering the carbohydrate structure, causing reduced availability of enzymes for hydrolysis. Partial solubility and complexity of the lignin makes it tough for degradation by microorganisms [12]. Still, with the diversity of associations, harmful organisms for wood require the breakage of aryl carbon bonds and aryl ether bonds requiring increasing the cost of production of specific enzymes or developing non-specific pathways for delignification. Additionally, lignin is less hydrophilic in nature than the polysaccharides helping to alter the permeability of cell wall by sealing it and enabling water transport through the vascular tissue [13]. The aromaticity of lignin lends itself to improving the heat stability of wood. Lignin nature has been broken with separated technical lignin, changing it into carbon fiber by using controlled increased temperature. While lignin is more heat stable but the structure of lignin is greatly obstructed by the thermal change and scientists should be advised that lignin structure can alter by processing at higher temperatures used in the production some thermoplastic materials. This type of modification comprised of polymerization, loss of hydroxyl group and synthesis of new acidic groups.

4. Sources of lignin

The sources through which we obtain lignin biomass are the extraction and the secondary treatments those having high impacts on its mechanical and physical properties. Lignin is derived from numerous sources like pulp, wood and paper, sugarcane and cereal straws using variety of pulping methods. The lignin content in woody plants from angiosperms and gymnosperms is the maximum with the order of 30–40% whereas other sources only comprise about 3–25% in terms of weight [14]. The pulp and paper sector harvest a large quantity of lignin with greater potential by future lignocellulosic biorefineries. Negligible lignin is recovered from grass leaves and solid waste in urban and rural areas where their lignin content is assessed to be less than 15%.

4.1. Wood

In the middle lamella, lignin comprises non-uniform thickness and in the primary and secondary cell walls depending on the plant species and the type of cell. Roughly 70% of the total lignin biomass of the cell wall is concentrated in the thick layer of secondary wall. Quantity and quality of lignin alter naturally among wood species with 19–28% in angiosperms and 24–33% in gymnosperms. Hardwood in angiosperms in general comprises more hemicellulose and low quantity of lignin than softwood in gymnosperms [15]. Not only the quantity of lignin varies between hardwoods and softwoods, their concentration also varies in the location within a tree.

The immature wood has higher concentration of lignin than latewood. Lignin content is also flexible within different populations of plants in the same genus. The average concentration of lignin ranges 25% in *Pinus monticola* to 30% in *Pinus palustris* within the genus Pinus.

Lignin in angiosperms is syringyl-guaiacyl type and in gymnosperm is naturally guaiacyl with limited p-hydroxyphenyl lignin in both forms. Both structures are different arising from the linkages that can occur during polymerization. Guaiacyl lignin can experience coupling reactions at the five positions of the phenylpropane unit and this delivers a substantial place for cross-linking and branching reactions, that particularly occur during delignification process. The occurrence of syringyl units makes angiosperm (hardwood) lignin more readily detached during pulping process by limiting lignin forming condensed structures without the open methoxy position in the syrginyl units. In compression-wood lignin, it is difficult enough to hydrolyze as it comprises a higher amount of condensed p-hydroxyphenyl units [16].

4.2. Pulp and paper industry

Species containing hardwood and softwood, and some types of once a year plants, have marketable interest as a basis of cellulose fibers to produce board and paper products. Lignin is not easy to separate in a native form from plant material. The paper and pulp industry is primarily commercial way of lignin, the delignification process, though, modifies lignin to various grades. In technical fiber liberation processes, like sulfite or alkaline pulping, vast amounts of lignin are liquefied as alkali lignin and lignosulphonates, respectively. Large quantity of lignin was made available every year from the paper and pulp industry as byproducts of the delignification process. These sulfite or sulfate lignins have fluctuating levels of covalently bonded sulfur ensuing in the polymer with different characteristics that the original lignin [17]. Typically, impurities in lignin including low molecular weight sugars and resin acids are removed through distillation method. Kraft lignin is typically purified by Kraft black spirits, which are complex assortments of fibrous materials and dissolved organics such as hemicelluloses, lignins, acids, sugars, and resins and inorganic salts such as ash [18]. Modern technology in increasing lignin recovery from black liquor using carbon dioxide acidification that has been transported to industry and has manufactured a readily accessible dry lignin dust stream [19].

4.3. Sugarcane bagasse

Sugarcane bagasse is the leftover fiber after sugars have been removed. As an agro-industrial residue, *Saccharum officinarum* (sugarcane bagasse) is another source of lignin biomass. It is a byproduct of the sugarcane industry with almost 32–34% cellulose, 19–24% hemicellulose, 25–32% lignin, 6–12% extractives and 2–6% ashes [20, 21]. Composition of sugarcane bagasse is close to the other plant cell walls. Each class of plants, grasses, gymnosperm and angiosperm manufacture higher lignin content in one type of the phenylpropane repentance. Sugarcane bagasse lignin comprises a higher quantity of H-type lignin, p-hydroxyphenyl, and result in lower methoxy content than softwood and hardwood lignins [22]. Approximately 250–280 kg of bagasse are produced from processing ton of sugarcane which roughly produced 54 million tons of bagasse every year [23]. Now, a large quantity of bagasse is burnt as a low-grade fuel for recovery of energy and a limited quantity is being used to make pulps, board materials and composites.

http://dx.doi.org/10.5772/intechopen.73284

5. Agricultural residues

The benefits to use agricultural residues are threefold for economic, environmental and technological results related to a green economy. Dissimilar from pulps of wood, pulps of agricultural can get by means of more ecologically benign pretreatment and bleaching methods [24]. In classical wood pulping methods, maximum bleaching of pulp takes place using chlorine-based chemicals or chlorine whereas straw is treated with slight additions of chlorine-free chemicals that results without the production of toxic chemicals. Additionally, agricultural residues comprise generally a more permeable structure and a lower lignin concentration than woody plants which enhances their pulping procedures. Agricultural residues may be rice and wheat straws which are discussed in this section. Rice straws also called as cereal-straw are additional source of lignin production. Total rice straw production of the word, *Oryza sativa*, is 525 million tons per year.

6. Chemical structure of lignin

The lignocellulosic material is a compound of biopolymers with cellulose intertwined (35–83% dry weight), hemicellulose (0–30%), lignin (1–43%) and some extra complexes (xylose, arabinose, tannin, etc.) [25]. In woody plants, lignin plays a key role in a construction and strength to cell walls, changeable fluid movement by wary in contradiction of biochemical straining by struggling enzymatic breakdown of other complexes [26]. Its structure includes phenylpropane components, linked from three aromatic alcohol ancestors, p-coumaryl, coniferyl and sinapyl alcohols. The phenolic administrations that started from these monolignols are said p-hydroxyphenyl (H, from coumaryl alcohol), guaiacyl (G, from coniferyl alcohol) and syringyl (S, from sinapyl alcohol) moieties. Overall the biological lignification way, the monolignols

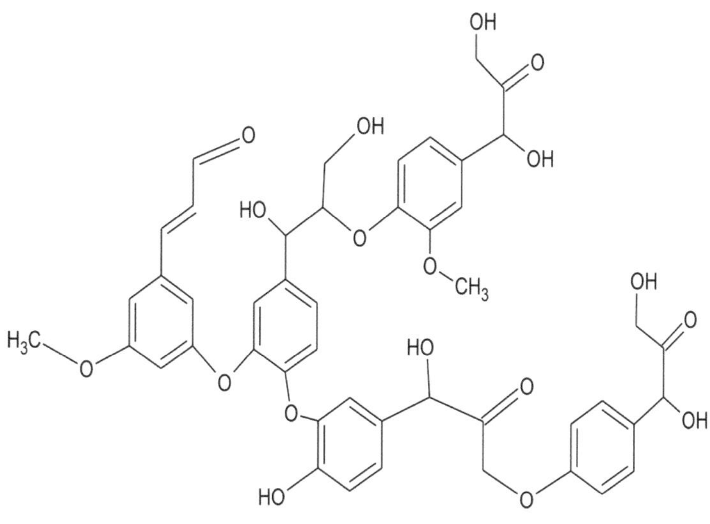

Figure 1. Chemical structure of lignin.

components are connected together by means of radical link reactions [27] to synthesize a complex 3D molecular structure that comprises a great variety of bonds with naturally around 50% ether linkages [9]. Lignin composition and content are predisposed by the environment andalso by the species. Hardwood lignins comprise principally of G and S units and traces of H units [26], however softwood lignins majorly comprise G units, with low levels of H units. Grasses lignins—monocots—incorporate G and S units at analogous levels and more H units than dicots. Grounded on complete lignin structure projected by Adler in 1977, lignin is documented as a majorly branched polymer with a diversity of functional groups: aliphatic and phenolic hydroxyls, carboxylic, carbonyl and methoxyl groups. The chemical structure of lignin is drawn in the given **Figure 1**. The profusion of the chemical sites suggest different potentials for chemical changings and recommends that lignin might play a vital character as a new chemical predominantly in the development of supramolecular structure and aromatic chemicals [26].

7. Extraction process and their resulting technical lignin

Physical, chemical and biochemical treatments are used for the extraction of lignin from other cellulosic materials. The botanical sources as well as the pulping source which is also called as the delignification process have great effect on the final structure, purity and corresponding properties of the lignin [28]. Cleavage of ester and ester linkages is the basis of the pulping process and the resulted technical lignin are different from the plant lignin. In this part, our main focus is to recover the lignins by the different extraction processes which are commercially available. Sulfur and sulfur-free processes are the two main categories which are shown in **Figure 2**. Differences in chemical structures effects the lignin reactivity like for the various modification [24].

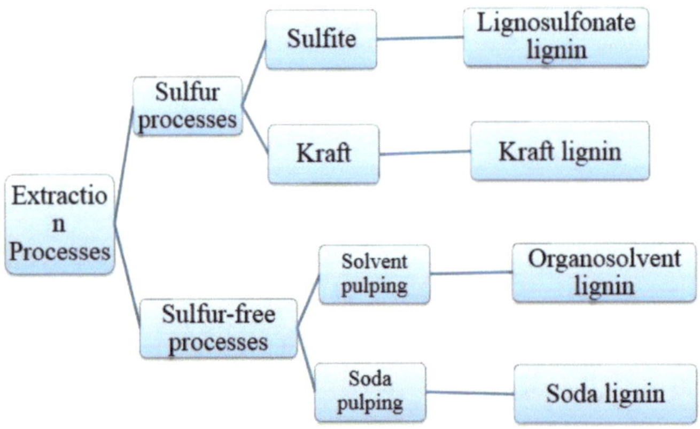

The meal of wood weighted 25 g after extraction was kept in a percentage liter beaker, three bottles tight with the reflux condenser, through dipping funnel and a nitrogen jar. In the water bath flask heated for 10–15 min, oil was attached to pump that attached and the solvent having

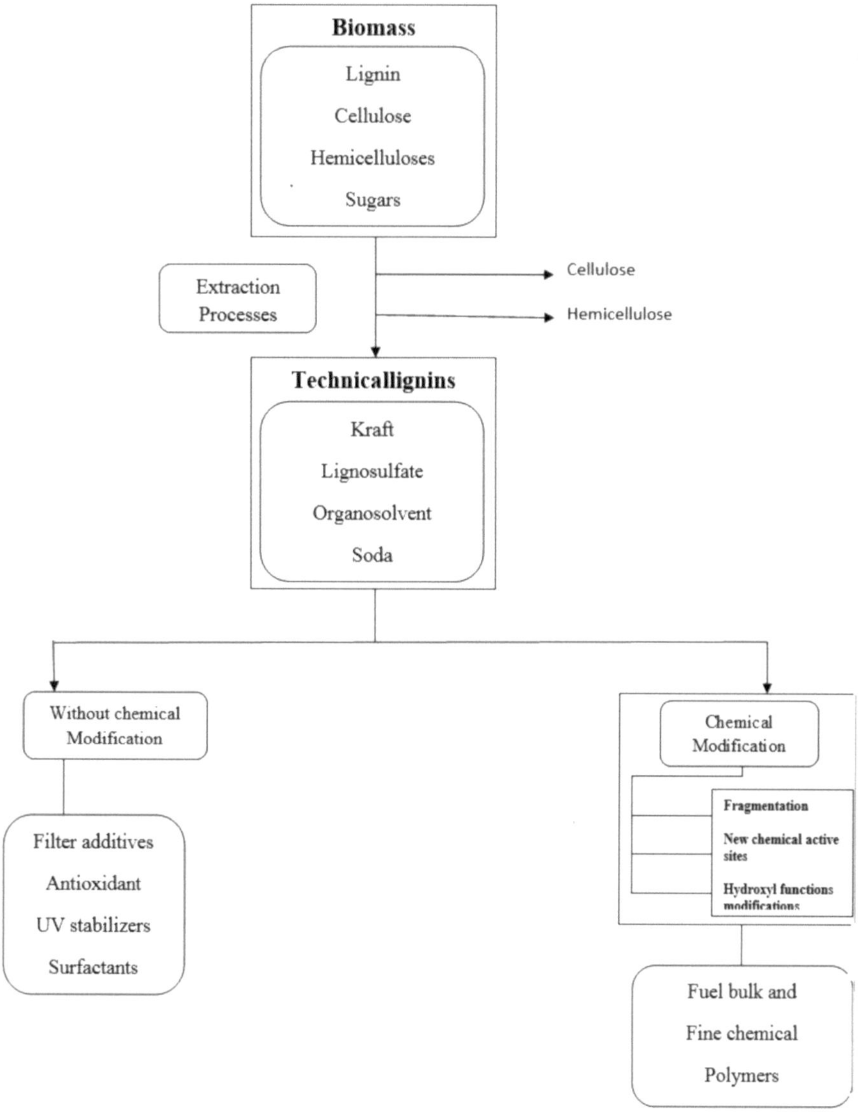

Figure 2. Isolation of lignin process.

46 g HCl as dry gas is poured by the dropping funnel. Heat the reaction mixture and gently refluxed for changing ages of time of 1–8 and 48 h beneath the nitrogen air [29]. For the analysis of sample, the use of aspen wood reaction mixture was homogenized utilizing a stirrer. Reaction mixture cooled, filtered while nitrogen enclosed and the washed accumulation by using filter by an extra process comprising similar dissolvable. Remaining pulp was dehydrated until the constant weight come and examined for methoxyl concentration. Then filter the solution and filtrate was killed with excess of Na_2CO_3 and allowed to remain overnight. We supposed to reduce the weight of liquid layer by using nitrogen spill [30].

8. Lignin isolation from the spruce wood

In preceding trials to reason about the fairness of dissimilar dissolvable constituents, originally separated spruce wood prospered in the dissolvable medium having 0.2 N (solution by weight), HCl dry gas and nitrogen air refluxed for 4 h time duration.

The solvents selected were dioxane-methanol and chloroform-methanol as suggested by dioxane-water, Schuerch and dioxane. Than filter the reaction mixture, and the filtrate, after equilibrium with Na_2CO_3, was concentrated to a little capacity. The lignin was encouraged by compelling this focus, as a fine stream, under the superficial of 1% sodium sulfate solution. The lignin was expelled by the process of centrifugation and decantation also the mother mixers were thoroughly removed with ether [31]. The leftover wood supper, the lignin, and the ether-solvent concentrate were dried to steady weight and examined for methoxyl content. The procedure of isolation of lignin from spruce wood is given in **Figure 2**. It was examined that the alkylation reaction and disclosed as being because of the etherification of benzyl liquor gatherings or the re-etherification of benzyl ether gatherings [32].

There is an impressive lessening in the aggregate accessible lignin items when contrasted with those obtained either by the one-stage 2.5 h or the one-stage 4.0 h extraction. The synthetic idea of these items detached by the progressive stepwise technique changed significantly, demonstrating the lignin substance [30]. It must be called attention to that every wood deposit was dried first noticeable all and phosphorus pentoxide in vacuum desiccator, and it is possible

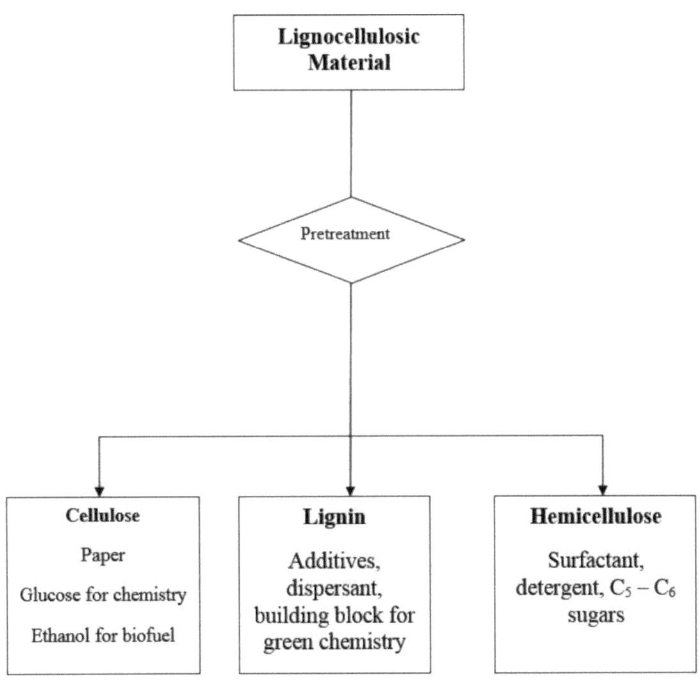

Figure 3. Digestion of lignocellulosic material.

that amid this time condensation or oxidation happens that diminishes the simplicity of ensuing isolation. The subsequent reaction items were considered some time recently, the lignin recouped by precipitation into 1% sodium sulfate, and the ether-dissolvable oils removed from the fluid filtrate [33].

The benefit of utilizing fluid dioxane over dioxane alone is demonstrated by the fundamentally expanded yield and n-methoxy substance of the disengaged lignin as shown in **Figure 3**. Doubtlessly, in the watery framework, some sugar or other non-methoxyl-containing parts that show up with the lignin in the unadulterated dioxane extraction has been isolated and show up rather in the ether remove offering ascend to the moderately bring down methoxyl substance of this portion.

9. Lignin as antioxidant

Utilization of the biomass product which is obtained at the technological cycle is the concept of biorefinery. In the chemical and biochemical processes lignin is separated as the byproduct from the lignocelluloses. It is an important component of the polymer system due to the good potential of recognition. About 98% of the total lignin is burned to improve the energy balance in the main process while only minute quantity of lignin (1–2%) is used to make the value-added products [34]. Lignin acts as binder, plasticizers, sorbent, composite of materials, additive for food preparation and polymeric carrier of biologically active agents. These are the traditional applications of lignin [35]. Lignin also acts as the antioxidants. Lignin acts as the stabilizers in that reaction which are induced by oxygen and its reactive species and also have the property to slow down the aging of biological systems and composites. The antioxidant property of lignin shows lots of applications in industry, healthcare and agriculture [36]. Due to high thermal and biological activity, lignin is used in place of those molecules which have lower molecular weight and also where the antioxidant activity of the single molecule is insufficient. For the composition and structure of lignin, polydispersity and chemical heterogeneity along with MMD is asked so that its application as the antioxidant can be determined. The efficiency of lignin as antioxidant is related to its structure and solubility [37]. It is difficult to allocate the lignin as antioxidant efficacy to specific organizational elements due to the complex structure of the lignin with comparison to synthetic antioxidants or polyphenols which are naturally defined like flavonoids and tannins [38]. For the antioxidant activity of the lignin, free phenolic hydroxyl groups and ortho-methoxy substitution in aromatic rings are essential. But some compounds like the carbonyl group in the side chain shows the negative effect on the antioxidant activity of the lignin. Radical scavenging activity of lignin is decreased due to various factors which are important in functioning of lignin-like high molecular weight, polydispersity and heterogeneity [39]. In situ, lignins are embedded in the complex network system which contains the different types of polymers including protein, macromolecules or molecules according to the biological region, isolated lignins and contained mixtures also have the ability to change the antioxidant activity of the lignins by changing the polarity like in the presence of polysaccharides which are more polar than the lignin and by decreasing the phenolic compounds concentration. Nowadays our aim is to investigate the novel type of technical lignin. BIOLIGNINTM, a product of biomass refinery technology which is recently developed (CIMV, France) [40] realizes the plant biomass fractionation and also three other compounds were obtained, celluloses, hemicelluloses and lignin, and the lignin itself is a commodity chemical and a raw material which is used for further processing. Fractionation of the technical lignins with the organic solvent which is a best way for obtaining the molecular mass distribution (MMD), component composition and for the production of value-added products [41]. Present work was done for the evaluation of the suitability of fractionation of BIOLIGNINTM, which acts as a tool form obtain the products which have a good antioxidant property specifically for polyurethanes (PU) which is a composite material on the basis of polyether's-polyols and aromatic isocyanates, for example, elastomers. So that is why BIOLIGNINTM was fractionated by using the two organic solvents which are of different polarity through sequential extraction [42]. Each fraction was characterized by the structure of lignin macromolecule and due to presence of the carbohydrates and other mixtures. These characteristics were then compared with the data of effect on lignin addition on model of thermos-oxidation destruction. Antioxidant efficiency of lignin fractions were indicated by PU elastomers [43].

10. Lignin potential

Only 2% of lignin is used for the combustion and energy production while around 50-million-ton lignin is produced every year by paper and the pulp sector [44]. Lignin structure, heterogeneity and the industrial processing costs for delignification are the main reasons for the limited utilization of the lignin [45, 46]. For the removal of these limitations, modification of several types is used which increases its chemical reactivity, decreases the brittleness of the lignin resultant polymers, regularity, increases its solubility in the organic diluents and progress the comfort of processing the lignin [47]. Lignin potential is considered important for the production of the additives, coating material and resins [48]. Synthetic polymers and the aromatic chemicals can be replaced with the new renewable aromatic compounds [49]. It is experimentally proved the high antioxidant capacity of the lignin and its applications in the polymer enhancement [50]. Lignin depolymerization can be used on large scale for the production of valuable chemicals due to the lignin macromolecule structure of the lignin with the many phenolic compounds [51]. Lignin is also used in the fabrication of composite material. Finally, it was showed that exterior grade particleboards which are made by the lignin-based adhesives can successfully met the requirements of international standards.

11. Application of lignin

The handiness of commercial lignosulfonates items come from their scattering, authoritative, complicated and emulsifying characteristics. Industry initially started to utilize lignins in the 1880s when lignosulfonates were utilized as a part of calfskin tanning and color showers. From that point forward, lignosulfonates have even discovered applications in nourishment items, filling in as emulsifying factory machine a creature encourage and as crude material in generation of vanillin [52]. Enthusiasm for the physiological part of bioactive mixes show in plants has expanded drastically finished the most recent decade and it is quite compelling in connection to human wellbeing. Notwithstanding customary application systems, cancer prevention agent is a huge use of lignin. Lignin is called a free radical borrower, and relaxes the reactions started by oxygen and type of free radicals [39]. Past researches in our study center have shown the cell support action of lignins got from sugar stick [53]. To arrange of negative impacts, it is significant to do soundings of possible eye and skin bothering. These categories of studies are compulsory for all type of chemicals [54], and are done with a change of the past test.

12. Antimicrobial properties of purified lignin

Antimicrobial possessions of the lignin are well known. Procedure of hard wood development is linked to potential harmfulness to microorganisms of phenolics deposited; this is due to natural durability of the wood species. Phenolic compounds are used in the food preservation process due to the presence of phenolic compounds in lignin which have the ability to inhibit the microbial growth with phenolic acids and flavonoids. Carvacrol and cinnamaldehyde, which are the phenolic monomers, are used in meat, rice, cheese and in fresh fruits and vegetables [54]. However, some applications are not of scientific consideration like the lignin has the potential to improve the intestinal health and animal welfare.

Dietary Alcell lignin reduced concentrations of *Escherichia coli*, when compared to antibiotic-free diet or containing one antibiotic, was suggested in the further research [55]. Main cause of food borne illness is basically the contamination of the intestinal Escherichia coli with the poultry carcases in the slaughter houses, and at the same time antibiotic resistant gene is spread in food chain though the poultry litter [56]. Litter Escherichia coli is also the pathogen of the cellulitis, which is the cause of the carcass condemnation at processing plants [57]. According to the new finding, the load of *Escherichia coli* can be reduced in the chicken intestine and in litter by the purified Alcell lignin and in this way, we can improve the safety of poultry products and also the cellulitis. Mechanism of lignin action is not well-defined. Polyphenolic compounds of lignin cause the cell membrane to damage and also cause the lysis of bacteria which also causes the release of cell contents. Carvacrol, thymol and cinnamaldehyde, a phenolic monomer, which also possess the antibacterial activity [58, 59]. Antimicrobial activity of the carvacrol and thymol is because it causes the integration of the cell membrane and cell content also releases. While cinnamaldehyde has the ability to penetrate in the cell membrane of bacteria which causes the reduction in the intracellular pH and causes the ATP depletion [60]. The antibacterial mechanistic actions are seen in many phenolic compounds.

13. Application of lignin in ethanol production

In the production process of biomass, ethanol lignin exists as in two forms; insoluble in residues and soluble in pretreatment solvent. The lignin which is soluble in solvent is restored by the process of washing and purification which removes the excess water and the extraction process is used for the insoluble lignin [61]. By the comparison of the lignin residues with the burning lignin for heat and power, high-value co-products of biomass ethanol are expected. It is a challenge to use the lignin residue at economical level for commercial process, but it is necessary for economic viability of such plants and it is beneficial to environment and also of high profit. Lignosulfonates is a commercial lignin, which is produced as a byproduct in paper and pulp industry. Lignosulfonates also acts as the emulsifiers in the animal field and in the production of vanillin used as a raw material. Lignin possesses antioxidant and antimicrobial properties as a complex phenolic polymer [39]. Different numbers of factors influence the chemical structure of the lignin which can be modulated be different means including botanical regions, condition of extraction from in situ network and the environmental conditions of growth. Delignification techniques consist of two processes: firstly, cleavage of the covalent lignin and then the solubilization of polymer fragments. Different techniques are used for the delignification of the lignin use of organic solutions, use of acid or alkali or several other parameters (time, temperature, etc.). These conditions affect the chemical structure of the lignin (molar mass, cross-linking, density and the functionality) and it's bioactive. Biomass ethanol as biofuel has many advantages including economic, environmental and social stability [62]. Corn stover is largely available that is why it is considered a good candidate for the production of ethanol. Ref. [63] defined that 0.25 kg ethanol can be made from 1 kg (dry basis) of corn stover, and it has been predictable that 0.34 kg ethanol can be formed from 1 kg (dry basis) of corn stover in the close future [64]. For the production of 150 million kg of ethanol, we have to consume 450 million kg of corn stover per year. It contains almost 18% lignin so per year almost 81 million kg of lignin will be generated [61].

14. Antioxidant activity in RBCs of human

Antioxidant capacity in red blood cells is evaluated by the inhibition of the human erythrocytes hemolysis which is induced by the AAPH (2,2-azobis (2-amidinopropane)) (a peroxyl radical initiator). With the help of vein puncture, blood samples were taken from different healthy individuals. Blood was centrifuged at 10,000 rpm for 10 min and washed with the help of phosphate buffer saline (PBF). After washing supernatant coat was removed carefully and then these washed red blood cells are suspended in the buffered saline. Hemolysis of erythrocytes was determined by the AAPH adjustment. Ref. [65] of a method defined by [66]. For the induction of hemolysis, 25 μl of aliquots of RBCs suspension were preserved at specific temperature in the presence of 100 mM AAPH. IC_{50} of the hemolysis is determined which was induced by the AAPH. By using the spectrophotometer at 540 nm, hemolysis was monitored. Inhibition of hemolysis which was obtained by AAPH is studied at the different concentrations of lignin. For all the lignin studies, inhibitory effect is concentration dependent, the higher the concentration, greater will be the inhibitory effect. Most of the antioxidant lignin, bagasse, has a 100% inhibition effect at the low concentration. For the IC_{50} values of the lignin, dose response curves are calculated [52].

It is dangerous to give these biological actions to specific structural components due the molecular difficulty of the lignin as related to the actions of chemically well-defined flavonoids and tannins [37]. Lignin can resist the viability and growth of effected cancer cells because it is a major component of dietary fiber and in this way, lignin can inhibit the enzymes which are related to the synthesis of superoxide anion radicals [67]. In a new study, it is examined that the lignins which are ready at high temperature, increased catalyst, longer reaction period and dilute ethanol displayed high antioxidant action [68]. It is compulsory to study the result of new properties on skin and eyes. We studied the lignins for skin and eye irritation. It is shown that all the lignins are non-irritants to skin and eyes after application of 0.2 and 0.5 g, respectively. None of the compounds showed irritation from mild to severe (i.e., severe erythema, oedema). So, these lignins are studied as non-irritants to skin and eye, however, more studies should be performed for using them in topical formulations [52].

15. Cell reinforcement movement

The examination of potential latest applications of lignin in many imaginations is of strange significance. Among these applications the use in the restorative ground would be of intrigue mainly because of their cancer stoppage agent capacity. The inhibitory result is fixation in need of for every one of the lignins studied, being the best inhibitory effect at the most astounding focus measured. Bagasse, the most cell support lignin, talented a 100% inhibitory effect at least fixations (from 100 g/ml). Amount reaction curves were determined for the lignins strained over a scope of focuses, allowing IC50 esteems to be acquired (fixation starting half limit on hemolysis started by AAPH) [68]. These qualities uncovered Bagasse as the lignin with most astounding cancer prevention agent action, on account of its low IC50 esteem and with a comparative incentive than those introduced by epicatechin. Heterogeneity as far as segment creation (starch admixtures) and polydispersity is the factor, which can diminish definitely the cancer prevention agent proficiency of segregated lignins [39]. Because of the sub-atomic multifaceted nature of lignin, it has been hard to allot these organic exercises to particular basic segments, contrasted with the exercises of synthetically characterized tannins and flavonoids [37]. As a noteworthy segment in dietary fiber, lignin can hinder the movement of compounds identified with the era of superoxide anion radicals and block the development and practicality of disease cells [67]. In a current paper, it has been shown that, as a rule, the lignins arranged at lifted temperature, longer reaction time, expanded impetus, and weakened ethanol demonstrated high cancer prevention agent action [68]. Additionally, considers are important to be led to comprehend the connection between substance structures of the lignins announced in this paper and their cancer prevention agent exercises [52].

16. Hydrogels application based on lignin

As observed in the previous section, the biorenewable polymer lignin can be applied for the hydrogel applications as a starting material. The development of hydrogels on the most current years has seen a rapid move from industrial to consistent polymers as one of the energetic segment [69]. The main attention for polymer classification based on hydrogels residue to their innate properties, like, biodegradability, biocompatibility, little poisonous value, eco-agreeableness and defenselessness to enzymatic lying. Biopolymer applications based hydrogels are in biomedical, water sanitization, mending frameworks, biomimetic platforms and medication transportation devices [54]. Between massive typical polymers, lignin is an interesting member in hydrogels due to its focal links as communicated before that makes it an elastic physical for common applications in numerous fields. There is an extraordinary set of quantity of researches obtainable on the consumption of lignin, for example, spin in hydrogel polymer applications. So, additionally, we must do undercover persona lignin kinds founded upon hydrogels together within potential applications in feature. Basic lignin or Kraft produces green polymer hydrogel [30]. The both procedures of lignin used as a part of this examination were limited from crushing alcohol and described UV/FTIR spectroscopy.

The hydrogels education was fused by the use of unite copolymerization technique. The idea of acrylamide and poly vinyl liquor wrapped to promote allowance of acrylamide monomer is done by join copolymerization reaction. These hydrogels were showed applying FTIR and UV spectroscopy. These hydrogels were then visible to water absorptive and deswelling education. It was detected by training that the basic lignin hydrogel had in height inflammation quantities, reduced aquatic take-up and inflammation charges when related with the hydrogel Kraft lignin l. The condensed deswelling performance was qualified to the ideal system construction of the basic lignin, while the height inflammation amount was qualified to the cooperative support within the wipe subdivisions and the mass grid. Mutually, types of hydrogels were secondhand to think about the impression of table salt (NaCl) on the interest size at normal temperature and distension at several grades. It was detected that the delay of sodium chloride brought about an insistent digestion size of the hydrogels and this action came because of the reduction in the weight alteration within the gel and the outer explanation [33]. Lignin created hydrogels were similarly decided using alcohol as the back matrix by a similar study. Dark alcohol is a combination of sugars and is one of the major automatic wastes that outcome from the mashing strategy [70]. The black alcohol in this research was learned from basic crushing of rice stubble.

The hydrogels were set up by artificial cross-connecting process. In this process, dual discrete cross-connecting methods, exactly, cross-connecting by expansion reaction and cross-connecting by fundamental polymerization were applied. In cross-linking by radical polymerization, ceric ammonium sulfate was used as creator within the vision of N,N-methylenebisacrylamide to do bond copolymerization of poly vinyl liquor (PVA) and polyacrylamide (PAAm). Anyhow, amid the cross-connecting by the progress reactions, the polymerization of hydrogels was confirmed without the originator. The hydrogels organized using the individual systems were labeled by FT-IR spectroscopy and examining electron microscopy (SEM). The hydrogels settled applying radical polymerization were happening to show an increased inflammation size, 60%, associated with 27.27% developed from the hydrogels that were set up by development response. NaCl effect on the digestion ability of these hydrogels was also educated at countless temperatures and pH. From investigation, it was hypothetical that lignin including dark alcohol can be used as green hydrogel experiences [5]. Heat touchy lignin constructed hydrogels had likewise been set by using the acidic destructive lignin. In this study hydrogels link copolymerization reply was used to set the hydrogels. The copolymerization response was done using acidic lignin and N-isopropyl acrylamide (NIPAAm) within the sight of N,N-methylenebisacrylamide

(MBBAm) as the cross-linker and H_2O_2 as the originator. The common hydrogels were later predictable for their warm, morphological and inflammation performs. The opening calculation in the hydrogels was noticed to rise by the development in the acidic lignin and was complete from the SEM slow. The thermogravimetric study of these hydrogels is showing their break down in the heat range of 400–410°C, although the difference review calorimetry (DSC) consider established to minor the basic explanation temperature (LCST) of 31°C. The point-by-point inspection of these hydrogels stated their temperature touchy countryside. Lignin founded hydrogels were equally approved aimed at their important usage as aerogel and cryogels [69].

In the field of material applications, the cell reinforcement action has been researched in the point of utilizing lignin as a characteristic substitute of engineered cancer prevention agents required for the detailing of polyolefins [71]. The similarity among lignin and polymer network could be effectively anticipated utilizing the dissolvability parameters figured after the Small, Van Krevelen and Hoy technique. A similar component ensuring polyolefins against radical oxidative debasement is significant for assurance of unsaturated fats. The utilization of regular cancer prevention agents to ensure lipids and sustenance has effective long history [72]. The capacity of lignin has prevented lipid peroxidation in human red platelets. Polymer bundling materials are collaborating with stuffed merchandise because of mass exchange between the two accomplices, which is impacting sustenance quality [73]. It is intriguing to make positive utilization of such a connection to guarantee a maintained conveyance of a dynamic substance, for example, the cancer prevention agent into the nourishment stuff. In that point, dynamic bundling is currently progressively utilized as a supply for dynamic substance conveyance [74]. Supplementing different procedures, for example, altered environment bundling, eatable covering or oxygen scroungers [75].

The goal of study is to research the potential of lignin to be utilized as a characteristic cancer prevention agent added substance in a dynamic sustenance bundling. The bundling framework picked was poly(lactide) (PLA) in light of its inexhaustible character. The incorporation of regular cancer prevention agents in PLA in the point of building up a dynamic material has just been proposed in writing [76]. Besides, a few investigations as of now exist indicating the practicality of consideration of lignin into PLA [77]. Two commercial salt lignins were picked due to their normal radical rummaging movement and their distinction in the free phenolic monomers content. PLA was mixed with lignin by twin screw expulsion and film tests were created by thermo-pressure. The impact of lignin amount and handling on the radical searching action was surveyed in an ethanol/water solution being a stimulant of greasy nourishments suggested by the European control for sustenance contact materials [78].

Assurance of lignin cancer preventive action is contradiction of DPPH free radical. Lignin cell reinforcement size remained examined by assessing free fundamental rummage impact. The DPPH radical probing exercises of organo-solvent lignin extraction and regular Kraft lignin were separated by minor modification technique as described by [79]. Quickly, a lignin test solution in 0.1 ml of dioxane-water (9/1, v/v) was added to 3.9 ml of a DPPH solution (25 mg/l in ethanol) as the free radical foundation and kept for 30 min at room temperature. The decline of the solution absorbance, because of proton donating action, was restrained at 517 nm (UV-2550, Shimadzu, Japan). The DPPH radical searching action was computed. Lignin cancer prevention agent capacity was contemplated by assessing free radical searching impact (DPPH). The outcomes showed the lignin acquired by acidic acid-water technique had the most astounding cell reinforcement movement [52]. Indeed, inquire about demonstrates the lignin has no unsafe impact on eyes and skin. In this manner, the high cell reinforcement capacity of lignin from the acidic acid-water strategy opens another application in corrective and topical plans.

17. Cytotoxicity of lignins

Possible applications of lignins in the corrective and pharmaceutical businesses expect them to be innocuousness. Potential skin bothering by a fixing or item is one of the different investigations embraced in the general evaluation process. Cell cytotoxicity measures are among the most well known in vitro bioassay techniques to anticipate the danger of an extensive variety of substances [80]. At present, couple of information exists on conceivable cytotoxic impacts of lignins. Thus, it is valuable to dispose of any undesirable impact. Moreover, it adds to describing their natural profiles. The cytotoxic impacts of the lignins were researched utilizing the cell layer uprightness of human keratinocyte HaCaT and murine fibroblast 3 T3 cells keeping in mind the end goal to foresee their skin disturbance potential. A colorimetric examine that measures the capacity of live cells to take-up unbiased red color was utilized. The 3 T3 nonpartisan red take-up examine was chosen since this test is prescribed by the US National Institute of Environmental Health Science (NIEHS) Interagency Coordinating Committee on the Validation of Alternative Methods (ICCVAM). The IC50 estimations of the lignins contemplated uncover that these mixes have cytotoxic impacts, however just at high focuses. Cytotoxic action increments with expanding presentation period with a focus subordinate reduction in the rate of impartial red take-up. The distinctions were not huge in all cases. The most cytotoxic compound was Curran 100 on HaCaT cells (no critical contrasts were found after 24, 48 and 72 h) and bagasse in the 3 T3 cell line [36]. The minimum cytotoxic lignin in both cell lines was lignosulphonate. Lignins appear to be more cytotoxic in the keratinocyte cell line, particularly after 24 h (there are critical contrasts between the items).

A decent relationship exists between the cytotoxic impacts on HaCaT cells and on 3 T3, particularly after 48 and 72 h ($r = 0.903$ and $r = 0.963$, separately; information not appeared). Contrasts between these two cell lines can, for this situation, be ascribed to contrasts in cell line affectability to the mixes considered. The cytotoxic information for the lignins were contrasted and that for epicatechin. The lignins are by and large similarly or less cytotoxic than epicatechin after 48 and 72 h of introduction, despite the fact that the 3 T3 cell line demonstrated a more harmful impact after 24 h presentation, on both HaCaT and 3 T3 cells (critical contrasts exhibited). The connection between potential cytotoxic properties and the cancer prevention agent capacity of lignins were likewise assessed. The most grounded cell reinforcement items were all in all likewise the most cytotoxic. In any case, the viable cancer prevention agent focuses are littler than the cytotoxic ones (cell reinforcement fixations are from 5- to 10-overlay lower) so the mixes are cell reinforcements at non-cytotoxic fixations. These informations are important to future applications, since they demonstrate that these lignins can be utilized over a powerful fixation run that is alright for the ordinary cell lines examined. A decent relationship was found among Mn and cytotoxic movement, particularly in the 3 T3 cell line. There was a reasonable pattern toward the heaviest lignin being the slightest cytotoxic [36].

18. Conclusion

It is examined that lignin extracts from remainder of biomass biofuel making displayed strong antioxidant properties and had little antimicrobial application. Commonly, the antimicrobial properties of lignin extracts were reliable with their antioxidant properties. Lignins and lignin-derivated complexes are not now used in the para-pharmaceutical or cosmetic productions. It has been advised that the antibacterial result of Kraft lignin is related with the huge antiradical property of its methanolic portion. Nevertheless, normal application arrangements, cancer inhibition negotiator is a massive use of lignin. Neither of the lignins exhibited any injurious conclusion on eyes and skin. The high antioxidant capacity of the lignins considered together with their protection when useful on eyes and skin, open new outlooks in topical formulations and their potential use in cosmetic. The huge heterogeneity of lignins describes that it as very significant to complete the comprehensive characterization of the structures and properties of the lignins examined. It will be possible to regulate their behavior in different applications, and the behaviors in which their structures stimulate their biological properties.

Author details

Zahed Mahmood[1]*, Muhammad Yameen[1], Muhammad Jahangeer[1], Muhammad Riaz[2], Abdul Ghaffar[1] and Irum Javid[3]

*Address all correspondence to: drzahiduaf2003@gmail.com

1 Department of Biochemistry, Government College University Faisalabad, Faisalabad, Punjab, Pakistan

2 Department of Allied Health Sciences, Sargodha Medical College, University of Sargodha, Sargodha, Punjab, Pakistan

3 Department of Biochemistry, Government College women University Faisalabad, Faisalabad, Punjab, Pakistan

References

[1] Haygreen JG, Bowyer JL. Forest products and wood science: An introduction. Forestry. 1998;**71**(1):79-80

[2] Gandini A. Polymers from renewable resources: A challenge for the future of macromolecular materials. Macromolecules. 2008;**41**(24):9491-9504

[3] Taylor G. Biofuels and the biorefinery concept. Energy Policy. 2008;**36**(12):4406-4409

[4] Pickett J, Anderson D, Bowles D, Bridgwater T, Jarvis P, Mortimer N, Poliakoff M, Woods J. Sustainable Biofuels: Prospects and Challenges. London, UK: The Royal Society; 2008

[5] Sarkanen KV, Ludwig CH. Liguins. Occurrence, Formation, Structure, and Reactions. New York: Wiley-Interscience; 1971

[6] Banoub J, Delmas GH, Joly N, Mackenzie G, Cachet N, Benjelloun-Mlayah B, Delmas M. A critique on the structural analysis of lignins and application of novel tandem mass spectrometric strategies to determine lignin sequencing. Journal of Mass Spectrometry. 2015;**50**(1):5-48

[7] Rodríguez JJ, Cordero T, Rodríguez-Mirasol J. Carbon Materials from Lignin and Their Applications. In Production of Biofuels and Chemicals from Lignin. Singapore: Springer; 2016. pp. 217-262

[8] Brunow G. Methods to reveal the structure of lignin. Biopolymers Online. 2005

[9] Chakar FS, Ragauskas AJ. Review of current and future softwood kraft lignin process chemistry. Industrial Crops and Products. 2004;**20**(2):131-141

[10] Hatakeyama H, Hatakeyama T. Lignin structure, properties, and applications. In: Biopolymers. Berlin, Heidelberg: Springer; 2009. pp. 1-63

[11] Lopes FJ, Silvério FO, Baffa DC, Loureiro ME, Barbosa MH. Determination of sugarcane bagasse lignin S/G/H ratio by pyrolysis GC/MS. Journal of Wood Chemistry and Technology. 2011;**31**(4):309-323

[12] Verma SR, Dwivedi UN. Lignin genetic engineering for improvement of wood quality: Applications in paper and textile industries, fodder and bioenergy production. South African Journal of Botany. 2014;**91**:107-125

[13] Ten E, Vermerris W. Functionalized polymers from lignocellulosic biomass: State of the art. Polymers. 2013;**5**(2):600-642

[14] Smolarski N. High-value Opportunities for Lignin: Unlocking its Potential. Paris: Frost & Sullivan; 2012. p. 15

[15] Bjornsson S. Advanced Control Methodology for Biomass Combustion (Doctoral dissertation)

[16] Novaes E, Kirst M, Chiang V, Winter-Sederoff H, Sederoff R. Lignin and biomass: A negative correlation for wood formation and lignin content in trees. Plant Physiology. 2010;**154**(2):555-561

[17] Sainsbury PD. Biocatalytic valorisation of lignin via genetic or chemical intervention of bacterial aromatic degradation pathways [doctoral dissertation]. University of Warwick

[18] Wallberg O, Jönsson AS, Wimmerstedt R. Fractionation and concentration of kraft black liquor lignin with ultrafiltration. Desalination. 2003;**154**(2):187-199

[19] Kouisni LA, Holt-Hindle P, Maki K, Paleologou MI. The lignoforce system: A new process for the production of high-quality lignin from black liquor. Journal of Science & Technology for Forest Products and Processes. 2012;**2**(4):6-10

[20] Sakdaronnarong C, Jonglertjunya W. Rice straw and sugarcane bagasse degradation mimicking lignocellulose decay in nature: An alternative approach to biorefinery. ScienceAsia. 2012;**38**(4):364-372

[21] Rezende CA, Ribeiro deAzevedo E, Polikarpov I, de Lima MA, Maziero P, Garcia W. Chemical and morphological characterization of sugarcane bagasse submitted to a delignification process for enhanced enzymatic digestibility. Biotechnology for Biofuels. 2011;**4**(1):54

[22] Doherty W, Halley P, Edye L, Rogers D, Cardona F, Park Y, Woo T. Studies on polymers and composites from lignin and fiber derived from sugar cane. Polymers for Advanced Technologies. 2007;**18**(8):673-678

[23] Canilha L, Chandel AK, Suzane dos Santos Milessi T, Antunes FA, Luiz da Costa Freitas W, das Graças Almeida Felipe M, da Silva SS. Bioconversion of sugarcane biomass into ethanol: An overview about composition, pretreatment methods, detoxification of hydrolysates, enzymatic saccharification, and ethanol fermentation. BioMed Research International. 2012;**2012**

[24] Clancy-Hepturn M. Agricultural residues: A promising alternative to virgin wood fiber. Issue in Resources Conservation, Briefing Series. 1998(1)

[25] Ralph J, Lundquist K, Brunow G, Lu F, Kim H, Schatz PF, Marita JM, Hatfield RD, Ralph SA, Christensen JH, Boerjan W. Lignins: Natural polymers from oxidative coupling of 4-hydroxyphenyl-propanoids. Phytochemistry Reviews. 2004;**3**(1-2):29-60

[26] Laurichesse S, Avérous L. Chemical modification of lignins: Towards biobased polymers. Progress in Polymer Science. 2014;**39**(7):1266-1290

[27] Hu TQ, editor. Chemical Modification, Properties, and Usage of Lignin. New York: Kluwer Academic/Plenum Publishers; 2002

[28] Vázquez G, Antorrena G, González J, Freire S. The influence of pulping conditions on the structure of acetosolv eucalyptus lignins. Journal of Wood Chemistry and Technology. 1997;**17**(1-2):147-162

[29] Crestini C, Pastorini A, Tagliatesta P. The immobilized porphyrin-mediator system Mn (TMePyP)/clay/HBT (clay-PMS): A lignin peroxidase biomimetic catalyst in the oxidation of lignin and lignin model compounds. European Journal of Inorganic Chemistry. 2004; **2004**(22):4477-4483

[30] Dong A, Fan X, Wang Q, Yu Y, Wang P, Yuan J, Cavaco-Paulo A. Changes on content, structure and surface distribution of lignin in jute fibers after laccase treatment. Journal of Natural Fibers. 2017:1-2

[31] Shioya T, Akiyama T, Yokoyama T, Matsumoto Y. Formation rate of benzyl cation intermediate from p-hydroxyphenyl, guaiacyl, or syringyl nucleus in acidolysis of lignin. Journal of Wood Chemistry and Technology. 2017;**37**(2):75-86

[32] Montgomery JR, Lancefield CS, Miles-Barrett DM, Ackermann K, Bode BE, Westwood NJ, Lebl T. Fractionation and DOSY NMR as analytical tools: From model polymers to a technical lignin. ACS Omega. 2017;**2**(11):8466-8474

[33] Perazzini R, Saladino R, Guazzaroni M, Crestini C. A novel and efficient oxidative functionalization of lignin by layer-by-layer immobilised Horseradish peroxidase. Bioorganic & Medicinal Chemistry. 2011;**19**(1):440-447

[34] Lora JH, Glasser WG. Recent industrial applications of lignin: A sustainable alternative to nonrenewable materials. Journal of Polymers and the Environment. 2002;**10**(1):39-48

[35] Huijgen WJ, Telysheva G, Arshanitsa A, Gosselink RJ, De Wild PJ. Characteristics of wheat straw lignins from ethanol-based organosolv treatment. Industrial Crops and Products. 2014;**59**:85-95

[36] Ugartondo V, Mitjans M, Vinardell MP. Comparative antioxidant and cytotoxic effects of lignins from different sources. Bioresource Technology. 2008;**99**(14):6683-6687

[37] Sakagami H, Hashimoto K, Suzuki F, Ogiwara T, Satoh K, Ito H, Hatano T, Takashi Y, Fujisawa SI. Molecular requirements of lignin–carbohydrate complexes for expression of unique biological activities. Phytochemistry. 2005;**66**(17):2108-2120

[38] KoŠíková B, Lábaj J, Gregorová A, Slameňová D. Lignin antioxidants for preventing oxidation damage of DNA and for stabilizing polymeric composites. Holzforschung. 2006;**60**(2):166-170

[39] Dizhbite T, Telysheva G, Jurkjane V, Viesturs U. Characterization of the radical scavenging activity of lignins—natural antioxidants. Bioresource Technology. 2004;**95**(3):309-317

[40] Delmas GH, Benjelloun-Mlayah B, Bigot YL, Delmas M. Functionality of wheat straw lignin extracted in organic acid media. Journal of Applied Polymer Science. 2011;**121**(1): 491-501

[41] Gosselink RJ, van Dam JE, de Jong E, Scott EL, Sanders JP, Li J, Gellerstedt G. Fractionation, analysis, and PCA modeling of properties of four technical lignins for prediction of their application potential in binders. Holzforschung. 2010;**64**(2):193-200

[42] Ropponen J, Räsänen L, Rovio S, Ohra-aho T, Liitiä T, Mikkonen H, van de Pas D, Tamminen T. Solvent extraction as a means of preparing homogeneous lignin fractions. Holzforschung. 2011;**65**(4):543-549

[43] Arshanitsa A, Ponomarenko J, Dizhbite T, Andersone A, Gosselink RJ, van der Putten J, Lauberts M, Telysheva G. Fractionation of technical lignins as a tool for improvement of their antioxidant properties. Journal of Analytical and Applied Pyrolysis. 2013;**103**:78-85

[44] De Wild PJ, Huijgen WJ, Gosselink RJ. Lignin pyrolysis for profitable lignocellulosic biorefineries. Biofuels, Bioproducts and Biorefining. 2014;**8**(5):645-657

[45] Vanholme R, Demedts B, Morreel K, Ralph J, Boerjan W. Lignin biosynthesis and structure. Plant Physiology. 2010;**153**(3):895-905

[46] Neutelings G. Lignin variability in plant cell walls: Contribution of new models. Plant Science. 2011;**181**(4):379-386

[47] Argyropoulos DS. U.S. Patent No. 9,340,426. Washington, DC: U.S. Patent and Trademark Office; 2016

[48] Vishtal AG, Kraslawski A. Challenges in industrial applications of technical lignins. BioResources. 2011;**6**(3):3547-3568

[49] Zakzeski J, Jongerius AL, Bruijnincx PC, Weckhuysen BM. Catalytic lignin valorization process for the production of aromatic chemicals and hydrogen. ChemSusChem. 2012; **5**(8):1602-1609

[50] Piazza GJ, Lora JH, Garcia RA. Flocculation of high purity wheat straw soda lignin. Bioresource Technology. 2014;**152**:548-551

[51] Horáček J, Homola F, Kubičková I, Kubička D. Lignin to liquids over sulfided catalysts. Catalysis Today. 2012;**179**(1):191-198

[52] Vinardell MP, Ugartondo V, Mitjans M. Potential applications of antioxidant lignins from different sources. Industrial Crops and Products. 2008;**27**(2):220-223

[53] Mitjans M, Vinardell MP. Biological activity and health benefits of lignans and lignins. Trends in Comparative Biochemistry & Physiology. 2005;**11**:55-62

[54] Tay CH, Pugh BT, Clough SR, Magee BH. Dermal irritation assessment of three benzene sulfonate compounds. International Journal of Toxicology. 2004;**23**(1):11-16

[55] Baurhoo B, Letellier A, Zhao X, Ruiz-Feria CA. Cecal populations of lactobacilli and bifidobacteria and *Escherichia coli* populations after in vivo Escherichia coli challenge in birds fed diets with purified lignin or mannanoligosaccharides. Poultry Science. 2007; **86**(12):2509-2516

[56] Nandi S, Maurer JJ, Hofacre C, Summers AO. Gram-positive bacteria are a major reservoir of Class 1 antibiotic resistance integrons in poultry litter. Proceedings of the National Academy of Sciences of the United States of America. 2004;**101**(18):7118-7122

[57] Kumor LW, Olkowski AA, Gomis SM, Allan BJ. Cellulitis in broiler chickens: Epidemiological trends, meat hygiene, and possible human health implications. Avian Diseases. 1998;**42**:285-291

[58] Lee KW. Essential oils in broiler nutrition [doctoral dissertation]. Uttercht University; 2002

[59] Bozin B, Mimica-Dukic N, Simin N, Anackov G. Characterization of the volatile composition of essential oils of some Lamiaceae spices and the antimicrobial and antioxidant activities of the entire oils. Journal of Agricultural and Food Chemistry. 2006;**54**(5):1822-1828

[60] Oussalah M, Caillet S, Lacroix M. Mechanism of action of Spanish oregano, Chinese cinnamon, and savory essential oils against cell membranes and walls of *Escherichia coli* O157: H7 and *Listeria monocytogenes*. Journal of Food Protection. 2006;**69**(5):1046-1055

[61] Dong X, Dong M, Lu Y, Turley A, Jin T, Wu C. Antimicrobial and antioxidant activities of lignin from residue of corn stover to ethanol production. Industrial Crops and Products. 2011;**34**(3):1629-1634

[62] Wu C, Chen F, Wang X, Kim HJ, He GQ, Haley-Zitlin V, Huang G. Antioxidant constituents in feverfew (*Tanacetum parthenium*) extract and their chromatographic quantification. Food Chemistry. 2006;**96**(2):220-227

[63] Farrell AE, Plevin RJ, Turner BT, Jones AD, O'hare M, Kammen DM. Ethanol can contribute to energy and environmental goals. Science. 2006;**311**(5760):506-508

[64] Lau MW, Dale BE. Cellulosic ethanol production from AFEX-treated corn stover using Saccharomyces cerevisiae 424A (LNH-ST). Proceedings of the National Academy of Sciences. 2009;**106**(5):1368-1373

[65] Abajo C, Boffill MÁ, del Campo J, Méndez MA, González Y, Mitjans M, Vinardell MP. In vitro study of the antioxidant and immunomodulatory activity of aqueous infusion of *Bidens pilosa*. Journal of Ethnopharmacology. 2004;**93**(2):319-323

[66] Miki M, Tamai H, Mino M, Yamamoto Y, Niki E. Free-radical chain oxidation of rat red blood cells by molecular oxygen and its inhibition by α-tocopherol. Archives of Biochemistry and Biophysics. 1987;**258**(2):373-380

[67] Pan X, Kadla JF, Ehara K, Gilkes N, Saddler JN. Organosolv ethanol lignin from hybrid poplar as a radical scavenger: Relationship between lignin structure, extraction conditions, and antioxidant activity. Journal of Agricultural and Food Chemistry. 2006;**54**(16): 5806-5813

[68] Thakur VK, Thakur MK. Recent advances in green hydrogels from lignin: A review. International Journal of Biological Macromolecules. 2015;**72**:834-847

[69] Davin LB, Lewis NG. Lignin primary structures and dirigent sites. Current Opinion in Biotechnology. 2005;**16**(4):407-415

[70] Samal SK, Fernandes EG, Corti A, Chiellini E. Hybrid polymeric composites based on polyethylene and lignin. International Journal of Materials and Product Technology. 2009; **36**(1-4):62-72

[71] Domenek S, Louaifi A, Guinault A, Baumberger S. Potential of lignins as antioxidant additive in active biodegradable packaging materials. Journal of Polymers and the Environment. 2013;**21**(3):692-701

[72] Berlinet C, Ducruet V, Brillouet JM, Reynes M, Brat P. Evolution of aroma compounds from orange juice stored in polyethylene terephthalate (PET). Food Additives and Contaminants. 2005;**22**(2):185-195

[73] Van Bree I, De Meulenaer B, Samapundo S, Vermeulen A, Ragaert P, Maes KC, De Baets B, Devlieghere F. Predicting the headspace oxygen level due to oxygen permeation across multilayer polymer packaging materials: A practical software simulation tool. Innovative Food Science & Emerging Technologies. 2010;**11**(3):511-519

[74] Antunez PD, Botero Omary M, Rosentrater KA, Pascall M, Winstone L. Effect of an oxygen scavenger on the stability of preservative-free flour tortillas. Journal of Food Science. 2012;**77**(1)

[75] Hwang SW, Shim JK, Selke SE, Soto-Valdez H, Matuana L, Rubino M, Auras R. Poly(L-lactic acid) with added α-tocopherol and resveratrol: Optical, physical, thermal and mechanical properties. Polymer International. 2012;**61**(3):418-425

[76] Ouyang W, Huang Y, Luo H, Wang D. Poly(lactic acid) blended with cellulolytic enzyme lignin: Mechanical and thermal properties and morphology evaluation. Journal of Polymers and the Environment. 2012;**20**(1):1-9

[77] Arvanitoyannis IS, Bosnea L. Migration of substances from food packaging materials to foods. Critical Reviews in Food Science and Nutrition. 2004;**44**(2):63-76

[78] Blois MS. Antioxidant determination by the use of a stable free radical. Nature. 1958;**181**: 1199-1200

[79] Sanchez L, Mitjans M, Infante MR, Vinardell MP. Assessment of the potential skin irritation of lysine-derivative anionic surfactants using mouse fibroblasts and human keratinocytes as an alternative to animal testing. Pharmaceutical Research. 2004;**21**(9): 1637-1641

[80] Ugartondo V, Mitjans M, Lozano C, Torres JL, Vinardell MP. Comparative study of the cytotoxicity induced by antioxidant epicatechin conjugates obtained from grape. Journal of Agricultural and Food Chemistry. 2006;**54**(18):6945-6950

Chapter 9

An Overview on the Use of Lignin and Its Derivatives in Fire Retardant Polymer Systems

Neeraj Mandlekar, Aurélie Cayla, François Rault,
Stéphane Giraud, Fabine Salaün,
Giulio Malucelli and Jin-Ping Guan

Additional information is available at the end of the chapter

http://dx.doi.org/10.5772/intechopen.72963

Abstract

Lignin is a highly abundant bio-polymeric material that constitutes cellulose one of major component in cell wall of woody plants. Alternatively, large quantity of lignin is yearly available from numerous pulping and paper industries; this is the key point that justifies its large use for industrial applications. Lignin could be one of the most essential and sustainable bio-resources as raw material for the development of environmentally friendly polymer composite. Owing to its huge chemical structure, lignin can provide additional functionality such as filler, reinforcing agent, compatibilizer, stabilizer, etc. In this study, the fire retardant functionality of lignin has been employed in polymeric materials. Due to high charring capability, lignin is effectively used as carbon source in combination with other flame retardants for designing the intumescent system for polymeric materials. Further in this, several articles related to lignin-based intumescent are reviewed and interesting work formulation as well as meaningful results achieved in the flame retardancy are discussed. More attention is given to the studies concerning the use of current intumescent systems for textile applications by means of coating on fabric/nonwoven and melt blending in bulk polymers.

Keywords: lignine, textile, flame retardancy

1. Introduction

Nowadays the growing interest about a green and sustainable development has contributed to recall attention to biomass and specifically to lignocellulosic feedstock as a promising, renewable and vast resource as chemical component, in this context, lignin represents a component of renewable raw material that is available in sufficient quantity and is not in direct competition with food production. After cellulose, lignin is the second most abundant polymer from biomass and the main one based on aromatic units [1–3]. It has been reported that worldwide more than 50 million tons of lignin is generated every year as a by-product from pulp and paper-making industries, however less than 2% is actually recovered for utilization as a chemical product and the rest is considered as waste and primarily burned for recovering energy [4, 5]. Therefore, the comprehensive utilization of lignin remains a big challenge. Many efforts have been done to increase the use of lignin into added value added applications [5–7].

In addition to the low-value lignin application, a wide diversity of high-value industrial applications has been industrially realized or demonstrated including uses as novel materials, polymeric, oligomeric and monomeric feed stock. Some of these opportunities, such as the use of lignin or its derivatives in animal feed additives, agriculture, construction, textile, oil drilling, binders, dispersants and composites, are today commercial realities but many others such as the production of carbon fibre precursors, the broad incorporation of lignin in synthetic polymeric blends and textiles or the production of BTX remain longer term opportunities with great value and market potential. In recent years, lignin has been exploited in various polymer composite applications, for example, stabilizing agents, lubricants, coatings, plasticizers, surfactants, carbon fibre, fire retardant (FR), etc. For example, Movil et al. [8] reported the generation of hydrogen through electrochemical oxidation of waste lignin from pulping mills for energy storage. Qin et al. [9] grafted hydrophilic side chains on the sulphonated alkali lignin with different molecular weights, and the resultant product could be used as dispersants for coal-water slurry. Thunga et al. [10] investigated lignin as a suitable precursor for carbon fibres. Yu et al. [11] modified lignin with phosphorus nitrogen molecules to improve thermal stability and fire retardancy of polypropylene.

In recent time, lignin has received great attention as bio-based flame retardant additive because of its high char yield after decomposition from its aromatic framework. It is investigated that the presence of lignin can effectively reduce the flammability of polymers such as polypropylene, PBS, ABS, PET, etc. Further, it is reported that thermal degradation of lignin generates very high amounts of char, around 35–40% [12]. It is well known that the ability to form char during the thermal degradation is a basic aspect of flame retardant intumescent systems. Upon heating, fire retardant (FR) intumescent materials form a foamed cellular charred layer, which protects the underlying material from the action of heat flux and flame. The proposed mechanism is based on the charred layer acting as a physical barrier, which slows down heat and mass transfer between the gas and the condensed phase [13, 14]. Generally, intumescent formulations contain an acid source, a carbon source and a blowing agent. Several studies have been demonstrated that compound like bio-based lignin act as carbon source blended with acidic source such as ammonium polyphosphate (APP), melamine polyphosphate (MPP), metal phosphinate, etc., can be used as intumescent systems for polymer to develop materials with enhanced fire retardant properties [15, 16].

In the pursuit of non-halogenated flame retardant textile, lignin-based intumescent could be considered an efficient system for textiles materials. To provide fire retardancy to textiles, it could be interesting to use lignin-based intumescent at different level of textile structure. In one approach, lignin could be incorporated with flame retardant additives in polymers before the processing of the fibres. In other approach, lignin-based fire retardant formulations could be used as surface treatments by the padding, layer by layer (LBL) and back-coating of on textile structures. Both these approaches are discussed in detail further in the literature.

http://dx.doi.org/10.5772/intechopen.72963

The aim of this review is to investigate recent researches based on lignin as fire retardant additive for imparting flame retardancy to polymer composites. Intumescent systems based on lignin also have been investigated for textile materials when (1) used in bulk polymer matrix, (2) applied as thin coating or layer for fabrics, to confer fire retardancy. In addition, context also provides overview on different types of lignin, extraction processes and key player of global lignin market. Further different existing and emerging applications of lignin have been discussed briefly. Finally, the future perspective of lignin-based intumescent for imparting flame retardancy is also pointed out.

2. Chemical structure

Lignin plays a major role in woody plants, adding strength and structure to the cell walls, controlling fluid flow and protecting against biochemical stresses by inhibiting enzymatic degradation of other components [17]. It usually contributes between 15 and 40 wt.% of woody plants dry matter [18]. Its chemical structure consists of phenyl propane units, originating from three aromatic alcohol precursors (monolignols) such as p-coumaryl, coniferyl and sinapyl alcohols. The phenolic substructures that originate from these monolignols are called p-hydroxyphenyl (**H**, from coumaryl alcohol), guaiacyl (**G**, from coniferyl alcohol) and syringyl (**S**, from sinapylalcohol) moieties (**Figure 1**). Lignin composition and content are influenced by the species and also by the environment. Hardwood lignins consist principally of **G** and **S** units and traces of **H** units, whereas softwood lignins mostly comprise **G** units, with low levels of **H** units [19]. The variability of lignin composition is much greater in hardwoods than in softwoods.

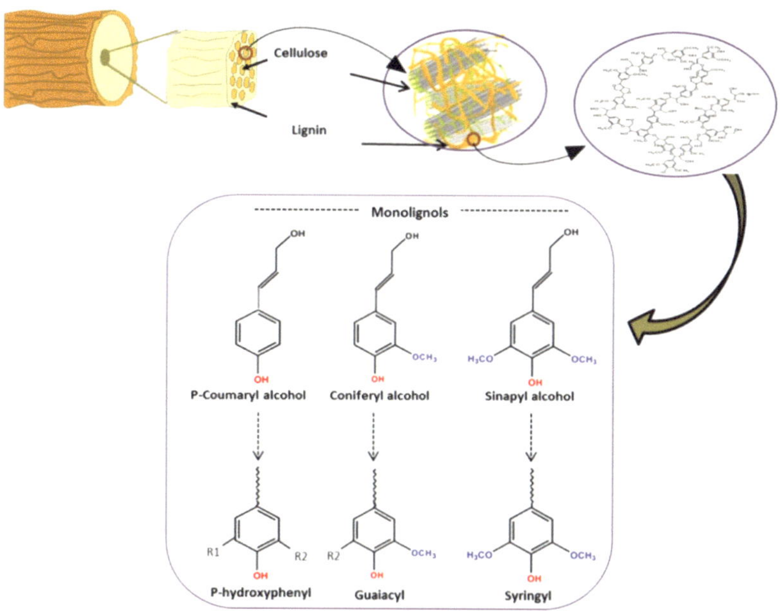

Figure 1. The three main precursors of lignin (monolignols) and their corresponding structures in lignin polymers [19].

Based on the first full lignin structure proposed by Adler in 1977 [20], lignin is recognized as a highly branched heteropolymer without any overall defined chemical structure (**Figure 2**). Lignin contains a variety of functional groups; the most common ones are aromatic methoxyl and phenolic hydroxyl, primary and secondary aliphatic hydroxyl and small amount of carbonyl group. The monomeric C9 unit is linked to form a polymer by C–O–C and C–C linkage. The most abundant lignin interunit linkage is the β-O-4 ether type of linkage, comprising about 50% of the interunit linkages in lignin (ca. 45% in softwoods and up to 60% in hardwoods). Other common lignin interunit linkages are α-O-4 type, resinol (β-β′), phenylcoumaran (β-5), 5-5′and 4-O-5 moieties (**Figure 2**). The number of these structures varies in different lignins.

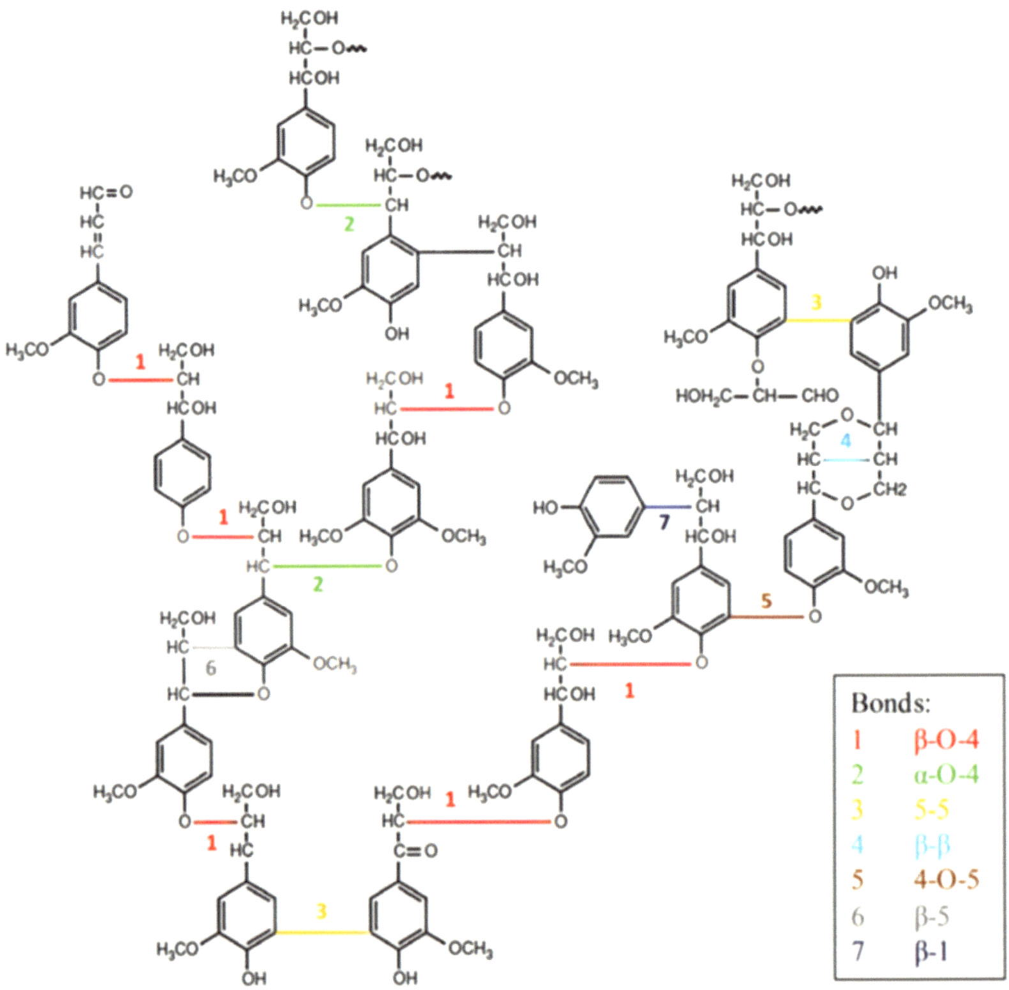

Figure 2. Main linkages in a softwood lignin.

3. Extraction processes of lignin and their resulting technical lignin

Lignin is extracted from the other lignocellulosic parts by physical and/or chemical and biochemical treatments. Technical lignin can be classified based on different point of view (**Table 1**). Common pulping processes are based on the cleavage of ester and ether linkages. It is worth noticing that resulted technical lignin differ considerably from the plant lignin. In this part, different commercially available extraction processes are discussed, which are used to recover technical lignins. **Figure 3** shows the classification into two main categories such as sulphur-bearing and sulphur-free processes.

3.1. Sulphur-bearing process

Sulphur lignin includes kraft and lignosulphonates lignin, which are primarily produced by pulp and paper industries and mainly correspond to the lignin extraction from the cellulose.

Lignin type	Scale	Chemistry	Sulphur content	Purity
Kraft	Industrial	Alkaline	Low	High
Lignosulphonate	Industrial	Acidic	High	Low
Organosolv	Industrial/Pilot	Acidic	Free	High
Soda	Industrial/Pilot	Alkaline	Free	Moderate
Hydrolytic	Industrial/Pilot	Acidic	Low/Free	Moderate

Table 1. Classification of technical lignins.

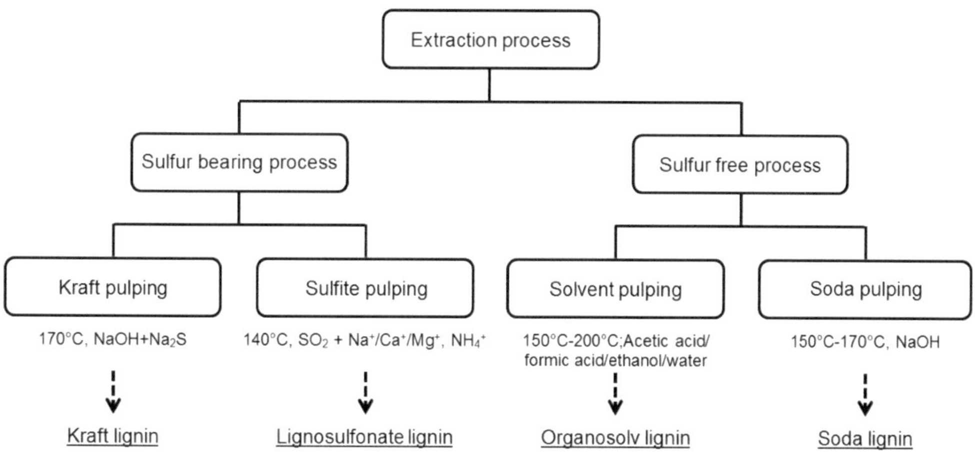

Figure 3. Different extraction processes to separate lignin from lignocellulosic biomass and the corresponding technical lignins.

3.1.1. Kraft process

The kraft process uses a mixture of chemicals including sodium hydroxide (NaOH) and sodium sulphide (Na_2S), known as white liquor, which breaks the bonds that link lignin to the cellulose. It basically converts wood or non-wood material into pulp and the generated black liquor that is then acidified to recover lignin. Considering the high sulphur environment used for kraft lignin extraction, it is quite surprising that the residual sulphur content is so low, typically below 2–3%. Moreover, it contains a high amount of condensed structures and a high level of phenolic hydroxyl groups [5], due to extensive cleavage of β-aryl bonds during cooking. The number-average molar mass (M_n) of kraft lignin is generally low, between 1000 and 3000 g mol^{-1} [21].

3.1.2. Sulphite process

The most abundant type of commercial lignin available is sulphite lignin because traditionally this was the most used type of pulping process. It is based on a cooking with an aqueous sulphur dioxide (SO_2) and a base calcium, sodium, magnesium or ammonium base. Lignosulphonates contain a considerable amount of sulphur in the form of sulphonate groups present on the aliphatic side chains. Lignosulphonates are water-soluble. They have a higher average molar mass than kraft lignin with a broad polydispersity index, around 6–8 [22]. However, they are generally contaminated by the cations used during pulp production and recovery. Their reactivity depends to some extent on the cation. Calcium- and ammonium-based products exhibit the lowest and the highest reactivity, respectively, while sodium- and magnesium-based lignosulphonates show a medium reactivity [23].

3.2. Sulphur-free process

Sulphur-free lignins are an emerging class of lignin products which have a low macromolecular size, after fractionation steps. The structure of these lignins is close to those of the native lignins. They show interesting properties that can make them an attractive source of low-molar mass phenol or aromatic compounds. Sulphur-free lignins can be divided into two main categories such as lignins from solvent pulping (organosolv lignin) and from alkaline pulping (soda lignin).

3.2.1. Organosolv process

Organosolv lignins are generally the purest ones, with the highest quality [24]. They show high solubility in organic solvents and practically insoluble in water, since they are very hydrophobic. They are recovered from the solvent by precipitation, which typically involves adjusting different parameters, such as concentration, pH and temperature. The most common organosolv processes are based on ethanol/water pulping (e.g., Alcell) and pulping with acetic acid, containing a small amount of mineral acid such as hydrochloric or sulphuric acid [25]. In addition, another extraction based on a mixture of formic acid, acetic acid and water was developed by CIMV Company (France). The lignin produced was called Bio-lignin© and is thought to be linear and have low molecular weight according to some published work [26, 27]. However, some of these isolation concepts, such as the Alcell process, using aqueous ethanol, are now being commercialized as a bio-refinery technology whereby the cellulose fraction is being used for ethanol production [28].

3.2.2. Soda process

Soda lignin originates from soda or soda-anthraquinone pulping processes. Soda-based cooking method are mainly used for annual plants such as straw, flax, bagasse, and to some extent hardwoods [29, 30] . Lignin extraction is based on hydrolytic cleavage of the native lignin but it results in a relatively chemically unmodified lignin compared to the others lignin types. A specific procedure for the precipitation of lignin in the black liquors of soda processes, which specially adapted from paper factories in the production of cellulose from annual plants or agricultural residual substances, is the Granit process, where the pH value of the liquor is lowered by acidification, typically with mineral acids. The recovery of spent cooking liquor from pulping of non-woods is not without problem, because of high content of silica which may co-precipitate with the lignin. So this lignin presents a low quality [31, 32]. In the meantime, companies have claimed to have solved this problem and to have obtained soda lignins with low ash and silica content [33].

Various alternative pulping processes such as hydrothermal, steam explosion, ionic liquids (ILs), enzymatic, etc. have been developed but failed for industrial or large-scale implementation due to several reasons. There are many efforts and new approaches searching for new technologies and enhancing the value of lignin and lignin products [34].

4. Global lignin market scenario

It is excepted that depending on the plant source and the corresponding structure, lignin and lignin derivatives are offered in quite different qualities for a multitude of applications. Isolated lignin is currently used in a wide range of low volume, niche applications such as concrete additive, animal feed, dye stuff and other applications (including cosmetics and absorbents). In addition to this, lignin is a key intermediate for the production of activated carbon, carbon fibres, vanillin, sorbents, plastic materials, phenol, binders and motor fuel. These applications have tremendous scope in the future market. There is an increasing demand for lignin in developing countries such as China due to the country's expanding industrial base [35].

Globally, major companies active in the lignin and lignin-based products industry, with top 10 players accounting for over 90% share of the global lignin market in 2015 [36]. The lignin market is highly concentrated, with top five players accounting for more than 85% share of the global market in 2015. Key players operating in the lignin market include Borregaard LignoTech, MeadWestvaco Corporation, Tembec Inc., Domtar Corporation, Domsjö Fabriker AB, Changzhou Shanfeng Chemical Industry Co. Ltd, Nippon Paper Industries Co. Ltd., UPM-Kymmene Corporation, Asian Lignin Manufacturing Pvt. Ltd., Lignol Energy Corporation, Green Value and Liquid Lignin Company (LLC).

At present, lignosulphonates accounted for the largest share of the global lignin market due to rising demand for lignin on the building and construction industry, are produced majorly by Borregaard (Norway) and Tembac (Canada), but there are also minor suppliers, for example, Nippon paper Industries (Japan), ALMT (India) and Domsjo AB. Kraft lignin is manufactured by MeadWestvaco (MWV) corporation (USA) and Domtar (Canada). Organosolv lignin are produced based on ethanol/water pulping (Lignol©, Canada) [28, 37], another extraction based on a mixture of formic acid, acetic acid and water (Bio-lignin©) was developed by CIMV Company (France) [38]. The degree of current utilization shows that at present only the potential of the lignin from sulphite pulping is utilized. The highest possible potential, however, has the lignin of the sulphate process due to the production volume.

The recent interest in lignin as renewable raw material feedstock is evidenced by the growing number of research documents containing the word "lignin application" which have been published in last 20 years (**Figure 4**). Especially during the last 10 years, more interest was developed on exploitation of lignin due to demand to promote "green concept". Several lignin market studies proposed high demand of lignin in forthcoming years. Zion Research has published a new report "Lignin Market" [39] for various applications; Global Industry Perspective, Comprehensive Analysis and Forecast 2014–2020. According to the report, global lignin market was valued at approximately USD 775 Million in 2014 and is expected to reach around USD 900 million in 2020, growing at a CAGR of slightly above 2.5% between 2015 and 2020. Global lignin market was dominated by the North America region followed by Europe. The lignin market in Europe region is expected to grow at a rapid pace in the years to come. The Europe region is followed by the Asia Pacific and Latin America, respectively. Asia Pacific is expected to be the fastest growing market for lignin during the forecast period

Most of the 70 million tons of lignin yearly occurring as a by-product in pulping is used for energy generation during sulphate pulping, because of the need to recover and recycle the inorganic cooking chemicals. For this reason, less than 2% of lignin and its by-products are materially utilized. Lignosulphonates accounted for the largest share of the global lignin market (1,000,000 tons per annum) [4, 40] due to rising demand for lignin on the building and construction industry. Demand for kraft lignin has increased (100,000 tap) due to growing demand for dye stuff in the paints and coatings and textile industries. Additionally, rising usage of kraft lignin as insecticides has contributed to the growth of global lignin market. Sulphur-free lignin is not yet commercialized in higher amounts (<5 kton).

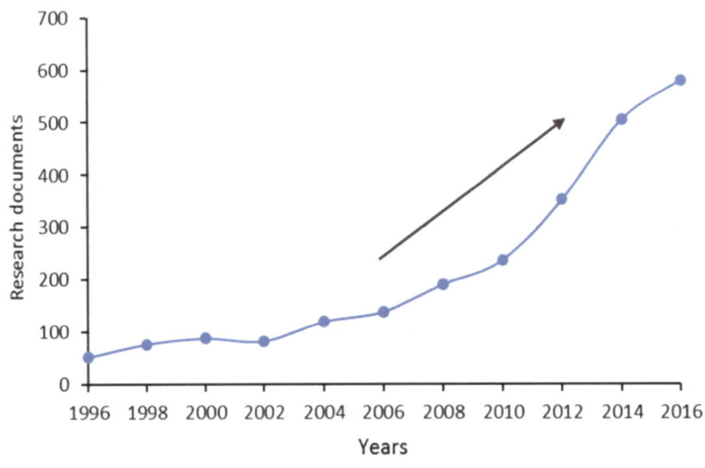

Figure 4. Research documents devoted to lignin applications in different areas published during 1996–2016 (Scopus® database with key word "lignin applications").

5. Industrial lignin applications

Lack of toxicity and versatility of lignin creates several potential industrial application routes. Stringent regulations, bulk availability, cost efficiency and growing need for bio-based and renewable chemicals are high-value lignin properties. Significant scope for diverse applications (**Figure 5**) principally segmented as power/energy, macromolecules and aromatics. Immense scope and opportunities to turn lignin into biofuel as companies are readily investing in the research and development for related applications. Overall, lignin applications can be divided into low or mid value and high-value applications. The low-value applications of lignosulphonates and sulphonated kraft lignins such as dust control, concrete admixtures and oil well drilling muds while most of the high-value applications target emerging applications, in which the chemical versatility of lignin could be fully utilized, or very large volume applications such as the production of BTX and other petrochemicals.

5.1. Low-value traditional applications

The traditional lignin applications include, primarily, lignin uses where lignin plays a replacement role for a relatively low-value chemical or material. As it was mentioned earlier, the largest current use of industrial lignin is fuel. However, there are several matured industrial

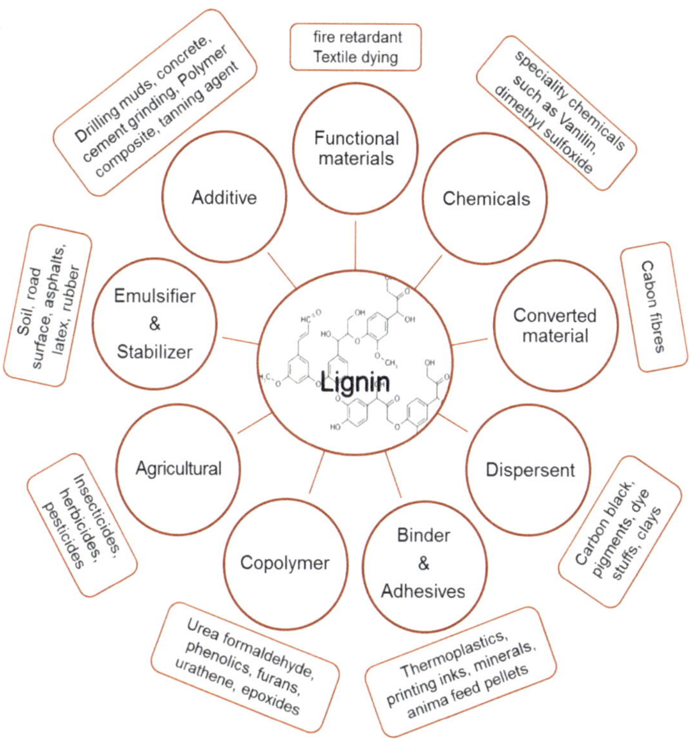

Figure 5. Current and potential applications of technical lignin.

lignin applications, constituting the bulk of the higher value lignin commercial chemical market, and this includes, primarily, lignosulphonates or sulphonated kraft lignins in large-medium-size markets such as additives for concrete admixtures, dust control, feed and food additives, dispersants, resin and binder compositions and oil well drilling. Examples of smaller market for technical lignins include carbon black, emulsifiers, water treatment, cleaning chemicals, leather tanning, battery expanders and rubber additives [41]. There are many reviews and literatures available, which cover traditional lignin applications [42] therefore, not discussed here.

5.2. High-value emerging lignin applications

From 1990 to 2010, several novel lignin applications have been proposed and described in the scientific and patent literature. Some of these novel applications have been demonstrated at larger than laboratory scale. For instance, the production of lignin-based carbon fibre (LCF) [10, 43, 44] is one of the brightest examples of successful lignin upgrading technology which has been scaled up to pilot scale. Another example of nontraditional lignin application is the use of upgraded technical lignins in polymers composites. For instance, depolymerization of lignins to produce valuable oxygenated aromatic compounds, and possibly olefins too, in replacement of petrochemicals [45, 46]. This section briefly summarizes these three relevant examples of emerging nontraditional lignin applications which in our opinion present the largest potential, in terms of volume and value, for commercialization of technical lignins in the future. Among all the emerging lignin applications, utilization of lignin's functionality in polymer composites is perhaps one with the large market potential in future lignin demand.

6. Fire retardant applications in polymer composite

For environmental concerns, halogen-free flame retardant (HFFR) has stimulated great attention in recent years due to the evolution of corrosive and toxic gases during the combustion of halogen-containing flame retardant materials [47]. The presence of lignin gives peculiar properties to the composite. Lignin can act as a stabilizer preventing polymer aging due to its antioxidant activity [48–50]. Lignin has the advantage of being a bio-based compound and can be effectively used as carbon source for the design of intumescent systems in combination with other flame retardant additives. Owing to the presence of highly aromatic structure lignin is able to produce a large amount of char residue upon heating at elevated temperature [12, 51], which reduces the combustion heat and heat release rate (HRR) of polymeric materials and thus enhances the fire retardancy. All forthcoming reviewed study have demonstrated that lignin-based intumescent could be a worthy alternative that represents a great advantage in comparison with the other proposed fire retardant system.

6.1. Lignin-based intumescence in synthetic polymers

Polypropylene is one of those most versatile polymers available with applications, both as a plastic and as a fibre. Its low density, low cost and easy processability makes it suitable for applications in automobiles, packaging and textile industry. However, PP has some drawbacks such as thermal stability, high flammability, thermal oxidation, etc. There are several ways to improve fire and thermal properties but incorporation of fillers into the polymer matrix has been extensively used to improve the mentioned properties. About decade ago in first study, A. De Chirico et al. [16] used lignin as charring agent with other traditional phosphorus flame retardants for improving the flame retardancy of polypropylene, each formulation containing 14 wt.% lignin and 6 wt.% of other flame retardants. TGA and cone calorimetry study showed that combination of lignin in intumescent formulation could prolong the combustion time, enhance the char residue and reduces the heat release rate during burning. Further, the same research group utilized direct lignin in polypropylene matrix to study thermal stability and fire performance of PP composite [52]. Polypropylene blends containing 5 and 15 wt.% of lignin was prepared by mixing in a screw mixer. Thermogravimetric analysis presented, increase in the thermal degradation temperature of the blends as a function of lignin content. The increase is more pronounced for the experiments carried out in air atmosphere, where the interactions between the PP and the charring lignin lead to the formation of a protective surface shield able to reduce the oxygen diffusion towards the polymer bulk. Another different study proposed by Yu et al. [11], wherein alkali lignin was modified by chemical grafting of flame retardant phosphorous and nitrogen-containing macromolecules, modified lignin (PN-lignin) blended with polypropylene polymer matrix to give enhanced fire property. Cone calorimetry study showed PN-lignin further reduces the peak heat release rate (−45%) and slows the combustion process, modified lignin (PN-lignin) also exhibits a much higher char-forming ability with a char of 61.4 wt.% (40.7 wt.% for lignin) at 600°C in N_2. Similar strategy is used by Liu et al. [53] to fabricate flame retardant wood plastic PP composite (WPC). Lignin was chemically modified by grafting of N-P then coordinated with metal ions. Wood-PP (WP/PP) and wood-PP-lignin (WP/PP/lignin) composites were fabricated via melt compounding using a rheomix at 180°C for 10 min. Flammability of samples was tested using UL94: the obtained results reported that adding 15 wt.% F-lignin resulted in a V-1 rating showing better FR properties than PP/WP composite. Compared with pure lignin incorporation of equal loading of F-lignin into WPC can further increase the thermal stability of PP/WP composite. Cone tests showed that optimum content of F-lignin in PP/WP shows 15 wt.%, which gave the best FR result by reducing PHRR, THR and MARHE. Moreover, smoke production was also reduced by 30 wt.%. The study by Acha et al. [54] present combination of lignin and jute fabric for impart fire retardancy and toughening to PP matrix. Different formulation of kraft lignin blended with PP in mixer at 180°C for 10 min and film were obtained by compression moulding and then jute fabric was sandwiched between the film and compression moulded PP at 180°C for 25 min. The blends showed higher degradation temperature compared to neat PP and with increased char residue. Char residue produced at degradation was directly related to lignin content in blend. With increasing lignin content time to ignition (TTI) decreases, but with increased char yield. Addition of jute fabric to PP leads to significant increase in stiffness of composite due to higher modulus of jute fibre. Jute fabric also improves tensile strength and impact behaviour. In different study Song et al. [55] investigated the synergistic effect of lignin for acrylonitrile-butadiene-styrene copolymer (ABS): improved flame retardancy was revealed with 20 wt.% of lignin. Cone calorimetry test demonstrated 20 wt.% lignin causing a 32% reduction in peak heat release rate. Further addition of styrene ethylene-co-butadiene styrene-grafted maleic anhydride (SEBS-g-MA) as in situ reactive compatibilizer for ABS/ lignin composite further reduces peak heat release rate (PHRR), indicating enhanced flame retardancy due to the improved interfacial adhesion. The *in situ* reactive compatibilization with SEBS-g-MA also contributes protective char layer formation. Thermal degradation and melting behaviour of PET was investigated by Canetti and Bertini [56]. Hydrolytic lignin was directly blended with PET thermoplastic using a single screw extruder. Thermal decomposition, melting behaviour and crystallinity of the resulting blend was studied by thermogravimetry, differential scanning calorimetry and X-ray diffraction, respectively. TG analysis results show that the introduction of lignin leads char formation and effective char layer (~27 wt.%) formed with 20 wt.% lignin content, evidenced of barrier effect. Further presence of lignin in blends promotes the crystallization process and induces a faster crystalline reorganization than that of the pure PET. The reduction in the entropy of melting due to a modification of the amorphous phase must be considered as responsible of the melting behaviour. Research group from China studies lignin-modified PU foam as a single component with excellent flame retardancy [57]. Corn straw lignin was first grafted with phosphate-melamine containing group to prepare a compound named as LPMC and then different wt.% ratio (5, 10, 15 and 20 wt.%) of LPMC were copolymerized with isocyanate to produce lignin-modified PU foam. TGA study showed initial degradation temperature of modified PU-LPMC foam that is slightly lower than that of pure PU, which accelerates intumescent action. In addition, the presence of LPMC could promote the generation of non-flammable gases during PU degradation, inhibiting the flame propagation and dehydration of PU to form compact char layer. Further char formation greatly increases with increasing LPMC content. Furthermore, flame retardancy assessed by limiting oxygen index (LOI) and UL-94 flame spread test illustrate, self-extinguishment and inhibition from melt dripping with small amount of LPMC and V-1 rating could be achieved for PU-$LPMC_{15}$ foam.

6.2. Lignin-based intumescence in bio-polymers

Recent years have seen the development of polymer made from renewable resources (bio-polymers). These materials represent an interesting alternative to traditional polymers due to their independence towards petroleum and, in some cases, they offer a different end of life scenario (bio-degradation). In previous years, bio-polymers were essentially used for short life products, for example, packaging. Now durable applications can be visualized. However, to penetrate some industrial sectors, functional properties, like thermal stability and flame retardant properties must be improved. Few studies have been proposed to use lignin as an additive for bio-polymers to modify flame retardant properties and thermal stability and to get a completely "green" material. In this regard Zhang et al. [58] reported a novel intumescent flame retardant (IFR) consisting of microencapsulated ammonium polyphosphate (MCAPP) and lignin to confer fire retardancy to poly lactic acid (PLA). The IFRs formulation was made up of MCAPP, lignin and organic modified montmorillonites (OMMTs). The ratio of MCAPP to lignin was fixed at 3:1, and the loading of OMMTs were kept at 0.5, 1.0 and 2.0 wt.%, the total content of IFR in PLA composites

http://dx.doi.org/10.5772/intechopen.72963

was kept at 23 wt.%. PLA composite containing 21 wt.% MCAPP/lignin and 2 wt.% OMMT showed the best LOI value of 35.3 and UL-94 V-0 classification, and exhibited a remarkable enhancement of the flame retardancy. For the same blend, the reduction of PHRR was 79% and that of THR was 60% as assessed in cone calorimetry test. In addition, the char residue analysis clearly showed that the incorporation of OMMTs can improve the char quality with a much more compact and continuous morphology. Alternatively, Reti et al. [59] studied lignin-based intumescent to provide flame retardant PLA. 10 wt.% kraft lignin and 30 wt.% ammonium polyphosphate (APP) were blended with PLA and compared with same composition of PLA/APP/starch and PLA/APP/PER systems. Further flammability properties were assessed by limiting oxygen index (LOI) and UL-94 vertical flame test. Blends containing lignin exhibited LOI value as high as 32 vol.%, slightly lower than starch containing blends (40 vol.%) but acceptable to achieve self-extinction. However, UL-94 test revealed that the presence of lignin and starch leads to superior fire properties and V-0 rating was achieved. Cone calorimeter results demonstrate that lignin-based formulation under test condition form an efficient intumescent protective char layer structure, which further reduces the heat release rate (HRR) besides, the observed 47% reduction in peak HRR is higher than that of the starch containing (PHRR, −41%) blend. However, the best cone test results were obtained for PLA/APP/PER system. In other study, Bertini et al. [60] exploited rice husk lignin, isolated by means of acidolytic (AL) and alkaline enzymatic (AEL) extraction methods. The isolation methods provides lignin samples with significant differences among their molecular, thermal and chemical features. Bio-composites of poly(3-hydroxybutyrate) (PHB) with AL and AEL lignin were prepared by weight ratios of 97.5/2.5, 95/5, 90/10 and 85/15%. The morphological, structural and thermal characteristics of the bio-composites were extensively studied. TG analysis of the PHB-AL bio-composites showed an enhancement of the thermal resistance, being the thermal degradation process shifted to higher temperatures. The increase of thermal stability was observed as a function of the lignin amount in PHB-AL bio-composites. Further, Ferry et al. [61] used lignins (alkali and organosolv) as flame retardant to improve the fire behaviour polybutylene succinate (PBS) bio-polyester. Lignin was successfully modified by grafting of macromolecular phosphate molecule. 20% loading of modified lignin in PBS matrix, alkali lignin significantly reduces heat release rate, promotes thick charring behaviour and was proved to be more interesting than organosolv lignin due to the release of sulphur dioxide during decomposition. A study by Liu and coworker [62] explains modification of alkali lignin by grafting of phosphorus-nitrogen-zinc-containing macromolecule and blending of modified lignin (PNZn-lignin) with bio-degradable polybutylene succinate (PBS) to impart flame retardancy. Incorporation of PNZn-lignin can effectively increase to some extent the thermal stability of PBS composites by gradual shift of T_{max}. Adding 10 wt.% of PNZn-lignin, PHRR and THR are remarkably reduced by 51 and 68%, respectively. Meanwhile, the AMLR and TSR are also decreased, respectively, by 54 and 55% as compared with pure PBS. Impressively, the char residue is enhanced from 9 wt.% for pure PBS to 55 wt.% with 10 wt.% of PNZn-lignin content. The prescence of Zn (II) ions leads to a compact, thick and strong char layer and makes PBS to show reduced flammability and smoke release. **Tables 2** and **3** collect the most significant results based on these attempts.

Polymer	Formulation	Flame retardancy	Ref.
PP	20 wt.% total add-on (14 wt.% lignin + 6 wt.% $Al(OH)_3$/PVA/MeP/AHP/APP)	Synergism with other FRs, longer combustion time and reduced PHRR with phosphate FRs	[16]
PP	Lignin 5 and 15 wt.% add-on	Higher char yield with increased lignin content, protective char layer formed	[52]
PP/WPC	Up to 20 wt.% total add-ons of chemically functionalized lignin and wood	PHRR decreased by 9%, THR by 25%, AMLR by 19% and char residue increased by 30%	[53]
PP	Lignin up to 20 wt.% and jute fabric	Char yield increases with increasing lignin content which leads to improved FR properties	[54]
PP	Soda lignin grafted with N-P containing macromolecule up to 30 wt.%	PHRR reduction 45%, char residue 21% increased, compact char layer formation	[11]
ABS	Wheat straw alkali lignin up to 20 wt.% with SEBS-g-MA up to15 wt.%	PHRR reduction 32%, protective char layer formation increased	[55]
PET	Hydrolytic lignin up to 20 wt.%	Formation of protective char layer, degradation temperature increases	[56]
PU	Corn straw lignin grafted phosphate melamine compound up to 20 wt.%	Improvement in thermal stability, increased charring, self-extinguishment, inhibition of melt dripping and smoke production	[57]

Table 2. Lignin-based fire retardant system to improve fire retardancy.

Polymer	Formulation	Flame retardancy	Ref.
PLA	Alkali lignin + MCAPP 23 wt.% with OMMTs 0.5, 1 and 2 wt.%	LOI 35%, PHRR reduction 79%, UL94 V0 classification, THR reduction 60%	[58]
PLA	Kraft lignin/starch 10 wt.% + APP 30 wt.%	PHRR reduced to 47%, UL94 V0 classification LOI improved to 32%, protective layer formed	[59]
PHB	Lignin isolated by alkali enzymatic (AEL) and acidolytic (AL) process up to 15 wt.%	PHB-AL bio-composite showed higher thermal resistance than PHB-AEL bio-composite	[60]
PBS	Alkali lignin 10 wt.% + macromolecular phosphate-modified alkali lignin 10 wt.%	Significant drop of PHRR and THR promotes thick charring behaviour	[61]
PBS	Alkali lignin grafted with N-P-Zn containing macromolecule (PNZn-lignin) up to 10 wt.%	Reduction in PHRR 51% and TSR 55%, char residue enhanced by 54% from pure PBS	[62]

Table 3. Lignin-based fire retardant system to improve fire retardancy of biopolymers.

7. Fire retardant textile applications

The development of flame retardant material is very crucial research topic and even more when it is related to textile material (**Figure 6**). Textiles are everywhere in our day today life and serve not only as clothes, curtains, bedding, seat covers and carpets but also in public transports and cinema halls. Different countries like UK, US, Canada, Australia and some EU countries have some stringent flammability requirement for textiles used in home appliances, public transport and public places. These textiles are made of natural or synthetic fibres show

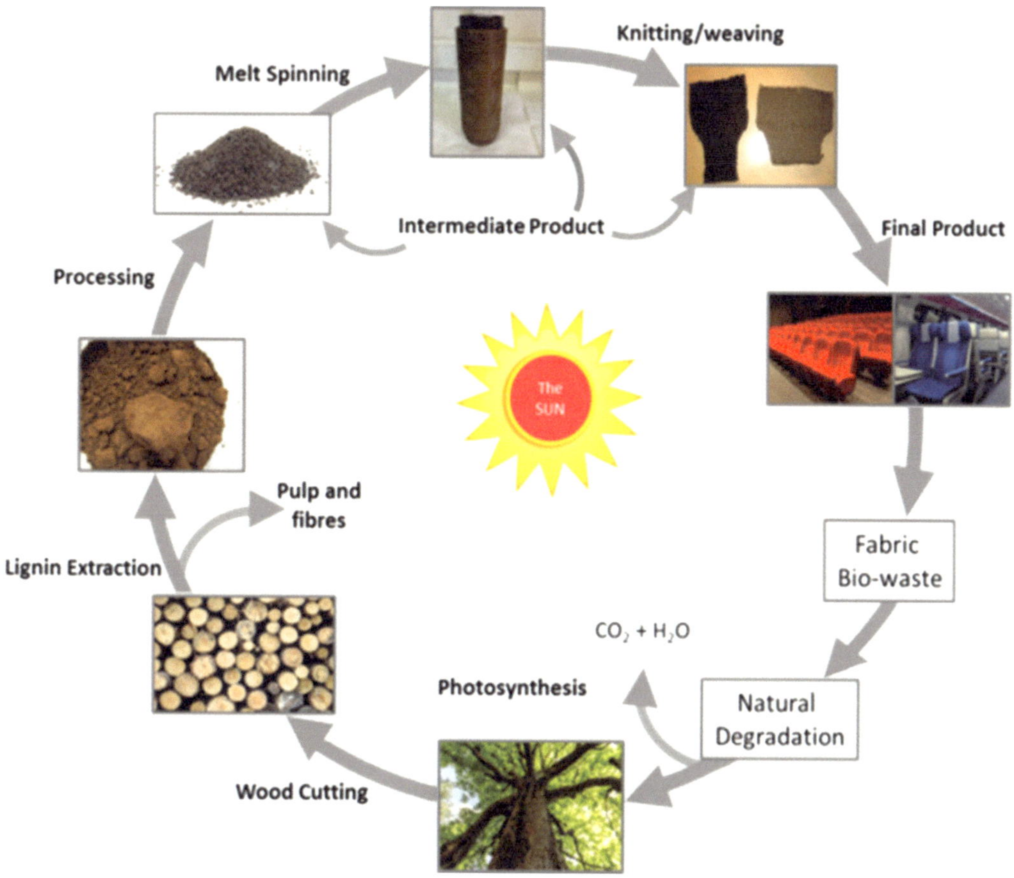

Figure 6. Life cycle of bio-based textile.

high burning rate due to their organic nature and the high specific surface further dripping effect make it more hazardous material. Therefore, a constant demand on fire protection is observed and even more since the toxicity of halogenated additives was highlighted [63]. As previously emphasized in this article, in the recent years researchers in the field of fire retardancy of polymeric substrates have geared up to find a sustainable approach to reduce the carbon foot print. In this connection, efforts have been extended, especially from the last 10 years, to use different eco-friendly bio-based extracted plant resources and other natural resources due to the desire to promote a "green concept". In this approach, a few researchers have started exploratory work to use a sustainable way to make the materials fire retardant and came out with good research findings where they used different bio-macromolecules as fire retardant alternatives instead of synthetic chemicals. The exciting results are documented in the scientific published articles [64–66]. In this direction, Alongi et al. used DNA bio-molecule to impart flame retardancy to cotton fabric [67]. Very recently, Basak et al. reported use of different bio-molecule such as banana pseudostem sap (BPS), spinach juice and hydrophobin for improving thermal stability and subsequent fire retardancy [68]. In another study, Wang et al.

developed novel eco-friendly chicken feather-based protein bio-molecule for imparting fire retardancy to cotton fabric [69]. In a different approach, the use of bio-based ingredients for the formulation of intumescent flame retardant system (IFR) has been proposed. Few studies proposed to replace the acid source by bio-based component: more specifically, fumaric acid and phytic acid were proposed as suitable candidates to develop intumescent system with PLA [70]. Bio-based carbonizing agents such as starch [71], cyclodextrin [72] and chitosan [73] have been proposed and showed high efficiency. Among those entirely bio-based carbonizing agents, lignin has also been considered by some researchers due to its abundance in nature and high charring capacity. Thus, very recently Cayla et al. have prepared fire retardant PLA fibres containing IFR composed of lignin and ammonium polyphosphate (APP) [74].

Two different well-known methods are discussed to confer flame retardancy to textile materials. These approaches are evidently easier to process as compared to synthesizing fire retardant fibres by copolymerization.

7.1. Fire retardant coating on textile surface

The first approach, which is widely used to provide low flammability to textiles, concerns the surface treatments such as padding, layer by layer (LBL) and back-coating of fire retardant intumescent formulation on textile structures during the finishing step. In the field of textile coating market, the standard formulation based on antimony III oxide (ATO) and brominated hydrocarbons, notably decabromodiphynyl ether (decaDBE) and hexabromocyclododecane (HBCD) are most commercially successful flame retardant formulations due to their low cost and durable flame retardancy effect. However, during the last 20 years environmental concerns have been raised to reduce their usage. The challenge of replacing these system by phosphorus-nitrogen-containing fire retardant system have been investigated [75, 76]. Consequently, phosphorous- and nitrogen-based fire retardant chemicals became dominant in the market because some of the phosphorous-/nitrogen-based species have high wash durability (even more than 50 washes). As a result, durable flame retardants based on the composition of phosphorous, nitrogen-like Tetrakis phosphonium chloride salt [proban® process] and N-alkyl phosphopropionamide [pyrovatex® and similar processes] derivatives are widely used commercially [77]. In fact, these chemical treatments bring stiffness in fabric due to high add-on%, are hazardous due to toxic chemical release, are expensive also because of a larger use of chemicals. Challenges associated to developments of flame retardant textile have been highlighted by Horrocks et al. in detail [77]. From that extensive review, it is quite clear that the major challenges for the flame retardant researchers are to develop more cost effective, environmental friendly and sustainable fire retardant chemicals. The first attempt to apply a bio-based intumescent coating to textile was carried out by Srikulkit et al. in 2006 [78], depositing multilayer of polyelectrolyte consisting of chitosan and poly(phosphoric acid) on silk. The results showed that an assembly consisting of 60 bilayers promotes the formation of a highly thermally stable char. Alternatively Laufer et al. [79] proposed fully bio-based intumescent based on phytic acid and chitosan applied to impart fire retardancy and reduced cotton PHRR by 50%. In the same scenario Alongi et al. reported DNA as an effective intumescent flame retardant for cotton fabrics as it contains all three components typical of an IFR formulation [67, 80]. Further by combining DNA with chitosan, the resulting assemblies (10 and 20 Bls) applied to cotton extinguished flame in horizontal flame spread tests and under a 35 kW/m^2 heat flux, produced a significant decrease of cotton PHRR and THR (−41 and −32%, respectively) [81]. Among those bio-based IFR components, lignin has also been considered as carbonizing agent by some researchers due to its abundant availability as raw material from pulp and paper making industries. In this context, Reti et al. [82] showed that intumescent films based on lignin are able to confer flame retardancy to nonwovens composed of hemp/wool. Films are elaborated with the formulations PLA/APP/LIG used as a potential flame retardant treatment on nonwoven. Intumescent film of about 250 μm thickness was applied on nonwoven using moulding press at 180°C. Horizontal and vertical flame spread measurements have shown no combustion when PLA FR films are coated on nonwovens. Self-extinguishable materials were obtained. In cone calorimetry, nonwoven covered with PLA/APP/Lignin film increased TTI by 50%, nonwoven treated with these coatings have around 40% of PHRR reduction. In the same scenario, work was performed in our lab as PET fabric was coated with intumescent solution [83]. Intumescent flame retardant coating based on lignin and APP in polyurethane (PU) were prepared by dispersing lignin (10 wt.%) and APP (10 wt.%) in PU solution. The dispersions were applied onto PET fabrics using a K Control Coater with a specific threaded rod to obtain a wet film of 100 μm at 2 m s^{-1}. The coated textile was then cured at 155°C for 4 min for crosslinking. Fire test performed using cone calorimetry with 25 kW/m^2 revealed that intumescent coating of APP/lignin/PU on PET fabric shows good result with increasing the number of coating layers. Reduction of peak heat release rate (PHRR) of ~59% was observed with eight coating layers.

7.2. Melt blending and spinning with fire retardant additives

Apart from the use of back coating, which may be applied to any textile comprising any type of fibres, flame retardant can also be applied to fusible fibre forming polymer during processing stage. There are two established viable ways to embed a flame retardant component in polymeric textiles. The first method deals with synthesizing fire retardant fibres by copolymerizing a FR component in thermoplastic macromolecular chain, where the flame retardant properties are firmly anchored in the fibre such as Trevira® CS a fire retardant copolyester fibres [84]. Despite their interesting fire performance, processing these materials is a costly and challenging task which often stops at laboratory scale, besides melt dripping issue is also addressed to PET. Therefore, there is a need for fire retardants that reduces melt dripping and promote the char formation. The second method deals with melt blending of FR additives and polymers. This method is discussed further in this review. Phosphorus-based compounds are widely investigated as FR additive in intumescent system by melt blending to impart flame retardancy to bulk polymer. Most of studies have been carried out on plates and composites. This section is not aimed to review all of them, as two detailed studies have been already published [85, 86]. In this context efforts have been extended, especially from the last 10 years, to use different eco-friendly bio-based extracted plant resources and other natural resources due to the desire to promote a "green concept". Thus, the formulation of intumescent flame retardant (IFR) systems may be based on bio-based acidic source such as phytic acid [87], metallic phytates [64], different bio-based carbonization agents like chitosan [73], cyclodextrin [72] and starch [71] exploiting melt blending processes. Among bio-based carbonizing agents, lignin has also been considered as an effective char former due to its highly aromatic structure. As reviewed in a previous section, lignin has been studied as a bio-based carbon source in intumescent system for various thermoplastic polymers to develop thermally stable and flame retardant composites.

As observed for bulk polymer, the combination of lignin with conventional flame retardants may enhance the overall flame retardant properties [16, 52]. Very recently, Cayla et al. [74] have used the same melt blending and spinning approach to develop flame retardant textile of PLA by using kraft lignin (LK) as carbon source with ammonium polyphosphate (APP) flame retardant additive, blended prior to the extrusion process. Several blends based on PLA/LK/APP were prepared to determine the possibility of melt spinning. Kraft lignin at 5 and 10 wt% with same proportion of APP was blended with PLA. TGA study showed that the combination of PLA/LK/APP exhibits single stage decomposition and degradation of PLA/LK/APP composites starts at a lower temperature than PLA. UL94 vertical flame tests performed on fabric samples showed that all the PLA/LK/APP ternary composites achieved V-0 rating with lowering the sample burning length. The combustion behaviour of fabric specimen was studied by cone calorimetry at heat flux 25 kW/m^2. The presence of lignin only in PLA does not improve the fire properties, but the combination of PLA/LK/APP exhibits good enhancement of fire retardant properties: in fact, t_{ign} is closer to that of the neat PLA and overall THR & MARHE values are lower than 56% and 43%, respectively. Besides, the combination of LK/APP allows increasing the charring effect that leads to the maximum residue at the end of combination.

In the same line, another work presents [88] intumescent formulations of lignin and APP with bio-based Polyamide 11 (Rilsan® PA11) prepared and extruded at 220°C. Blends were prepared by taking different weight ratios of intumescent formulation, keeping as high as 20 wt.% the total add on. TGA analysis shows an increase of char residue in the lignin-containing blends. Sheets of the different formulations were tested under the cone, using a heat flux of 25kWm^{-2}. Based on cone calorimetry results some of selected formulations were spindrawn by melt spinning to produce multifilament and finally knitted textile structures were preferred to woven fabrics. Cone calorimetry tests performed on the fabrics containing 10 wt.% lignin and 10 wt.% of OP1230 showed reduction of PHRR and THR of about 50 and 30%, respectively and an increase of time to ignition. In another work by Gaan et al. [89], flammability of cellulose-based textiles obtained from cotton fibre (Cot-Cell) and peat fibres (P-Cell), this latter is rich source of lignin (about 30 wt.%) are studied. Fibres are pretreated to make them a similar contraction. Both fibres were knitted in single jersey pattern. In a first step, flammability of knitted textile was evaluated for their flammability using the vertical fabric strip, BKZ-VB Swiss standard test. To further improve the flammability, a series of phosphoramidate was synthesized and Cot-Cell and P-Cell textile were treated with the FR compound normal by soaking in solution followed by drying in air. TGA and pyrolysis combustion flow calorimeter (PCFC) were used as screening tools.

8. Conclusion and future perspectives

The pursuit of sustainable development are the driving forces for increasing the interest in the use of material derived from renewable and bio-based sources. As indicated in this review, lignin is a promising component as carbon source in intumescent flame retardant systems due to its high charring capacity and abundancy in nature. In recent years, novel emerging

http://dx.doi.org/10.5772/intechopen.72963

applications of lignin have been explored: among them, it is worthy to mention the production of chemicals by depolymerization, manufacturing of lignin-based carbon fibres (LCF) and their use in polymer composites are the examples some niche areas. As presented in this review, the highly aromatic structure of lignin makes it suitable candidate as char former (carbon source) for intumescent system to improve flame retardant properties of polymer. Further, lignin-based intumescent have been studied extensively in bulk polymer including a wide range of synthetic thermoplastic and with bio-based polymers. In the same scenario, few researcher extended the lignin utilization to develop flame retardant textile. Furthermore, lignin-based intumescent are applied to textile by mean of textile coating and to bulk polymers through melt blending. Very few literature studies are available related to lignin as additive for intumescent systems in combination with other flame retardant to use as intumescent coating material for textiles. On the other side, very recently lignin-based intumescent systems have been used in bulk polymer to develop flame retardant PLA multifilament fibres and textile structures. Therefore, lignin-based intumescent are a possible way to develop 'green' flame retardant materials that could design the formulation of an intumescent system for future. The number of increasing scientific articles involving lignin is the indication of the growing lignin utilization in various niche areas. As far as flame retardancy is concerned, we expect more and more scientific and industrial research exploring flame retardancy specifically for textile applications.

In summary, there are many interesting application investigated: some are novel applications but for this purpose lignin quality standard must be developed, because at present available lignin types differ substantially in composition, structure, purity, etc., depending on the plant source, extraction process and many other factor. It is generally excepted that intumescent systems appeared to be very efficient when used as coatings, regardless of the underlying substrate since no interference may occur, and their efficiency can be assured. In contrast, the situation becomes more difficult when lignins are employed in bulk; in fact, at processing temperature chemical and physical interactions may occur. At the same time, lignins must be highly compatible with the polymer matrix; certainly, a homogenous distribution and uniform dispersion are mandatory. On the other hand, the possibility to create an expanded char layer that can protect polymer during combustion is primarily depends on coherency and consistency of the char layer; in fact, it must be mechanically strong and compact, hence strongly increasing its efficiency.

Author details

Neeraj Mandlekar[1,2], Aurélie Cayla[1,2]*, François Rault[1,2], Stéphane Giraud[1,2], Fabine Salaün[1,2], Giulio Malucelli[3] and Jin-Ping Guan[4]

*Address all correspondence to: aurelie.cayla@ensait.fr

1 Univ. Lille Nord de France, Lille, France

2 ENSAIT, GEMTEX, Roubaix, France

3 Politecnico di Torino, Alessandria, Italy

4 Soochow University, China

References

[1] White JF, Werpy T, Petersen G, Holladay JE, Program USD of EO of the B, Aden A, et al. Top Value Added Chemicals From Biomass: I. Results of Screening for Potential Candidates from Sugars and Synthesis Gas [Internet]. Department of Energy Report. U.S. Department of Energy [Office of] Energy Efficiency and Renewable Energy; 2004

[2] Cazacu G, Pascu MC, Profire L, Kowarski AI, Mihaes M, Vasile C. Lignin role in a complex polyolefin blend. Industrial Crops and Products. 2004;**20**(2):261-273

[3] Sahoo S, Seydibeyoğlu MÖ, Mohanty AK, Misra M. Characterization of industrial lignins for their utilization in future value added applications. Biomass and Bioenergy. Oct 2011;**35**(10):4230-4237

[4] Gosselink RJA, de Jong E, Guran B, Abächerli A. Co-ordination network for lignin—standardisation, production and applications adapted to market requirements (EUROLIGNIN). Industrial Crops and Products. 2004;**20**(2):121-129

[5] Laurichesse S, Avérous L. Chemical modification of lignins: Towards biobased polymers. Progress in Polymer Science. 2014;**39**(7):1266-1290

[6] Thakur VK, Thakur MK, Raghavan P, Kessler MR. Progress in green polymer composites from lignin for multifunctional applications: A review. ACS Sustainable Chemistry and Engineering. 2014;**2**:1072-1092

[7] Naseem A, Tabasum S, Zia KM, Zuber M, Ali M, Noreen A. Lignin-derivatives based polymers, blends and composites: A review. International Journal of Biological Macromolecules. 2016;**93**:296-313

[8] Movil O, Garlock M, Staser JA. Non-precious metal nanoparticle electrocatalysts for electrochemical modification of lignin for low-energy and cost-effective production of hydrogen. International Journal of Hydrogen Energy. 2015;**40**(13):4519-4530

[9] Qin Y, Yang D, Guo W, Qiu X. Investigation of grafted sulfonated alkali lignin polymer as dispersant in coal-water slurry. Journal of Industrial and Engineering Chemistry. 2015;**27**:192-200

[10] Thunga M, Chen K, Grewell D, Kessler MR. Bio-renewable precursor fibers from lignin/polylactide blends for conversion to carbon fibers. Carbon NY. 2014;**68**:159-166

[11] Yu Y et al. Functionalized lignin by grafting phosphorus-nitrogen improves the thermal stability and flame retardancy of polypropylene. Polymer Degradation and Stability. 2012;**97**(4):541-546

[12] Li B, Zhang X, Su R. An investigation of thermal degradation and charring of larch lignin in the condensed phase: The effects of boric acid, guanyl urea phosphate, ammonium dihydrogen phosphate and ammonium polyphosphate. Polymer Degradation and Stability. 2002;**75**(1):35-44

[13] Chen C, Gu X, Jin X, Sun J, Zhang S. The effect of chitosan on the flammability and thermal stability of polylactic acid/ammonium polyphosphate biocomposites. Carbohydrate Polymers. 2017;**157**:1586-1593

[14] Feng J et al. Syntheses of metallic cyclodextrins and their use as synergists in a poly(vinyl alcohol)/intumescent flame retardant system. Industrial & Engineering Chemistry Research. 2013;**52**(8):2784-2792

[15] Réti C, Casetta M, Duquesne S, Bourbigot S, Delobel R. Flammability properties of intumescent PLA including starch and lignin. Polymers for Advanced Technologies. Jun 2008;**19**(6):628-635

[16] De Chirico A, Armanini M, Chini P, Cioccolo G, Provasoli F, Audisio G. Flame retardants for PP based on lignin. Polymer Degradation and Stability. 2003;**79**:139-145

[17] Boerjan W, Ralph J, Baucher M. Lignin B iosynthesis. Annual Review of Plant Biology. Jun 2003;**54**(1):519-546

[18] Lin SY, Dence CW. Methods in Lignin Chemistry. Vol. 53. Berlin Heidelberg: Springer; 1989

[19] Doherty WOS, Mousavioun P, Fellows CM. Value-adding to cellulosic ethanol: Lignin polymers. Industrial Crops and Products. 2011;**33**(2):259-276

[20] Adler E. Lignin chemistry? Past, present and future. Wood Science and Technology. 1977;**11**(3):169-218

[21] Marton J, Marton T. Molecular weight of kraft lignin. TAPPI Journal. 1964;**47**(8):471-476

[22] Vishtal A, Kraslawski A. Challenges in industrial applications of technical lignins. BioResources. 2011;**6**(3):3547-3568

[23] El Mansouri N, Pizzi A, Salvadó J. Lignin-based wood panel adhesives without formaldehyde. Holz als Roh- und Werkst. Feb 2007;**65**(1):65-70

[24] Sannigrahi P, Ragauskas AJ, Miller SJ. Lignin structural modifications resulting from ethanol organosolv treatment of Loblolly pine. Energy and Fuels. 2010;**24**(1):683-689

[25] Shukry N, Fadel SM, Agblevor FA, El-Kalyoubi SF. Some physical properties of acetosolv lignins from bagasse. Journal of Applied Polymer Science. Jul 2008;**109**(1):434-444

[26] Kham L, Le Bigot Y, Delmas M, Avignon G. Delignification of wheat straw using a mixture of carboxylic acids and peroxoacids. Industrial Crops and Products. 2005;**21**:9-15

[27] Delmas M. Vegetal refining and agrichemistry. Chemical Engineering and Technology. 2008;**31**(5):792-797

[28] Lignol Innovations Ltd. | Sustainable Development Technology Canada. [Online]. Available from: https://www.sdtc.ca/en/organizations/lignol-innovations-ltd. [Accessed: Feb 13, 2017]

[29] Saake B, Lehnen R, Saake B, Lehnen R. Lignin. In: Ullmann's Encyclopedia of Industrial Chemistry. Weinheim, Germany: Wiley-VCH Verlag GmbH & Co. KGaA; 2007

[30] Rodríguez A, Sánchez R, Requejo A, Ferrer A. Feasibility of rice straw as a raw material for the production of soda cellulose pulp. Journal of Cleaner Production. 2010;**18**(10):1084-1091

[31] Lora J. Chapter 10 – Industrial commercial lignins: Sources, properties and applications. In: Belgacem MN, Gandini A, editors. Polymers and Composites from Renewable Resources ABT-M. Amsterdam: Elsevier; 2008. pp. 225-241

[32] Gellerstedt G, Henriksson G. Chapter 9 – Lignins: Major sources, structure and properties. In: Belgacem MN, Gandini A, editors. Polymers and Composites from Renewable Resources ABT-M. Amsterdam: Elsevier; 2008. pp. 201-224

[33] Gosselink RJA et al. Analytical protocols for characterisation of sulphur-free lignin. Industrial Crops and Products. 2004;**19**:271-281

[34] Adorjan I, Zhou Z, Jaaskelainen AS, Influence of carbonyl groups on the brightness reversion of eucalyptus kraft pulp. European Workshop on Lignocellulosics and Pulp, Vienne, Austria. 2006

[35] Market Research Store. Global Lignin Market – Industry Analysis, Size, Growth, Trends, Statistics, Segment and Forecast 2014-2020. 2015. [Online]. Available from: http://www.marketresearchstore.com/report/lignin-market-for-z36379

[36] Mordor Intelligence. Global Lignin Products Market – Segmented By Product Type, Source, Application, and Geography – Trends And Forecasts (2015-2020). 2016. [Online]. Available from: http://www.mordorintelligence.com/industry-reports/global-lignin-products-market-industry?gclid=CjwKEAjwrIa9BRD5_dvqqazMrFESJACdv27GeNb_zoHSY_-SAKQPrE1UG3nOjo7Qm6uvQ2yaFtiO5xoCNwjw_wcB.

[37] Wang M, Leitch M, Xu C. Synthesis of phenol-formaldehyde resol resins using organosolv pine lignins. European Polymer Journal. 2009;**45**(12):3380-3388

[38] Delmas G-H, Benjelloun-Mlayah B, Bigot Y Le, Delmas M. Biolignin TM based epoxy resins. Journal of Applied Polymer Science. 2013;**127**:1863-1872

[39] Global Lignin Market – Industry Analysis, Size, Growth, Trends, Statistics, Segment and Forecast 2014-2020. 2015. [Online]. Available from: http://www.marketresearchstore.com/report/lignin-market-for-z36379

[40] von Lignin SN. [Online]. Available from: https://mediathek.fnr.de/downloadable/download/sample/sample_id/285/ [Accessed: Feb 13, 2017]

[41] Agrawal A, Kaushik N, Biswas S. Derivatives and applications of lignin – An insight. The Scitech Journal. 2014;**1**:30-36

[42] Glasser WG, Northey RA, Schultz TP, editors. Lignin: Historical, Biological, and Materials Perspectives. Vol. 742. Washington, DC: American Chemical Society; 1999

[43] Maradur SP, Kim CH, Kim SY, Kim BH, Kim WC, Yang KS. Preparation of carbon fibers from a lignin copolymer with polyacrylonitrile. Synthetic Metals. 2012;**162**(5-6):453-459

[44] Baker DA, Rials TG. Recent advances in low-cost carbon fiber manufacture from lignin. Journal of Applied Polymer Science. Oct 2013;**130**(2):713-728

[45] Pandey MP, Kim CS. Lignin depolymerization and conversion: A review of thermochemical methods. Chemical Engineering and Technology. Jan 2011;**34**(1):29-41

[46] Galkin MV, Samec JSM. Lignin valorization through catalytic lignocellulose fractionation: A fundamental platform for the future biorefinery. ChemSusChem. Jul 2016;**9**(13):1544-1558

[47] Segev O, Kushmaro A, Brenner A. Environmental impact of flame retardants (persistence and biodegradability). International Journal of Environmental Research and Public Health. 2009;**6**(2):478-491

[48] Gregorová A, Košíková B, Moravčík R. Stabilization effect of lignin in natural rubber. Polymer Degradation and Stability. 2006;**91**(2):229-233

[49] Pouteau C, Dole P, Cathala B, Averous L, Boquillon N. Antioxidant properties of lignin in polypropylene. Polymer Degradation and Stability. 2003;**81**:9-18

[50] Pucciariello R, Villani V, Bonini C, D'Auria M, Vetere T. Physical properties of straw lignin-based polymer blends. Polymer (Guildf). 2004;**45**(12):4159-4169

[51] Hu J, Xiao R, Shen D, Zhang H. Structural analysis of lignin residue from black liquor and its thermal performance in thermogravimetric-Fourier transform infrared spectroscopy. Bioresource Technology. Jan 2013;**128**:633-639

[52] Canetti M, Bertini F, De Chirico A, Audisio G. Thermal degradation behaviour of isotactic polypropylene blended with lignin. Polymer Degradation and Stability. 2006;**91**:494-498

[53] Liu L, Qian M, Song P, Huang G, Yu Y, Fu S. Fabrication of green lignin-based flame retardants for enhancing the thermal and fire retardancy properties of polypropylene/wood composites. ACS Sustainable Chemistry & Engineering. 2016;**4**(4):2422-2431

[54] Acha BA, Marcovich NE, Reboredo MM. Lignin in jute fabric–polypropylene composites. Journal of Applied Polymer Science. 2009;**113**:1480-1487

[55] Song P, Cao Z, Fu S, Fang Z, Wu Q, Ye J. Thermal degradation and flame retardancy properties of ABS/lignin: Effects of lignin content and reactive compatibilization. Thermochimica Acta. 2011;**518**(1-2):59-65

[56] Canetti M, Bertini F. Influence of the lignin on thermal degradation and melting behaviour of poly(ethylene terephthalate) based composites. E-Polymers. 2009;**49**:1-10

[57] Zhu H et al. Preparation and characterization of flame retardant polyurethane foams containing phosphorus–nitrogen-functionalized lignin. RSC Advances. 2014;**4**(98):55271-55279

[58] Zhang R, Xiao X, Tai Q, Huang H, Yang J, Hu Y. The effect of different organic modified montmorillonites (OMMTs) on the thermal properties and flammability of PLA/MCAPP/lignin systems. Journal of Applied Polymer Science. 2013;**127**(6):4967-4973

[59] Reti C, Casetta M, Duquesne S, Bourbigot S, Delobel R. Flammability properties of intumescent PLA including starch and lignin. Polymers for Advanced Technologies. 2008;**19**:628-635

[60] Bertini F, Canetti M, Cacciamani A, Elegir G, Orlandi M, Zoia L. Effect of ligno-derivatives on thermal properties and degradation behavior of poly(3-hydroxybutyrate)-based biocomposites. Polymer Degradation and Stability. 2012;**97**(10):1979-1987

[61] Ferry L, Dorez G, Taguet A, Otazaghine B, Lopez-Cuesta JM. Chemical modification of lignin by phosphorus molecules to improve the fire behavior of polybutylene succinate. Polymer Degradation and Stability. 2015;**113**:135-143

[62] Liu L, Huang G, Song P, Yu Y, Fu S. Converting industrial alkali lignin to biobased functional additives for improving fire behavior and smoke suppression of polybutylene succinate. ACS Sustainable Chemistry & Engineering. 2016;**4**(9):4732-4742

[63] Darnerud PO. Toxic effects of brominated flame retardants in man and in wildlife. Environment International. 2003;**29**:841-853

[64] Costes L et al. Metallic phytates as efficient bio-based phosphorous flame retardant additives for poly(lactic acid). Polymer Degradation and Stability. 2015;**119**:217-227

[65] Basak S, Samanta KK, Chattopadhyay SK. Fire retardant property of cotton fabric treated with herbal extract. Journal of Textile Institute. Dec 2015;**106**(12):1338-1347

[66] Malucelli G et al. Biomacromolecules as novel green flame retardant systems for textiles: an overview. RSC Advances. Sep 2014;**4**(86):46024-46039

[67] Alongi J et al. Intrinsic intumescent-like flame retardant properties of DNA-treated cotton fabrics. Carbohydrate Polymers. 2013;**96**(1):296-304

[68] Basak S, Ali SW. Sustainable fire retardancy of textiles using bio-macromolecules. Polymer Degradation and Stability. 2016;**133**:47-64

[69] Wang X, Lu C, Chen C. Effect of chicken-feather protein-based flame retardant on flame retarding performance of cotton fabric. Journal of Applied Polymer Science. 2014;**131**:n/a-n/a

[70] Bourbigot S, Fontaine G, Duquesne S. Intumescence for the flame retardancy of polylactide, 21th BCC Conference – Recent Advances in Flame Retardancy of Polymeric Materials, Stamford, États-Unis, Mai. 2010

[71] Wang J, Ren Q, Zheng W, Zhai W. Improved flame-retardant properties of poly(lactic acid) foams using starch as a natural charring agent. Industrial and Engineering Chemistry Research. 2014;**53**:1422-1430

[72] Feng J-X, Su S-P, Zhu J. An intumescent flame retardant system using β-cyclodextrin as a carbon source in polylactic acid (PLA). Polymers for Advanced Technologies. Jul 2011;**22**(7): 1115-1122

[73] Xiao Y, Zheng Y, Wang X, Chen Z, Xu Z. Preparation of a chitosan-based flame-retardant synergist and its application in flame-retardant polypropylene. Journal of Applied Polymer Science. 2014;**131**(19):1-8

[74] Cayla A, Rault F, Giraud S, Salaün F, Fierro V, Celzard A. PLA with intumescent system containing lignin and ammonium polyphosphate for flame retardant textile. Polymers. 2016;**8**:331-346

[75] Wang MY, Horrocks AR, Horrocks S, Hall ME, Pearson JS, Clegg S. Flame retardant textile back-coatings. Part 1: Antimony-halogen system interactions and the effect of replacement by phosphorus-containing agents. Journal of Fire Sciences. Jul 2000;**18**(4):265-294

[76] Horrocks AR, Wang MY, Hall ME, Sunmonu F, Pearson JS. Flame retardant textile back-coatings. Part 2. Effectiveness of phosphorus-containing flame retardants in textile back-coating formulations. Polymer International. Oct 2000;**49**(10):1079-1091

[77] Horrocks AR. Flame retardant challenges for textiles and fibres: New chemistry versus innovatory solutions. Polymer Degradation and Stability. 2011;**96**(3):377-392

[78] Srikulkit K, Iamsamai C, Dubas ST. Development of flame retardant polyphosphoric acid coating based on the polyelectrolyte multilayers technique. Journal of Metals Materials and Minerals. 2006;**16**(2):41-45

[79] Laufer G, Kirkland C, Morgan AB, Grunlan JC. Intumescent multilayer nanocoating, made with renewable polyelectrolytes, for flame-retardant cotton. Biomacromolecules. Sep 2012;**13**(9):2843-2848

[80] Alongi J et al. DNA: a novel, green, natural flame retardant and suppressant for cotton. Journal of Materials Chemistry A. 2013;**1**(15):4779

[81] Carosio F, Di Blasio A, Alongi J, Malucelli G. Green DNA-based flame retardant coatings assembled through Layer by Layer. Polymer (Guildf). 2013;**54**(19):5148-5153

[82] Réti C, Casetta M, Duquesne S, Delobel R, Soulestin J. Intumescent biobased-Polylactide films to flame retard nonwovens. Journal of Engineered Fibers and Fabrics. 2009;**4**(2):33-39

[83] Giraud S, Rault F, Rochery M, Gacquere L. Use of bio-based carbon source to develop intumescent flame retardant polyurethane coating for polyester textile fabric. 14th European Meeting on Fire Retardant Polymers, FRPM13. Lille, France. 2013

[84] Black S. The development of Trevira CS (comfort and safety). Journal of the Society of Dyers and Colourists. Oct 2008;**108**(9):372-373

[85] Alongi J, Han Z, Bourbigot S. Intumescence: Tradition versus novelty. A comprehensive review. Progress in Polymer Science. 2014;**51**:28-73

[86] Richard Horrocks A. Textile flammability research since 1980 – Personal challenges and partial solutions. Polymer Degradation and Stability. 2013;**98**(12):2813-2824

[87] Cheng X-W, Guan J-P, Tang R-C, Liu K-Q. Phytic acid as a bio-based phosphorus flame retardant for poly(lactic acid) nonwoven fabric. Journal of Cleaner Production. 2016;**124**:114-119

[88] Cayla A, Rault F, Giraud S, Salaün F, Campagne C, Fierro V. Interest of wood wastes (lignin) as flame retardant fillers in bio-based polymers (PA11 and PLA). Eurofillers and Polymer Blends, Montpellier, France. 2015

[89] Salmeia AK, Jovic M, Ragaisiene A, Rukuiziene Z, Milasius R, Mikucioniene D, et al. Flammability of cellulose-based fibers and the effect of structure of phosphorus compounds on their flame retardancy. Vol. 8. In: Polymers. 2016

Chapter 10

Utilisation of Lignins in the Bioeconomy: Projections on Ionic Liquids and Molecularly Imprinted Polymers for Selective Separation and Recovery of Base Metals and Gold

Peter Papoh Ndibewu and Pierre Gerard Tchieta

Additional information is available at the end of the chapter

http://dx.doi.org/10.5772/intechopen.72542

Abstract

A brief review has been herein done of technologies involved in the exploitation of lignin, in order to provide an introduction to the subject from the perspective of a fast technologically advancing economy. Lignocellulosic materials and biomass have historically been utilised from since time memorial, but a new conversation is emerging on the role of these materials in modern bioeconomies. This new discourse needs to help us understand how technologies for managing and processing lignocellulosic materials both as biosynthetic moieties, biogenic wastes or simply renewable biopolymer—both established and novel—should be deployed and integrated (or not) to meet developmental requirements of the sustainability paradigm. The world is caught in the middle of green technology advocating for more and more focus on renewable sources of manufacturing raw materials and that of the molecularly imprinted/synthesised or genetically engineered ones. The utilisation of lignins (from renewable sources) in both the industry as the base for the formulation of ionic liquids (with yet a wider industrial applications), and also is a potential scaffold material for functionally modified imprinted polymers (LCIPS) for the selective recovery of base metals and gold, respectively, evidently incorporates the bioeconomy aspirations.

Keywords: lignins, ionic liquids, polyelectrolytic polymers, gold, bioeconomy

1. Introduction

Lignins [1–10], in particular, and in general lignocellulosic materials (LCMs) [11–16] and biomass [17–20], respectively, have historically been utilised since time memorial, but a new conversation is emerging on the role of these materials in modern bioeconomies [21–25]. Due to the nature of the molecule(s) as a novel commodity for many interesting natural and manufactured products [26–29], a modern bioeconomy [30–33] is not simply a rerun of former ones. This new discourse needs to help us understand how technologies [34–38] for managing and processing lignocellulosic materials both as biosynthetic moieties [39–49], biogenic wastes [50–58] or simply renewable biopolymer [59–69]—both established and novel—should be deployed and integrated (or not) to meet developmental requirements of the sustainability paradigm [70–75]. More so, in a closed-loop and resource-security agendas that undoubtedly sits behind the bioeconomy aspirations now being voiced by many schools of thought in many countries and regions of a so fast technologically advancing world [76–87]. The world of which is caught in the middle of green technology [88–92] advocating for more and more focus on renewable sources [93–98] of manufacturing raw materials and that of the molecularly imprinted/synthesised or genetically engineered ones [99–102].

The utilisation of lignins [103–106] (from renewable sources) in both the industry as the base for the formulation of ionic liquids (with yet a wider industrial applications) and as a potential scaffold material for functionally modified imprinted polymers (LCIPS) [107–114] for the selective recovery of base metals and gold, respectively, evidently incorporates the bioeconomy aspirations. Further, one new trend in lignin research is to steer biosynthetic pathways towards the biosynthesis of molecules that, upon incorporation into the lignin polymer, will improve lignin degradation, has been echoed by some researchers [5].

The name for lignin was derived from the Latin word for "wood" [94]. Lignin is a complex biopolymer based on 4-hydroxyphenylpropanoids [8] and is synthesised by almost all terrestrial plants [115] and also in some algae [116]. It is known to be the second most abundant natural polymer on Earth, just after cellulose [94]. It is one major structural component of higher plants' cell wall [94], contributing significantly to the recalcitrance of the woody biomass [117] that provides mechanical structure and resistance to environmental stress and microbial decay [94]. Lignin, as a second most abundant biopolymers present in terrestrial ecosystems is reported to contain organic carbon 30%, approximately [6]. Structurally, lignin in its native state is defined as protolignin [118], and can be defined as an amorphous, polyphenolic material arising from an enzyme-mediated dehydrogenative polymerisation of three phenylpropanoid monomers. That is, namely monolignols, namely coumaryl, coniferyl, and sinapyl alcohols, which corresponds, respectively, to p-hydroxyphenyl (H), guaiacyl (G), and syringyl (S), forming the lignin structures (**Figure 1**) [118].

As seen in the lignin biosynthesis (**Figure 1**), it can act as building blocks in phenylpropanoid pathway, and thus helps in precipitation of lignin in the presence of acidic solutions [119, 120].

The lignin structure consists of three dimensional cross-linked macromolecules, hereunder illustrated (**Figure 2**) depicts the three sub-divisions of alcohols namely: sinapyl, coniferyl and p-coumaryl alcohol (**Figure 1**), with various lignin residues constructed from the three monolignols as illustrated in **Figure 3**.

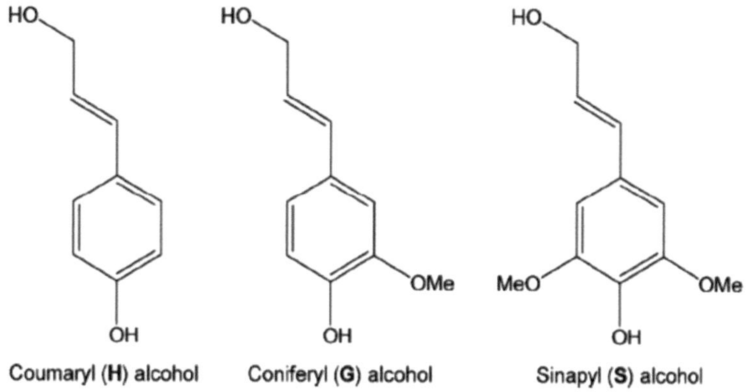

Figure 1. Phenylpropanoid precursors in lignin biosynthesis [118].

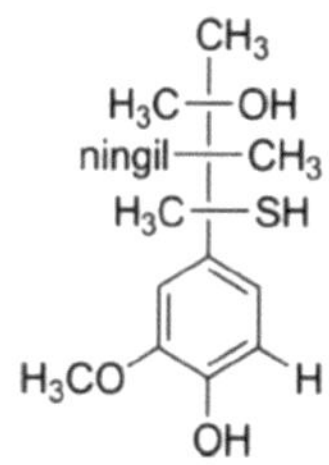

Figure 2. Structure of lignin [1].

Lignin can be considered essential "glue" that holds three polymers (lignin, cellulose, and hemicellulose) together in the plant cell wall (**Figure 2**) and is linked to polysaccharides, creating stable lignin-carbohydrate cross-linkages [121]. This is one of the main reasons for preventing selective separation of the biomass components in a biorefinery process [121]. Currently, it is understood that lignin, cellulose, and hemicellulose are not just individual units in a plant cell wall but are intimately interrelated and associated.

The chemical structure of cellulose has been clearly elucidated as a polysaccharide consisting of a linear chain of $\beta(1 \rightarrow 4)$ linked D-glucose units, from several hundred to many thousands (**Figure 5A**) [94]. Cellulose is embedded in the primary cell wall as protection (**Figure 4**), in fact, its regular structure makes it strong and resistant to most hydrolysis treatments [121].

On one hand, the cellulose polymers are arranged in microfibrils that are organised in fibrils; combined with cellulose fibres, which are responsible for the fibrous nature of lignocellulosic biomass [11, 12, 16] cell walls. On the other hand, hemicellulose is a branched polysaccharide composed of several different heteropolymers, such as arabinoxylans and glucomannans [12].

It is composed by a mixture of different pentoses (xylose and arabinose) and hexoses (glucose, mannose and galactose), respectively, containing C_5 and C_6 atoms (**Figure 5B**). It should be noted that hemicelluloses in deciduous woods (hardwoods) and agricultural herbaceous products like wheat straw, corn stover and switch grass predominantly consist of D-xylose units (xylans),

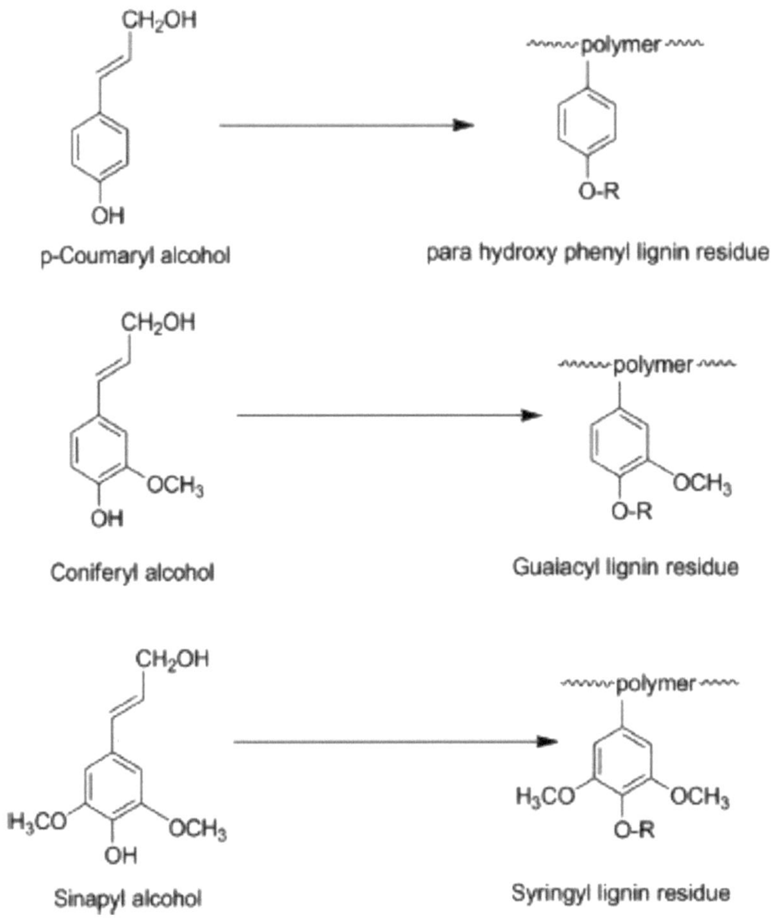

Figure 3. Structure of lignin residues from the three monolignols [1].

whereas coniferous (softwood) hemicellulose is mainly made up of D-mannose (mannans), L-arabinose (arabinans) and D-galactose (galactans) [123]. While they make up 30% of the cell wall material, xylans are the main hemicelluloses in hardwood, and predominate in annual plants and cereals making. Its structure is less regular than cellulose, and as such its resistance is also lower, while its role in the cell wall is to bind together different cellulose fibres together.

According to Huang [124], lignocellulosic biomass can be roughly classified into virgin biomass (e.g. trees and grasses), agricultural and forestry waste biomass (e.g. straws, sugarcane bagasse and paper mill discards) and energy crops (e.g. *Arundo donax* and switch grass). Some examples of different biomass type composition are shown in **Table 1**.

Lignocellulose that refers to the plant biomass accounts for more than 60% of the total biomass [11]. Authors [125, 126] clarify that this includes, but not limited to, the wood residues (sawdust and paper mill discards), grasses, wastepaper, agricultural residues (straw, stover, peelings, cobs, stalks, nutshells, non-food seeds and bagasse), domestic wastes (lignocellulose garbage and sewage), food industry residues and municipal solid wastes (MSWs). These

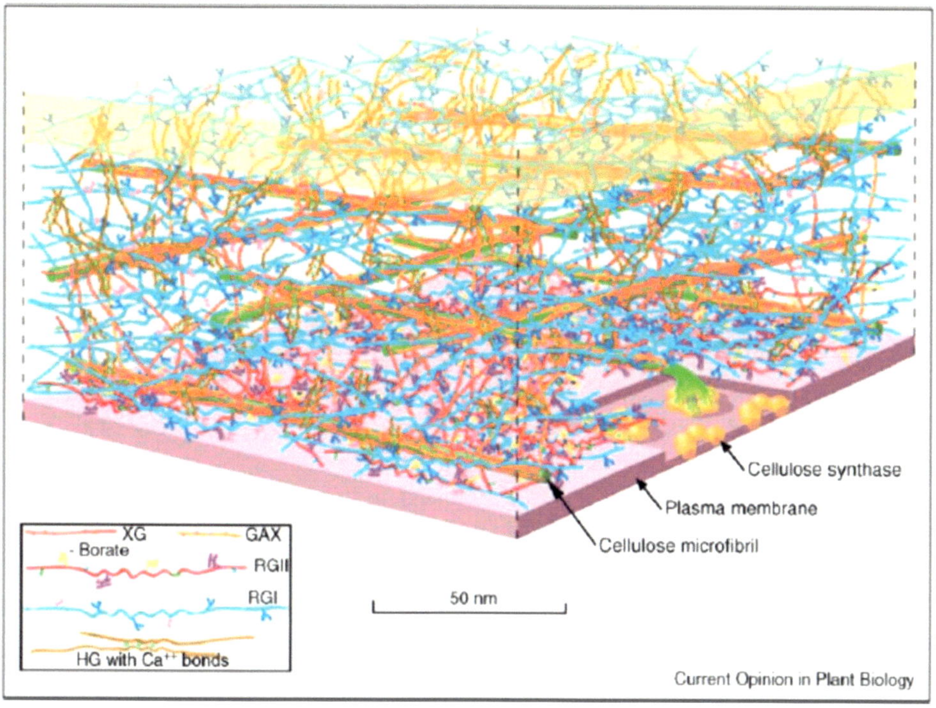

Figure 4. Schematic representation (Adapted from [8]) depicting the complexity of the plant cell wall [122].

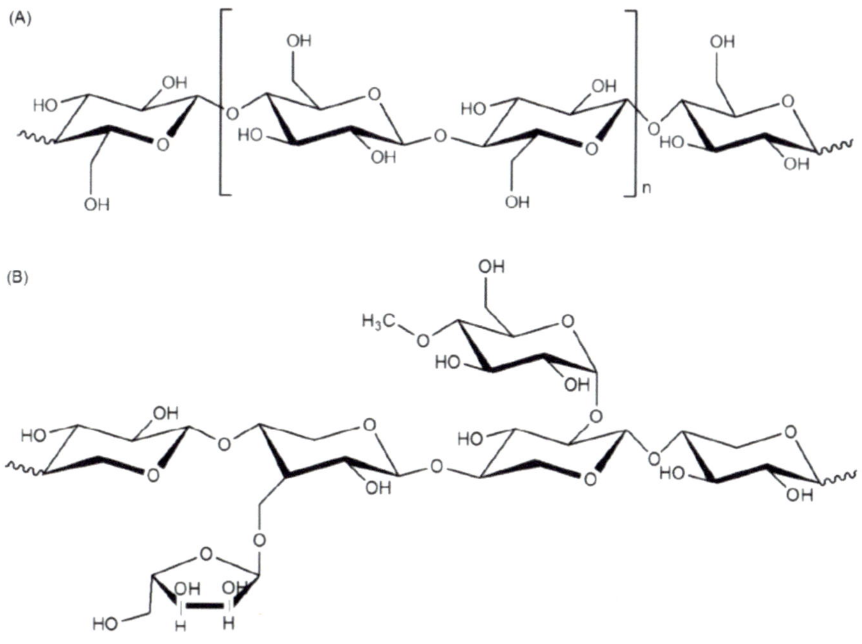

Figure 5. Structure of cellulose (A) and hemicellulose (B) [94].

composition constitutes the bulk of lignocellulosic materials (LCMs) of some importance in bioeconomic transformation [21]. In their review, Bilal et al. [11] provided an excellent detailed compositional profile of various previously reported LCMs herein summarised in **Table 2**. It is reported in the literature [94] that lignocellulosic biomass is essentially the inedible component of a plant, mainly composed of carbohydrate (cellulose and hemicellulose) and phenolic (lignin)-based polymers, chemically bounded together. From its composition viewpoint, these three main components represent from 85 to 95% by weight of the overall biomass; while the remaining part includes other components such as proteins, silica, fatty acids, pectin and salts.

Lignin samples have been fully characterised [127] with respect to lignin and sugar content and the amount of hydroxyl and carboxyl groups elucidated (**Table 3**).

It is a well-known fact that biomass was the main source of energy and materials prior to the industrial revolution. But following the exploitation of fossil fuels during the 1900s, biomass utilisation had declined sharply [52], except for a few cases that is worth mentioning. For example, during and after the aforementioned period, biomass usage was constant or even on the increased in the timber and fibre industries [128], during the 1930s agricultural surplus in the USA [129] and during World War II petroleum shortages in Europe [130]. As one of the twenty-first century human goal to reduce society's climate footprint and other environmental burdens, there is now a renewed interest in biomass, sparked by the aim to achieve a more secure supply of resources and to encourage the bioeconomy [21, 25, 30, 32]. Unfortunately, activities designed to achieve those aims are at the centre of a battlefield where recent food versus fuel and deforestation debates reveal that the source of the biomass being exploited is an important consideration. Thus, attention should be paid to the effective exploitation of renewable materials (unavoidable organic and biogenic wastes), including lignins and LCMs that might otherwise go unused or potentially be landfilled or cause disposal problems. The biotransformation and/or green treatment of lignins and LCMs with a focus on making new ionic liquids and molecularly imprinted polymers for selective separation and recovery of base metals and gold will result in tremendous gains in the bioeconomy loop.

The Dutch government's initiative in the 1920s of composting municipal solid waste (MSW) for land-expansion in Europe apparently was sparked by the concept of bioeconomy and these efforts was later followed by much wider interest in utilising MSW [131] much more sustainably. In recent times, many nations have designed very attractive strategies or roadmaps with clear settings on how they plan to approach the bioeconomy in their own respective countries

	Cellulose (% of d.m.)	*Hemicellulose (% of d.m.)*	*Lignin (% of d.m.)*	*Other (% of d.m.)*
Corn Stover	38	26	19	17
Soybean	33	14	-	53
Wheat Straw	38	29	15	18
Rye Straw	31	25	-	44
Switchgrass	37	29	19	15
Miscanthus	43	24	19	14
Forage Sorghum	34	17	16	33
Sweet Sorghum	23	14	11	52
Bagasse	40	30	20	10

Note: d.m. = dry matter

Table 1. Composition of main lignocellulosic biomasses [124].

http://dx.doi.org/10.5772/intechopen.72542

Lignocellulosic materials	Carbohydrate composition (% dry wt.)		
	Lignin	Cellulose	Hemicellulose
Agricultural residues	5–15	37–50	25–50
Banana waste	14	13.2	14.8
Bagasse	23.33	54.87	16.52
Barley hull	19	34	36
Barley straw	6.3–9.8	36–43	24–33
Bamboo	23	49–50	18–20
Corn straw	8.2	42.6	21.3
Corn cobs	15	45	35
Corn stover	19	38	26
Cotton seed hairs	0	85–95	5–20
Coffee pulp	15.6–19.1	33.7–36.9	44.2–47.5
Douglas fir	15–21	35–48	20–22
Eucalyptus	29	45–51	11–18
Grasses	10–30	25–40	25–50
Horticultural waste	36	34.5	28.6
Hardwood	18–25	40–55	24–40
Olive tree biomass	19.1	25.2	15.8
Jute fibers	21–26	45–53	18–21
Leaves	0	15–20	80–85
Nut shells	30–40	25–30	25–30
Newspaper	18–30	40–55	25–40
Oilseed rape	14.2	27.3	20.5
Oat straw	10–15	31–35	20–26
Poplar wood	10–21	45–51	25–28
Pulp and paper sludge	16	23.4	8.6
Pine	23–29	42–49	13–25
Rice Straw	18	32.1	24
Rice husk	15.4–20	28.7–35.6	11.96–29.3
Sugar cane bagasse	20	42	25
Sweet sorghum	21	45	27
Softwood	25–35	45–50	25–35
Sponge gourd fibers	15.46	66.59	17.44
Sorted refuse	20	60	20
Solid cattle manure	2.7–5.7	1.6–4.7	1.4–3.3
Swine waste	NA	6	28
Sugar beet	NA	5	5.5
Tamarind kernel	NA	10–15	55–65
Winter rye	16.1	29–30	22–26
Wheat straw	16–21	29–35	26–32
Water-hyacinth	3.55	18.4	49.2
Wheat bran	8.3–12.5	10.5–14.8	35.5–39.2

Table 2. Compositional analysis of representative common lignocellulosic feedstocks [11].

Sample number	Raw material	Lignin type (commercial name)	Sample code	Lignin content (%)	COOH (mmol/g)	Phenolic OH (mmol/g)	Total sugars (%)
1	Hemp	Soda	He	Nd	2.1	1.58	2.4
2	Sisal	Soda	Si	Nd	1.18	2.29	7.7
3	Abaca	Soda	Ab	Nd	1.14	2.73	5.5
4	Straw	Soda	St	Nd	2.1	2.43	Nd
5	Jute	Soda	Ju	Nd	1.78	2.38	Nd
6	Flax	Soda	Fl-1	87.8	1.9	1.1	1.7
7	Softwood	Lignosulfonate (Borresperse 3A)	SW-LS-1	Nd	3.5	1.1	1.3
8	Softwood	Kraft (Indulin AT)	SW-Kr-1	90.0	2.5	1.8	2.06
9	Softwood	Lignosulfonate (Wafex P)	SW-LS-2	Nd	1.2	1.1	24.5
10	Flax	Soda (Bioplast)	Fl-2	87.8	1.9	1.1	1.7
11	Flax	Soda, oxidised	Fl-ox	92.1	1.8	0.9	1.6
12	Softwood	Kraft (Curan 100)	SW-Kr-2	88.6	1.7	0.8	2.26
13[a]	Softwood	Soda, precipitation high pH	SW-SF-1	64.7	Nd	Nd	1.77
14	Softwood	Kraft	SW-Kr-3	96.8	2.0	2.5	0.71
15	Mixed hardwoods	Organosolv (Alcell)	HW	96.5	0.78	2.4	0.32
16	Softwood	Soda, precipitation low pH	SW-SF-2	99.2	1.4	1.0	0.65
17	Softwood	Kraft (Curan 2711P)	SW-Kr-4	80.3	2.7	1.6	2.58

Nd, not determined.
[a] Lignin 13 contains about 40% ash.

Table 3. Chemical characteristics of lignin samples [127].

or regions. From thence, there are undoubtedly new opportunities in exploiting biogenic wastes, of which lignins and lignocellulosic materials are some of the important prospect. This desire to valorise wastes and to holistically move waste treatment (and resources) up the waste hierarchy in an eco-friendly manner should contribute to a renewed vitality that is likely to nurture the bioeconomy [52]. Well, the valorisation of wastes happening so quickly and catalysed by technological and industrial developments, is an emerging subject going beyond energy applications. A good selection of recent ideas in the field of biotransformation of waste biomass into high value biochemicals has been exemplified by Brar et al. [132]. Typically, there is interest in generating higher value substances and products from biogenic waste, such as pigments, phenolics and nutrients, and so captivating can be the make new ionic liquids and molecularly imprinted polymers for the selective separation and recovery of base metals and gold, for industry applications and environmental pollution controls.

This chapter briefly reviews many of the key technological ingredients in the bioeconomy that are being explored for the utilisation of lignins and lignocellulosic materials (LCMs), as well as a brief description of biogenic wastes as a resource. Creating a categorical scheme for describing these renewable resources towards their potential to be explored in the manufacturing of a new range of ionic liquids (e.g. polyelectrolytic polymers) and bio-based imprinted polymers (e.g. molecularly imprinted polymers, MIPs) for selective separation and recovery of base metals and gold is increasingly an interesting avenue of research. Today's world of which is caught in the middle of green technology advocating for more and more focus on renewable sources for manufacturing raw materials is attempting desperately to meet up with the aspirations of bioeconomy. In this regard, the discussion herein presented in this chapter articulates well this new sustainability paradigm focusing on lignins and lignocellulosic materials. The utilisation of lignins (from renewable sources) in both the industry as the base for the formulation of ionic liquids (with yet a wider industrial applications) and as a potential scaffold material for functionally modified imprinted polymers (LCIPS) for the selective recovery of base metals and gold, respectively, evidently incorporates the bioeconomy principles. Further, one new trend in lignin research is to steer biosynthetic pathways towards the biosynthesis of molecules that, upon incorporation into the lignin polymer, will improve lignin degradation [5]. Due to the nature of the molecule(s) as a novel resourceful commodity for many interesting natural and manufactured products, a modern bioeconomy in this niche area may not simply be a rerun of former ones. This discourse spelled out in this chapter, although not very critically, needs to help understand how technologies for managing and processing lignins and lignocellulosic materials both as biosynthetic moieties, biogenic wastes or simply renewable biopolymers—both established and novel—should be deployed and integrated (or not) to meet developmental requirements of the so much emphasised sustainability paradigm. More so, in a closed-loop and resource-security agendas that undoubtedly sits behind the bioeconomy aspirations now being voiced by many schools of thought in many countries and regions of the so fast technologically advancing world. This is the foremost reason that many countries, including international agencies at large, are working towards ensuring sustainable development [54].

2. The bioeconomy

To understand the term bioeconomy [22], two main schools of thoughts have conceptualise it and have tried to explain profoundly the aspirations behind the concept, irrespective from its origin [30]. The first school of thought widely publicised in the literature [133–136] defines the concept based on the understanding that bioeconomy will rely on renewable biomass instead of finite fossil inputs for the production of a wide range of value-added products such as food, feed, bio-based products and bioenergy. Secondly, proponents in the second school of thoughts, also available from the literature [23, 136, 137] advocates that these products will be produced in biorefineries following a cascade principle in order to maximally valorise the available biomass. Giurca and Metz [138] articulates that, from these two conceptualised definitions, the bioeconomy can be understood as a cross-sectoral concept, where different sectors (e.g. food, feed, chemistry, energy, fuel and pharmaceutical sectors) are expected to cooperate in order to derive products from renewable biological resources stemming from agricultural-, forest- and fishery-sectors [22, 23]. What this entails is that initially, biomass is processed into high value products such as chemicals and pharmaceutical active ingredients, and the residues are then used in downstream processes for lower value applications until a minimum of waste remains at the end of the process [139–141]. Some other co-workers [23, 137] has summarised the principle of the bioeconomy to be one which can thus be considered a collection of sectors and subsectors (e.g. food, feed, chemistry, energy, fuel and pharmaceutical sector), all working in conjunction to derive products from renewable biological resources originating from agriculture, fisheries and forestry. In this context, lignins and lignocellulosic materials, considered renewable and resourceful agricultural wastes are good candidates for consideration in further research, development and application. Another author [52] stated that "bioeconomy" shares a lot in common with the aims of "sustainability" and the use of biomass and renewables. This author exemplifies that in this regard, bioeconomy could be understood as an economy in which food, chemicals, energy and materials largely come from plants and animals—essentially bio-based. Thus, in agreement with the fundamental principle of bioeconomy emphasising an economy established on biomass instead of fossil fuels, signalling a significant shift in socio-economic, agricultural, energy and technological systems since from its inception in the twentieth century. This alignment of the objectives of bioeconomy with those of sustainability may be able to create new directions and opportunities for meeting some of the requirements for sustainability [52]. As both scholars and policy-makers strongly assert that innovation plays a central role in ensuring a sustainable bioeconomy transition, both on the supply side, citing the need for new technologies and products, and on the user side, emphasising the consumption and waste patterns [22, 142–144], it must then be globally agreed that a multitude of different renewable resource-based technologies and innovation structures would have to come into play under the broad concept of "bioeconomy" [144] for this to happen. Technology Innovation Systems (TIS) perspectives [145, 146] can be regarded as a catalyst for the technology and innovation back-up.

In conclusion, it is widely ascertained, particularly in agreement with the views of both policy scholars and makers and scholars that the transition towards a bioeconomy is increasingly as one of the primary ways to reduce our dependency on fossil resources. However, research to provide the very complete myriad of economically viable technologies and biotransformation solutions of the total biomass (including lignins and lignocellulosic materials) remains scarce. This can be unarguably true of the socio-economic research on the transition towards the bioeconomy at the firm-level. This chapter serves a multiple purpose of creating the spark for more focussed research (lignin-biotransformation, lignin-based ionic polyelectrolytic polymers for industry usage, e.g. food, chemical and pharmaceutical, lignin-imprinted polymers for applications in gold and precious metals recovery and environmental remediation, just to name a few).

3. Lignins structure and functional properties

3.1. Structure

3.1.1. Lignin biosynthesis and distribution

Comprehensive approaches of the lignin biosynthetic pathway have been described in the literature [8]. It is common knowledge that the distribution and family expansion of the lignin biosynthetic genes in a few non-wood plants (rice, poplar and Arabidopsis, for example) exhibit a comparative genome expressive of the biosynthesis of lignin. Apparently, the complete lignin biosynthetic pathway first appeared in the lower plant (the moss). More so, the correlation between the expansion of lignin biosynthetic gene families and the substrate diversity is highly significant ($p < 0.05$). Furthermore, these authors observed that lignin biosynthetic genes expansion within the different families in terms of was very rapid. This observation could be a suggestion of a possible effective translation of transgenic lignin modification strategies from Arabidopsis (used as model species) to a more relevant plant with genetically differentiated biomass feedstocks only if it occurred between closely related species. A general reaction scheme for lignin biosynthesis is shown in **Figure 6** [8].

Chemically, lignins are synthesised via oxidative coupling between monomers of p-hydroxycinnamyl alcohols and other related molecules [6, 10, 147]. As illustrated in **Figure 6**, lignin is derived from three predominant p-hydroxycinnamyl alcohols, the monolignols: p-coumaryl, coniferyl and sinapyl alcohols. It is presumed [148] that all the enzymes required for monolignols biosynthesis are known, namely: phenylalanine ammonia lyase (PAL), cinnamate

4-hydroxylase (C4H), 4-coumarate-CoA ligase (4CL), shikimate hydroxycinnamoyl transferase (HCT), coumarate 3-hydroxylase (C3H), caffeoyl-CoA 3-O-methyltransferase (CCoAOMT), cinnamoyl-CoA reductase (CCR), cinnamyl alcohol dehydrogenase (CAD), ferulate 5-hydroxylase (F5H) and caffeic acid/5-hydroxyferulic acid O-methyltransferase (COMT). According to Vanholme et al. [5], the main biosynthetic route towards the monolignols p-coumaryl, coniferyl and sinapyl alcohol, adapted from Boerjan et al. [6, 10] is presented in **Figure 7**.

Obviously (**Figure 8**), the main building blocks of lignin are the hydroxycinnamyl alcohols (or monolignols), coniferyl alcohol and sinapyl alcohol, with typically minor amounts of p-coumaryl alcohol [149]. From the literature [150], biosynthetic pathway for the monolignols shows that they are synthesised from Phe through the general phenylpropanoid and monolignol-specific pathways. Whereas, Phe is derived from the shikimate biosynthetic pathway in the plastid [150]. Maybe, it is worth mentioning here that certain enzymes of the lignin biosynthetic pathway, like the cytochrome P450 enzymes cinnamate 4-hydroxylase (C4H), p-coumarate 3-hydroxylase (C3H), and ferulate 5-hydroxylase (F5H), are typically membrane proteins, presumably active at the cytosolic side of the endoplasmic reticulum [151, 152].

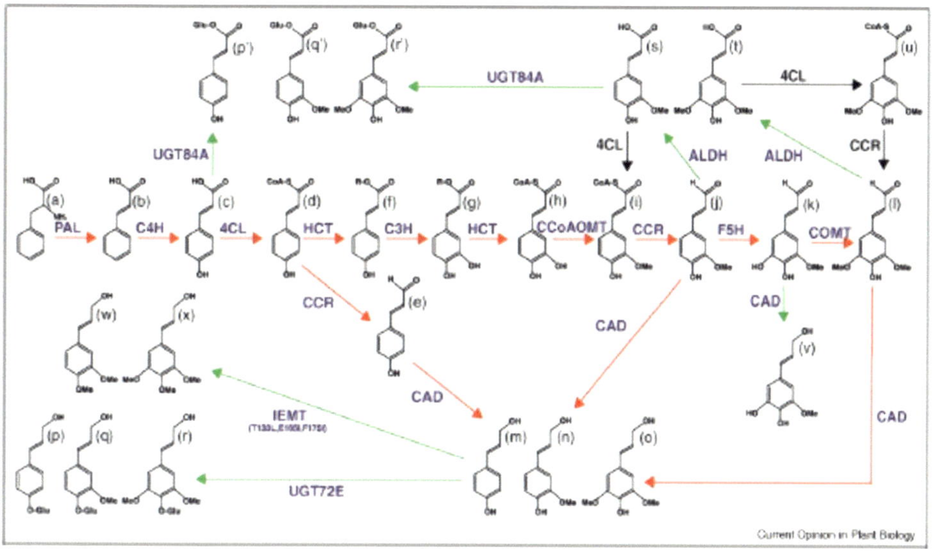

Figure 6. A schematic depiction of the known lignin biosynthesis pathways. 4CL, hydroxycinnamate-CoA/5-hydroxyferuloyl-CoA ligase; ALDH, hydroxycinnamaldehyde dehydrogenase; C3H, 4-hydroxycinnamate 3-hydroxylase; C4H, cinnamate 4-hydroxylase; CAD, hydroxycinnamyl alcohol dehydrogenase; CCoAOMT, S-adenosyl-methionine caffeoyl-CoA/ 5-hydroxyferuloyl-CoA-O-methyltransferase; CCR, hydroxycinnamoyl-CoA:NADPH oxidoreductase; COMT, caffeate O-methyltransferase; F5H, ferulic acid 5-hydroxylase; HCT, hydroxycinnamoyl-CoA shikimate/quinate hydroxycinnamoyl transferase; IEMT (T133 L, E165I, F175I), monolignol 4-O-methyltransferase (isoeugenol 4-O-methyltransferase carrying T133 L, E165I, F175I mutations); OMT, S-adenosyl-methione-caffeate/5-hydroxyferulate-O-methyltransferase; PAL, phenyl ammonia lyase; UGT84A, UDP-glucose: hydroxycinnamate glucosyltransferases; UGT72A, UDP-glucose:hydroxycinnamate 4-O-glucosyltransferase-like; a, phenylalanine; b, cinnamic acid; c, p-coumaric acid; d, p-coumaroyl-CoA; e, p-coumaraldehyde; f, p-coumaroyl shikimic acid/quinic acid; g, caffeoyl shikimic acid/quinic acid; h, caffeoyl-CoA; i, feruloyl-CoA; j, coniferaldehyde; k, 5-hydroxyconiferaldehyde; l, sinapaldehyde; m, p-coumaryl alcohol; n, coniferyl alcohol; o, sinapyl alcohol; p, p-coumaryl alcohol 4-O-glucoside; q, coniferyl alcohol 4-O-glucoside (coniferin); r, sinapyl alcohol 4-O-glucoside (syringin); s, ferulic acid; t, sinapic acid; u, sinapoyl-CoA; v, 5-hydroxy-coniferaldehyde; w, 4-methoxy-coniferyl alcohol; x, 4-methoxy-sinapyl alcohol; p0, p-coumaroyl glucose-ester; q0, feruloyl glucose-ester; r0, sinapoyl glucose-ester [8].

However, it remains unknown whether the other pathway active enzymes are also part of the metabolic complexes at the endoplasmic reticulum, although metabolic linkages has been shown between phenylalanine ammonia-lyse (PAL) and C4H [153, 154]. After synthesis, lignin monomers are exported into the plant cell wall where they will diffuse until they are oxidised by laccases or peroxidases and polymerised within the secondary plant cell wall [8]. The possible modes of export mechanism of the monolignols are illustrated in **Figure 7**.

Notwithstandingly the several proposed mechanisms (**Figure 8**), the exact mechanism of monolignol export from the cytosol to the apoplast still remains relatively unknown. The current up-to-date hypothesis suggests the exportation of free monolignols into the apoplast via plasma

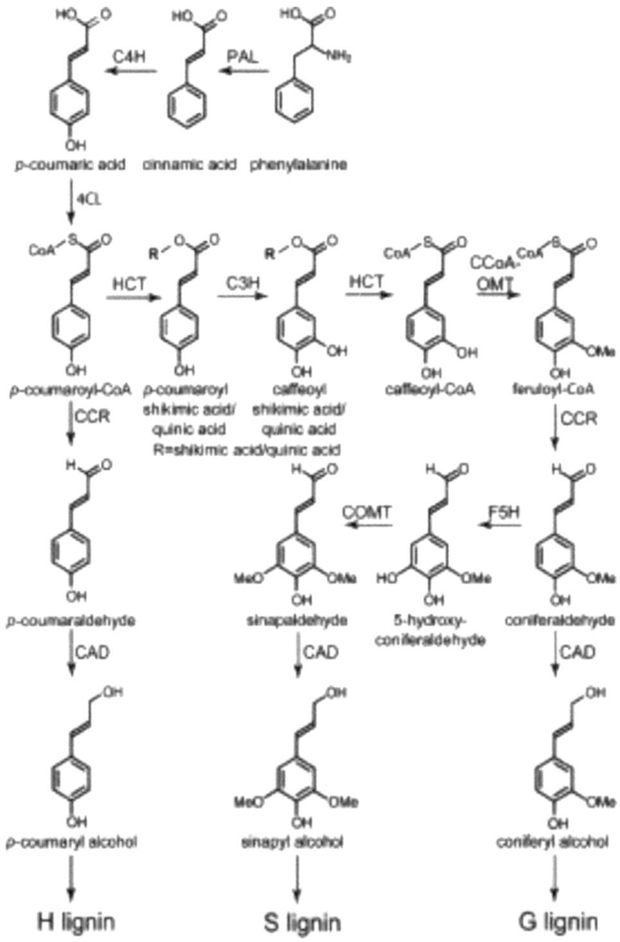

Figure 7. The main biosynthetic route towards the monolignols p-coumaryl, coniferyl, and sinapyl alcohol [6, 10]. PAL, phenylalanine ammonia-lyse; C4H, cinnamate 4-hydroxylase; 4CL, 4-coumarate: CoA ligase; C3H, p-coumarate 3-hydroxylase; HCT, p-hydroxycinnamoyl-CaA: Quinate/Shikimate p-hydroxycinnamoyl transferase; CCoAOMT, caffeoyl-CoA O-methyl transferase; CCR, cinnamoyl-CoA reductase; F5H, ferulate 5-hydroxylase; COMT, caffeic acid O-methyltransferease; CAD, cinnamyl alcohol dehydrogenase. Adapted from Vanholme et al. [5].

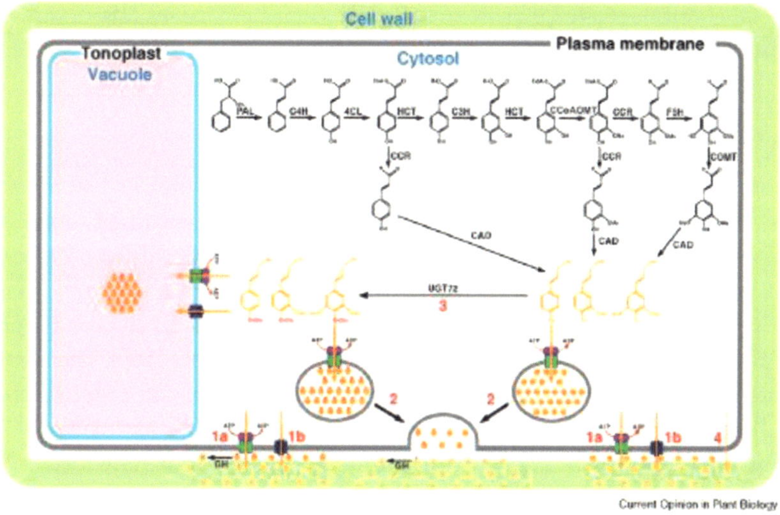

Figure 8. Schematic representation depicting lignin biosynthesis and possible modes of export mechanism of the monolignols: (1a) and (1b) protein-mediated transport mechanisms at the plasma membrane: (1a) active transporter (primary and secondary transport system) and (1b) facilitator protein; (2) secretion via vesicular export; (3) export of glycosylated monolignols (via a plasma membrane transporter or via vesicular export); (4) passive diffusion through the plasma membrane [8].

membrane transporters [155]. This is evidenced by the fact that feeding radiolabelled phenylalanine to lignifying cells did not show any vesicular accumulation of labelled monolignols [8, 155].

The most abundant renewable resource containing aromatic structures is lignin [156–158]. Lignin is the generic term for a large group of aromatic polymers resulting from the oxidative combinatorial coupling of 4-hydroxyphenylpropanoids [6, 10, 147]. Predominantly, these polymers are deposited in the thickened walls of secondary cells, resulting in their rigidity and making them impervious. Apart from the natural development via biosynthetic pathways and deposition of lignin, induced biosynthesis upon various biotic and abiotic stress conditions (e.g. wounding, pathogen infection, metabolic stress and perturbations in cell wall structure) have also been reported [159, 160]. As mentioned in the introduction, the protective characteristic of lignin, firmly protecting the cell wall polysaccharides from microbial degradation, and as a result imparting resistance to decay (recalcitrance), it thus presents one of the most important limiting factors in the conversion of plant biomass to pulp or biofuels. Extraction of lignin from plant biomass is a costly process. In recent times, research efforts are aimed at designing plants (bioengineering) that can either deposit less lignin or produce lignins that are more amenable to chemical degradation [161–163].

To conclude this section, one most recalls that the distribution of lignin within the cell walls is not uniform [149]. Zhou et al. [164] ascertained that the concentration of lignin in the middle lamella and the primary wall is higher than the concentration in the secondary wall. Nonetheless, it is reported in the literature [149] that the majority of the total amount of lignin present in the plant (75–85%) is located in the secondary wall, owing to its considerable larger volume. Structurally, the amount of lignin present in the plant varies from species to species, and range from 20 ± 4% in hardwoods to 28 ± 3% in softwoods and herbaceous angiosperms; monocots are less lignified (15 ± 4%) [121, 165].

3.1.2. Structure-dependent functional properties of lignins

Lignin (**Figure 1b**) is a highly abundant, low-cost biopolymer exhibiting phenylpropane units as a typical structural feature. This molecule is a major component of wood and annual plants. Lignin produced mainly as a waste product in the pulp industry on a scale of 40–50 million tons per year worldwide. Whereas, a considerable amount of lignin is burned so to recover as downstream chemicals from the pulping process, as well as provide process heat, only 2% of all lignins are utilised for rather low-value applications [166]. This picture has led to the general perception that lignin is just a waste product of the pulping process. As a result, this has restrained the research for value-added applications for lignin and lignin-based materials for quite some time. However, intensive research is now focused on seeking higher value products from lignin [167]. Traditionally, lignin has been derived from the kraft and sulphite processes that are ubiquitous in the pulp and paper industry from since time memorial [167]. It has been reported [166, 168, 169] that over 50 million tonnes are produced annually worldwide.

The structure of native remains unclear despite intensive studies [167] over several years. However, Kubo and Kadla [170] have reported several proposed structures purely based on the analysis of degradation products. This has been possible through the identification of the dominant linkages between the phenylpropane units and their abundance, as well as the abundance and frequency of certain functional groups [167](**Figure 9a** and **b**).

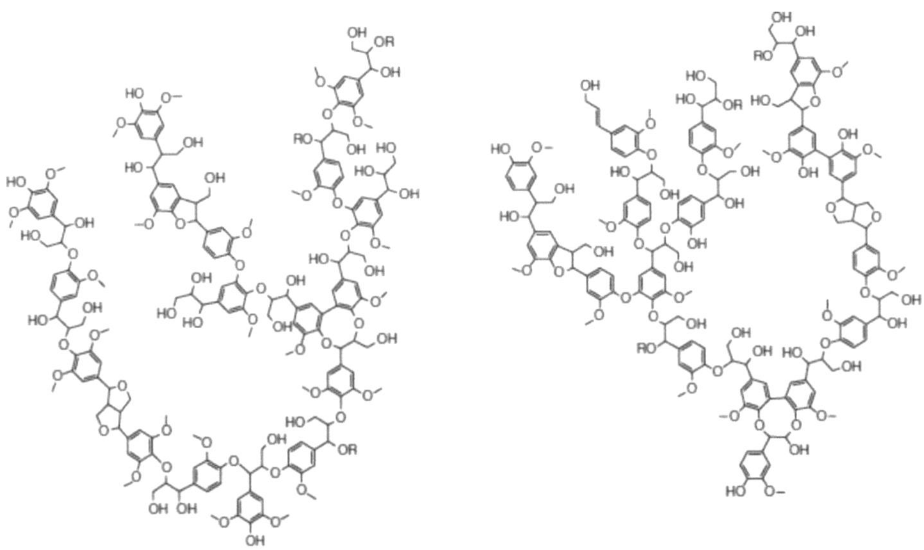

(a) (b)

Figure 9. (a) Hardwood native lignin; R = H, alkyl, aryl [170] (adapted from [167]); (b) softwood native lignin; R = H, alkyl, aryl [170] (adapted from [167]).

The functional groups (R = H, alkyl and aryl) have some great impact on the reactivity of the lignin molecule. It should be recalled that the lignin molecule mostly contains methoxyl groups, phenolic and aliphatic hydroxyl groups, and a few terminal aldehyde groups [167], though most of the phenolic hydroxyl groups are not available because of the fact that they are occupied in interunit linkages.

3.1.3. Characterisation of lignins

Chemically, lignin samples have been characterised with respect to lignin and sugar content and the amount of hydroxyl and carboxyl groups (**Table 4**).

Structurally, the characterisation of lignin can be well illustrated by the work of Li and Ge [171]. These co-workers utilised the byproduct of a non-wood material for the extraction of lignin. The plant material investigated was sugar cane bagasse from where lignin was extracted via acidification of black liquor prior to characterisation. Firstly, the authors extracted sugar cane bagasse lignin (EBL) and then modified it by oxidation, hydroxymethylation and sulphonation into a water-soluble lignosulphonate (EBL-M) as a final product for characterisation. Techniques employed successfully were infrared spectroscopy (IR), ultraviolet-violet spectrophotometry (UV), gel-permeation chromatography (GPC) and elemental (C, H, O, N and S) analysis. A typical IR spectra of extracted bagasse liquid (soluble aqueous fraction) is shown in **Figure 10**.

The characterisation of molecular weight and sulphonic group's content of lignosulphonates were determined by employing GPC and an automatic potentiometric titrator, respectively.

According to Li and Ge [171], the relative contents of moisture, ashes and pure lignin content of extracted bagasse lignin, can be measured as follows: an appropriate amount of raw EBL is weighed and placed in an oven and heated to 105°C in air. The corresponding mass loss would

Sample number	Raw material	Lignin type (commercial name)	Sample code	Lignin content (%)	COOH (mmol/g)	Phenolic OH (mmol/g)	Total sugars (%)
1	Hemp	Soda	He	Nd	2.1	1.58	2.4
2	Sisal	Soda	Si	Nd	1.18	2.29	7.7
3	Abaca	Soda	Ab	Nd	1.14	2.73	5.5
4	Straw	Soda	St	Nd	2.1	2.43	Nd
5	Jute	Soda	Ju	Nd	1.78	2.38	Nd
6	Flax	Soda	Fl-1	87.8	1.9	1.1	1.7
7	Softwood	Lignosulfonate (Borresperse 3A)	SW-LS-1	Nd	3.5	1.1	1.3
8	Softwood	Kraft (Indulin AT)	SW-Kr-1	90.0	2.5	1.8	2.06
9	Softwood	Lignosulfonate (Wafex P)	SW-LS-2	Nd	1.2	1.1	24.5
10	Flax	Soda (Bioplast)	Fl-2	87.8	1.9	1.1	1.7
11	Flax	Soda, oxidised	Fl-ox	92.1	1.8	0.9	1.6
12	Softwood	Kraft (Curan 100)	SW-Kr-2	88.6	1.7	0.8	2.26
13[a]	Softwood	Soda, precipitation high pH	SW-SF-1	64.7	Nd	Nd	1.77
14	Softwood	Kraft	SW-Kr-3	96.8	2.0	2.5	0.71
15	Mixed hardwoods	Organosolv (Alcell)	HW	96.5	0.78	2.4	0.32
16	Softwood	Soda, precipitation low pH	SW-SF-2	99.2	1.4	1.0	0.65
17	Softwood	Kraft (Curan 2711P)	SW-Kr-4	80.3	2.7	1.6	2.58

Nd, not determined.
[a] Lignin 13 contains about 40% ash.

Table 4. Chemical characteristics of lignin samples [127].

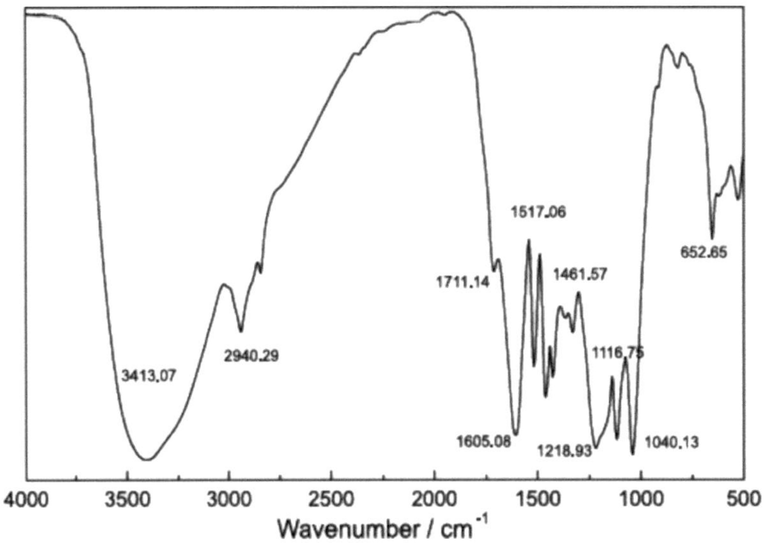

Figure 10. FT-IR spectrum of EBL-M [171].

be assigned as moisture. Another sample is placed in a muffle furnace and heated to 800°C in air. The remaining amount would be accounted as ashes. The pure lignin amount is subtracted from the moisture and ashes from the total amount of the sample. An example of the physico-chemical properties of extracted bagasse lignin (EBL) is shown in **Table 5**.

3.1.4. Trends and research avenues

From the literature, several reviews [37, 38, 172–174] have been published to cover various aspects of lignin, especially on processes aiming at its industrial applications and valorisation [149].

Extensive research has been conducted in the past decades in both the improvement of lignin extraction processes and the elucidation of its structure [127]. Such research efforts have been extended to cover works on the characterisation of chemical reactivity and functional properties, as well as the development of new applications of lignin [127]. In this regard, several co-workers [175–177] have shown that lignin is a versatile molecule endowed with multiple properties such as antioxidant (i.e. radical scavenger), UV-absorption, antifungal and antibiotic

Properties	EBL
Lignin source	Sugar cane bagasse pulping
Appearance	Light brown powder
Moisture / wt.%	5.2
Ashes / wt.%	2.9
Pure lignin / wt.%	91.9

Table 5. Properties of lignin extracted from sugar cane bagasse (EBL) [171].

activity. From the above, it is suggestive in agreement with the viewpoint of Boeriu et al. [127] that to enhance the industrial use of lignins, there is need for a continuous supply of lignin products with constant quality as related to purity, chemical composition and functional properties, meaning research is still to play central role to be able to meet the above goals.

As a conclusion and in alignment with the thoughts conceptualised by MacFarlane et al. [167], trends and research avenues focusing on lignins can be summarised within the following timeframes.

Firstly, economic studies must find a large market for lignin from the ubiquitous kraft process in the short term. The role that research can play in during this timeline is to intensify to improve on the quality and applications of kraft lignin as resins and adhesives. Penetration into this market is considered inevitable by Stewart [178].

It is predicted in the medium term that more lignin from unsulphonated sources will become available. As this lignin is likely to come abundantly from biorefinery platforms, owing to its uniqueness in properties compared to other feedstocks with a potential to uplift small-scale availability, research would have to stretch in that direction in the medium term. Initially, the resin and adhesives market may act as a catch-all for this more reactive lignin [167]. According to the same author, a product niche of high value and sufficient market size must be found for each lignin source, citing hydrophobic organosolv lignins as an example to hold the potential to add value to resins used for water resistant particleboard. Platform chemicals are an attractive medium-term market.

Finally, in the long term, biorefineries will be tuned to produce lignin for specific materials of high value, and lignin is expected to yield greater incomes than the other components of wood [167].

3.2. Renewable platform chemicals

3.2.1. Lignosulphonates

Lignosulphonate (**Figure 11**) is the sulphonated lignin that has been removed from wood by sulphite pulping. Hardwood lignosulphonate and softwood lignosulphonate are obtained from waste pulping liquor concentrate by the Howard process [179] after stripping and recovery of the sulphur [149]. They exhibit monomer molecular weights of 188 Da, and 215–154 Da, respectively [149].

Up to now, lignosulphonates are the only lignin products obtained from a pulping process that have found a considerable range of applications up to now are lignosulphonates (MacFarlane et al., 2013). These materials are endowed with excellent properties that, thus, are used as binders, emulsifiers and dispersants for a great variety of materials; and are non-hazardous. Good examples as binders are their common use for commodities like coal briquettes, ceramics, briquetting of mineral dust, and the production of plywood or particle boards, owing to their cheap cost. In addition, their ability to retain moisture and suppress dust makes them a useful tool for construction works, gravel roads, airports and sports facilities [180]. This material is an anti-settling agent that can prevent lumping, making lignosulphonates excellent

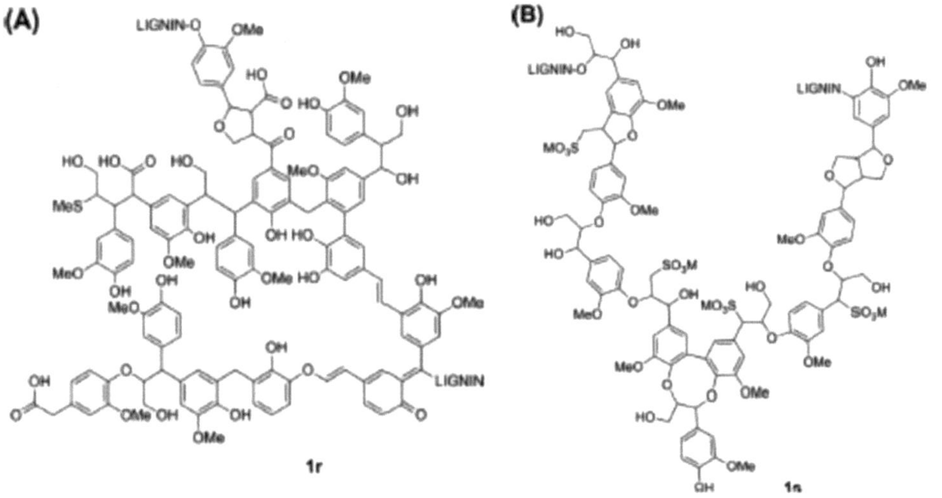

Figure 11. Structure of kraft lignin (A) and lignosulphonate (B) [149].

binders used in concrete mixtures, ceramics [180], gypsum-board production and for leather tanning. Lignosulphonates provide flowability and plasticity to cement [180]. These properties lends lignosulphonates as a replacement for more expensive materials that provide set retardation such as superplasticizers, gluconates and gluconic acid. In the wet-process, Portland cement mills lignosulphonates are utilised to increase the solids content of the raw slurries. It is stated [37, 38] that the demand for lignin-based concrete additives are on the increase with a market value worth as much as US $1.05–1.32/L as an aqueous solution. Nadif et al. [181] has also reported that sulphur-free lignin such as soda lignin has also been shown to improve flowability of mortar. It is known that lignosulphonates can also stabilise emulsions of immiscible fluids like asphalt emulsions, pesticide preparations, pigments and dyes [167]. Due to their low toxicity, they can be used as binders in animal feed and thereby improve the feed properties of pellets. In addition, lignosulphonates show the ability to keep micronutrients in solution which is useful for micronutrient transport or as a cleaning and decontaminating agent in water and soils [7, 181].

Apart from the usual (traditional) use of lignosulphonates as energy source and in leather tanning, these materials can now also be applied for polymer reinforcement, or as sequestrants (i.e. antioxidants). It has been suggested [176] that lignins can be applied for stabilisation of food and feed, due to their antioxidant and antifungal properties. Also, anti-carcinogenic and antibiotic activities of lignins have been reported [176]. When depolymerised, lignins can be used for the synthesis of valuable chemicals, such as vanillin.

3.2.2. *Lignin oligomers*

Chemically, lignin is viewed as a highly complex phenolic polymer, generally showing plant-specific compositions and linkage motifs [182, 183]. Very recent findings on milled wood lignin samples are suggesting, however, that lignin exists as linear oligomers. Presumably, strongly interactive in such a way that traditional analyses of the molecular weights are biased, thus, a suggestion of the existence of higher molecular weight polymeric units [184]. Apparently and to the best of the current knowledge, these lignin oligomers lack a defined primary structure, but rather represent random phenylpropanoid (C_9) polyphenols, mainly linked by arylglycerol ether bonds between phenolic para-coumaryl alcohol (2) (H-type), coniferyl alcohol (3) (G-type) and sinapyl alcohol (4) (S-type) units [185, 186]. Depending on the plant-type, one of the different lignin types dominants. Lignin of gymnosperms consists almost entirely of G-type lignin (G-lignin); dicotyledonous angiosperms produce a mixture of G- and S-type lignins (GS-lignin). All three types of lignin can be found in quantities in monocotyledonous lignin (GSH-lignin) [149].

For the moment, there are a multitude of practical considerations that preclude further use of lignin as a monomer feedstock [156, 158] but actual developments on the level of low-cost catalysis and separation technologies could stimulate its future use [187].

4. Treatments and technologies towards renewable chemicals recovery

4.1. Treatments and technologies

4.1.1. Chemical treatments an advanced thermochemical recovery

There are several thermochemical methods reported in the literature that have been studied in the recent past in order to depolymerise lignin and to convert it into value-added chemicals. To name a few, pyrolysis, gasification, hydrogenolysis, chemical oxidation and hydrolysis under supercritical conditions are the major methods to produce pyrolytic oil, syngas or phenols. **Figure 12** shows an overview of these methods and the resulting products thereafter the treatments.

Lignins from trees, plants and agricultural crops with different chemical composition and properties can be obtained by use of several extraction methods. These include the commercial chemical pulping processes (sulphite and kraft process) to produce lignosulphonates and kraft lignins as residue. Recently, commercialised alkaline pulping-precipitation process supply sulphur-free, free-flowing lignins. The challenge resides, however, in practically isolating pure lignin quantitatively from cell walls in an intact state. The lignin isolated by known methods (physical, chemical or enzymatic treatments) is a mixture of degraded or solubilised lignin from various unidentified morphological regions [127].

4.1.2. Specialties base chemicals recovery

Roopan (2016) has stated that lignin is a best essential source for bio-pesticides. This author collected lignin compounds from fluidized bed pyrolysis reactor and investigated it for utilisation in bio-oil production. Other authors [188] have also collected lignin, cellulose and hemicellulose macromolecules from waste disposables and purified the lignin without any contaminants of macromolecules for test against various toxi-micro-organisms of fungi and bacterial species like *Xanthomonas campestris*, *Pythium ultimum*, *Sclerotinia sclerotiorum*, *Clavibacter michiganensis*, *Rhizoctonia solani* and *Streptomyces scabies*.

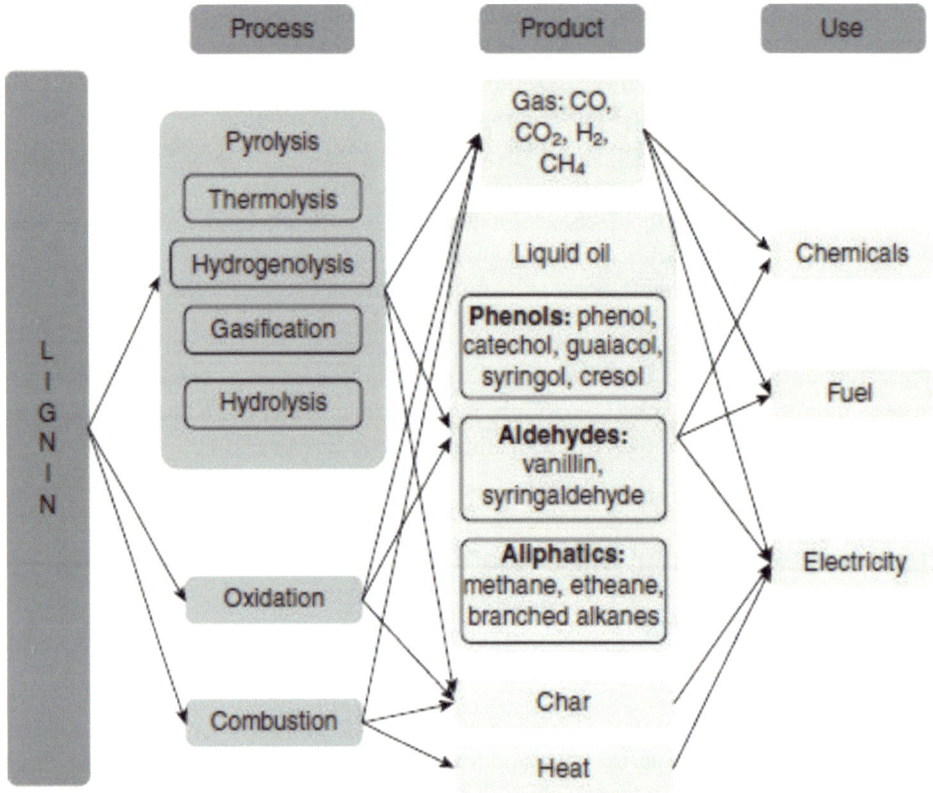

Figure 12. Major thermochemical lignin conversion processes and their potential products [167].

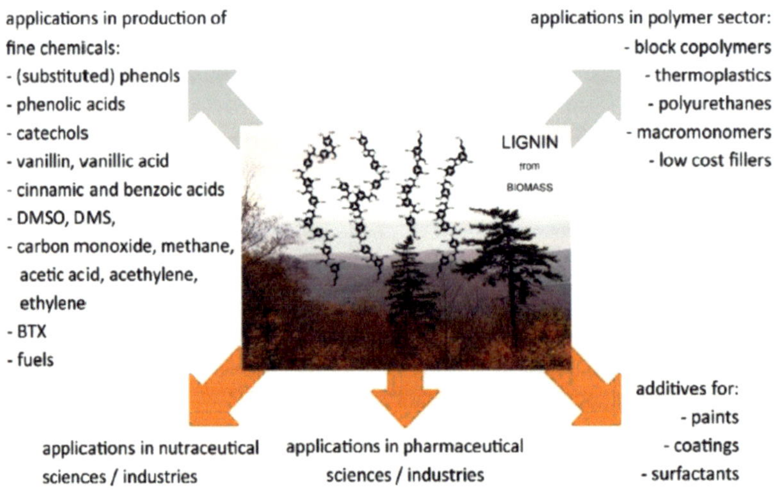

Figure 13. Existing and potential applications of lignin are summarised in [149].

Other specialty chemicals have been developed in recent times. Aromatic building blocks as platform chemicals in the field of polymer chemistry cannot are good examples, with scientific terms such as PET and bisphenol A becoming known to the general public. This is a result of their high production volumes, which were predicted to exceed 24 million tonnes [189] and 5.6 million tonnes [2], respectively, in 2015.

Since aromatic building blocks are to date being produced mainly from petroleum-based feedstocks, such as benzene, xylene and cumene, there is an ongoing quest for more sustainable and renewable sources for aromatic synthons evolving in both academia.

Llevot et al. [190] and Isikgor and Becer [191] and industry [192], using lignins. Some examples of bio-available aromatic systems include cardanol, originating from cashew nut shell liquid [193], hydroxymethylfurfural (HMF) obtained by carbohydrate dehydration.

Figure 13 summarises the speciality chemicals that can be obtained from lignins.

5. Separation and recovery of gold and other precious metals

5.1. Gold

5.1.1. Occurrence and fundamental properties of gold

Gold (Au) as a mineral usually occur in native forms associated with silver (Ag) along with copper (Cu), arsenic (As), iron (Fe), bismuth (Bi), tin (Sn) and zinc (Zn) [194]. The concentration of gold (Au) in the lithosphere is estimated to be 0.004 $\mu g\ g^{-1}$ in the rock, whereas in the soil it is about 0.001 $\mu g\ g^{-1}$ [195]. At the same time, its abundance in seawater and river is estimated at 0.05 $\mu g\ g^{-1}$ and 0.02 $\mu g\ g^{-1}$, respectively, much higher in the seawater and river [195]. Gold has been found in the atmospheric dust in a concentration as high as 440 mg g^{-1}, especially in the vicinity of gold mines [196]. Gold has a very high affinity for sulphur and often occurs as ultrafine solid solution inclusions in sulphide minerals [197]. The world distribution of gold is not with China as the leading country with 16.7% of the world production [198],

Precious metals such as gold (Au), silver (Ag), platinum (Pt), palladium (Pd), etc., are utilised in several manufacturing fields for various applications [180, 199]. These "noble" metals possess unique physical and chemical properties, thus, their high economic value.

Though the natural occurrence in the earth crust of precious metals and Au generally termed the platinum group elements (PGEs) is in very low abundance, their concentration and distribution in various types of rock formations have attracted enormous interest from miners in recent times. This is due to their rising demand in the technology sector owing to their applications as catalysts, biomedical tools (e.g. anticancer drugs), corrosion-resistant materials, high electrical conductivity materials, micro-electronics and also in jewellery manufacturing [200].

The fundamental properties of gold are indeed an interesting one. Gold (atomic number 79, mp 1065°C) is thence the noblest metal that is generally found in nature in the metallic state, and the only gold compounds that occur in a natural state are the telluride and stibnite species, $AuTe_2$ and $AuSb_2$ [202]. The metal is soft, yellow and malleable, ductile and resistant to attack by common chemicals and light [201].

Yannopoulos [194] has reported 19 isotopes (ranging from ^{185}Au to ^{203}Au) although only the ^{197}Au isotope is known to occur naturally. Of a great significance to the extractive metallurgist is the chemistry of gold complexes in aqueous solutions, and this is a far broader subject beyond the scope of this chapter. Traditionally, it has been accepted that gold complexes in aqueous solution can exist in one of two oxidation states, the aurous (+1) or the auric (+3), and forming Au^+ and Au^{3+}, and that all the gold complexes of hydrometallurgical interest fall into these two groups [202]. The electron configuration of the element in the ground state is [Xe] $4f^{14}5d^{10}6s^1$. From this is deduced the monovalent ion: [Xe] $4f^{14}5d^{10}$ and the trivalent ion: [Xe] $4f^{14}5d^8$. Thermodynamics predicts that neither aurous (Au^+) nor auric (Au^{3+}) cations will be stable in aqueous solution, but will be reduced by water to metallic gold. Gold does not dissolve in either nitric acid (an oxidising agent) or hydrochloric acid (a complexing ligand), but dissolves quite easily in a mixture of the two ($HCl:HNO_3$, 3:1), aqua regia [203]. This is because the oxidation potential for transition Au^o to Au^{3+} is as high as 9.225 eV [204].

Thus, many of the known thermodynamic characteristics of gold complexes can be rationalised, providing some basis for the prediction of their behaviour. Chlorine (Cl) and bromine (Br) can also dissolve gold to generate Au^{3+} and under these reducing conditions, Au^{3+} species can easily be disproportionately converted from Au^o to Au^+. In water, Au^+ and Au^{3+} ions occur in a hydrated form. Nicol et al. [202] states that a useful first generalisation about the stability of gold complexes relates to the nature of the ligand donor atom, that is, the atom in the ligand that is bonded directly to the gold. From this, both gold(I) and gold(III) are B-type metal ions, meaning that the stability of their complexes tends to decrease as the electronegativity of the ligand donor atom increases (i.e. as the tendency of the atom to attract electrons increases). This leads to stability orders such as $I^- > Br^- > Cl^- > F^-$, where the order of electronegativities is $F^- > Cl^- > Br^- > I^-$.

5.1.2. Selective separation and recovery of gold from various secondary sources

Mining efforts of precious metals (Au, Ag, Pt, etc.) are currently being hampered owing to chemical ambiguities of mineralization. The need for the development of simple, fast and reliable recovery procedures is important.

5.2. Precious metals

As the name suggests, this group consists of precious metals out of which Au and Ag were probably the first metals ever extracted on earth [205], whereas the platinum group metals such as Pt and Pd are increasingly finding their role in modern products especially as auto catalysts used in the emission control system of passenger vehicles. These elements are also used in jewellery. Au and Ag, both have high electrical conductivity which makes them highly desirable in modern electronics apart from their historical use as investment and jewellery.

The fact the modern society depends intensively on metals utilisation, is the justification that metals are the basis of our infrastructure and the technology, without which it is hard to imagine modern life. Metals have been generated as a result of geological events spanning over hundreds of millions of years. As such, metals are considered by some school of thoughts as non-renewable by nature as they which makes them a finite resource. This can be argued as true if metals, nevertheless, do not disappear after their consumption per se like some other resources like fuel minerals (e.g. oil, gas and coal due to their non-dissipative nature). Somehow, metals can be beneficiated by recycling them from urban mines, though with the product's long lifetimes and economic implications. Despite the above ascertainment, Bardi [205] and Prior et al. (2012) argue that despite the fact that the geological resources of metals are finite, their unprecedented extraction from the earth especially during the last two centuries has raised concerns regarding their long-term availability to meet the demand of future generations. Within the context of secure, uninterrupted and long-term availability of resources, a relatively new research field of resource criticality assessment got widespread popularity during the recent years [206].

Palladium (Pd) belongs to the precious metals group, that is, platinum group metals (PGMs), which are mainly used as catalysts in the automobile sector for pollution abatement. Other uses of PGMs are in fuel cells, petroleum refining, chemical industry, electronics, glass manufacturing, medical appliances, jewellery and as investment.

6. Concluding remarks

A brief review has been herein done of technologies involved in the exploitation of lignin, in order to provide an introduction to the subject from the perspective of a fast technologically advancing economy. Lignocellulosic materials and biomass have historically been utilised from since time memorial, but a new conversation is emerging on the role of these materials in modern bioeconomies. Due to the nature of the molecule(s) as a novel commodity for many interesting natural and manufactured products, a modern bioeconomy is not simply a rerun of former ones. This new discourse needs to help us understand how technologies for managing and processing lignocellulosic materials both as biosynthetic moieties, biogenic wastes or simply renewable biopolymer—both established and novel—should be deployed and integrated (or not) to meet developmental requirements of the sustainability paradigm, more so in a closed-loop and resource-security agendas that undoubtedly sits behind the bioeconomy aspirations now being voiced by many schools of thought in many countries and regions of the so fast technologically advancing world.

The world of which is caught in the middle of green technology advocating for more and more focus on renewable sources of manufacturing raw materials and that of the molecularly imprinted/synthesised or genetically engineered ones. **Figure 1** presents the main building blocks of lignin, from the hydroxycinnamyl alcohols (or monolignols) coniferyl alcohol and sinapyl alcohol, with typically minor amounts of p-coumaryl alcohol [6, 10].

The utilisation of lignins (from renewable sources) in both the industry as the base for the formulation of ionic liquids (with yet a wider industrial applications) and as a potential scaffold material for functionally modified imprinted polymers (LCIPS) for the selective recovery of base metals and gold, respectively, evidently incorporates the bioeconomy aspirations.

Further, one new trend in lignin research is to steer biosynthetic pathways towards the biosynthesis of molecules that, upon incorporation into the lignin polymer, will improve lignin degradation [5].

Acknowledgements

We thank the Departments of Chemistry (Faculty of Science) of both the Tshwane University of Technology (TUT), South Africa and the University of Douala, Cameroon, for support. The authors are grateful to the Nano-structured Materials Centre (CSIR) laboratory for their contribution in their earlier work performed in the characterisation of lignosulphonates and ionic polymeric-treated materials using the scanning electron microscopy (SEM).

The authors wish to thank Professor B.M. Mgangira (Council for Scientific and Industrial Research (CSIR), Built Environment, Pretoria, Republic of South Africa) and Professor R.I. McCrindle (Department of Chemistry of the Tshwane University of Technology, TUT) in co-authoring some of the published work on the treatment and applications of ionic and polyelectrolytic polymeric materials for the treatment of subgrade soil pavement.

Author details

Peter Papoh Ndibewu[1]* and Pierre Gerard Tchieta[2]

*Address all correspondence to: ndibewup@tut.ac.za

1 Department of Chemistry, Tshwane University of Technology, Pretoria, Republic of South Africa

2 Analytical Chemistry and Materials Science Laboratory, Department of Chemistry, Faculty of Science, University of Douala, Douala, Cameroon

References

[1] Roopan SMS. An overview of natural renewable bio-polymer lignin towards nano and biotechnological applications. A Review. International Journal of Biological Macromolecules. 2017;**103**:508-514

[2] García A, Spigno G, Labidi J. Antioxidant and biocide behaviour of lignin fractions from apple tree pruning residues. Industrial Crops and Products. 2017;**104**:242-252

[3] Jääskeläinen AS, Liitiä T, Mikkelson A, Tamminen T. Aqueous organic solvent fractionation as means to improve lignin homogeneity and purity. Industrial Crops and Products. 2017;**103**:51-58

[4] Gao L, Chen S, Zhang D. Advances in modifying lignin structures for largely enhancing high-lignin biomass saccharification. Process Biochemistry. 2017;**57**:175-180

[5] Vanholme R, Demedts B, Morreel K, Ralph J, Wout Boerjan W. Update on lignin biosynthesis and structure: Lignin biosynthesis and structure. Plant Physiology. 2010;**153**: 896-905

[6] Boerjan W, Ralph J, Baucher M. Lignin biosynthesis. Annual Review of Plant Biology. 2003;**54**:519-546

[7] Calvo-Flores FG, Dobado JA. Lignin as renewable raw material. ChemSusChem. 2010;**3**(11): 1227-1235

[8] Simmons BA, Loque D, Ralph J. Advances in modifying lignin for enhanced biofuel production. Current Opinion in Plant Biology. 2010;**13**:313-320

[9] Glennie DW, McCarthy JL. Chemistry of lignin. In: Libby CE, editor. Pulp and Paper Science and Technology. New York: McGraw-Hill Book Company, Inc.; 1962. pp. 82-107

[10] Boerjan W, Meyermans H, Chen C, Baucher M, Van Doorsselaere J, Morreel K, Messens E, Lapierre C, Pollet B, Jouanin L, Leplé JC, Ralph J, Marita JM, Guiney E, Schuch W, Petit-Conil M, Pilate G. In: Morohoshi N, Komamine A, editors. Molecular Breeding of Woody Plants. Amsterdam: Elsevier Science; 2001. pp. 187-194 (Chapter 23)

[11] Bilal M, Asgher M, Hafiz MN, Iqbal HMN, Hongbo Hu H, Zhang X. Biotransformation of lignocellulosic materials into value-added products—A review. International Journal of Biological Macromolecules. 2017;**98**:447-458

[12] Fortunati E, Luzi F, Puglia D, Torre L. Extraction of Lignocellulosic Materials From Waste Products. Terni, Italy: Civil and Environmental Engineering Department, UdR INSTM, University of Perugia; 2016

[13] Hu Y, Hao X, Wang J, Cao Y. Enhancing anaerobic digestion of lignocellulosic materials in excess sludge by bio augmentation and pre-treatment. Waste Management. 2016;**49**:55-63

[14] Andriania A, Tachibana S. Lignocellulosic materials as solid support agents for Bjerkandera adusta SM46 to enhance polycyclic aromatic hydrocarbon degradation on sea sand and sea water media. Biocatalysis and Agricultural Biotechnology. 2016;**8**:310-320

[15] Tzvetkov G, Mihaylova S, Stoitchkova K, Tzvetkov P, Spassov T. Mechanochemical and chemical activation of lignocellulosic material to prepare powdered activated carbons for adsorption applications. Powder Technology. 2016;**299**:41-50

[16] Lo SL, Huang YF, Chiueh TP, Kuan WH. Microwave pyrolysis of lignocellulosic biomass. The 8th International Conference on Applied Energy – ICAE2016. Energy Procedia. 2017;**105**:41-46

[17] Sansaniwala SK, Rosen MA, Tyagi SK. Global challenges in the sustainable development of biomass gasification: An overview. Renewable and Sustainable Energy Reviews. 2017; **80**:23-43

[18] Li Y, Zhou LW, Wang RZ. Urban biomass and methods of estimating municipal biomass resources. Renewable and Sustainable Energy Reviews. 2017;**80**:1017-1030

[19] Ferreira S, Monteiro E, Paulo Brito P, Vilarinho C. Biomass resources in Portugal: Current status and prospects. Renewable and Sustainable Energy Reviews. 2017;**78**:1221-1235

[20] Proskurina S, Heinimo J, Schipfer F, Esa Vakkilainen E. Biomass for industrial applications: The role of torrefaction. Renewable Energy. 2017;**111**:265-274

[21] Patermann C, Aguilar A. The origins of the bioeconomy in the European Union. New Biotechnology. 2017;**40**:20-24

[22] Lancker JV, Wauters E, Huylenbroeck GV. Managing innovation in the bioeconomy: An open innovation perspective. Biomass and Bioenergy. 2016;**90**:60-69

[23] McCormick K, Kautto N. The bioeconomy in Europe: An overview. Sustain. 2013;**5**: 2589-2608

[24] Schmid O, Padel S, Levidow L. The bio-economy concept and knowledge base in a public goods and farmer perspective. Bio-based and Applied Economics. 2012;**1**:47-63

[25] European Commission. Communication from the Commission to the European Parliament, the Council, the European Economic and Social Committee and the Committee of the Regions—Innovating for Sustainable Growth: A Bioeconomy for Europe. Brussels, Belgium: European Commission; 2012

[26] Quast K. Literature review on the use of natural products in the flotation of iron oxide ores. Minerals Engineering. 2017;**108**:12-24

[27] Costa R, Santos L. Delivery systems for cosmetics—From manufacturing to the skin of natural antioxidants. Powder Technology. 2017;**322**:402-416

[28] Bischoff T, Vogl CR, Ivemeyer S, Klarer F, Meier B, Hamburger M, Walkenhorst M. Plant and natural product based homemade remedies manufactured and used by farmers of six central Swiss cantons to treat livestock. Livestock Science. 2016;**189**:110-125

[29] Zhou J, Du G, Chen J. Novel fermentation processes for manufacturing plant natural products. Current Opinion in Biotechnology. 2014;**25**:17-23

[30] Ramcilovic-Suominen S, Pülzl H. Sustainable development e A 'selling point' of the emerging EU bioeconomy policy framework? Journal of Cleaner Production. 2017:1-11

[31] EC. 2012. Bio-economy Newsletter, February. http://ec.europa.eu/research/bioeconomy/press/pdf/120202_research_en.pdf [Accessed October 2015]

[32] EU (European Union). EU Presidency. En Route to the Knowledge-based Bioeconomy. Cologne Summit of the German Presidency, Cologne, Germany. 2007. http://www.bioeconomy.net/reports/files/koln_paper.pdf [Accessed: October, 2017]

[33] WCED (World Commission on Environment and Development). Report of the World Commission on Environment and Development: Our Common Future. 1987

[34] Toivanen H, Novotny M. The emergence of patent races in lignocellulosic biofuels, 2002–2015. Renewable and Sustainable Energy Reviews. 2017;**77**:318-326

[35] Bozell JJ, Petersen GR. Technology development for the production of biobased products from biorefinery carbohydrates—The US department of energy's "Top 10" revisited. Green Chemistry—Royal Society of Chemistry. 2010;**12**:539

[36] Lee S, Yoon B, Park Y. An approach to discovering new technology opportunities: Keyword-based patent map approach. Technovation. 2009;**29**:481-497

[37] Holladay JE, White JF, Bozell JJ, Johnson D. Top Value-added Chemicals from Biomass —Volume II? Results of Screening for Potential Candidates from Biorefinery Lignin [Internet]. Richland, WA; 2007. http://dx.doi.org/10.2172/921839 [Accessed 9 October, 2017]

[38] Holladay JE, Bozell JJ, White JF, Johnson D. Top Value Added Chemicals from Biomass: Results of Screening for Potential Candidate from Sugars and Synthesis Gas. Vol. 2. Pacific Northwest National Laboratory, US Department of Energy; 2007

[39] Angelini S, Cerruti P, Immirzi B, Scarinzi G, Malinconico M. Acid-insoluble lignin and holocellulose from a lignocellulosic biowaste: Bio-fillers in poly(3-hydroxybutyrate). European Polymer Journal. 2016;**76**:63-76

[40] Dutta S. Lignin Deconstruction: Chemical and Biological Approaches. Department of Chemical Engineering: National Taiwan University, Taipei, Taiwan; 2015

[41] Uzal-Novo E, Pomar F, Ros LVG, Espineira JM, Barcelo AR. Evolutionary History of Lignins. In: Department of Animal Biology, Plant Biology and Ecology, University of A of A Coruna, A Coruna. Spain: Spain & Department of Plant Biology, University of Murcia, Murcia; 2012

[42] Berthet S, Demont-Caulet N, Pollet B, Bidzinski P, Cezard L, Le Bris P, Borrega N, Herve J, Blondet E, Balzergue S, Lapierre C, Jouanin L. Disruption of LACCASE4 and 17 results in tissue specific alterations to lignification of *Arabidopsis thaliana* stems. The Plant Cell. 2011;**23**:1124-1137

[43] Bonawitz ND, Chapple C. The genetics of lignin biosynthesis: Connecting genotype to phenotype. Annual Review of Genetics. 2010;**44**:337-363

[44] Grabber JH, Lu F. Formation of syringyl-rich lignins in maize as influenced by feruloylated xylans and p-coumaroylated monolignols. Planta. 2007;**226**:741-751

[45] Aloni R, Feigenbaum P, Kalev N, Rozovsky S. Hormonal control of vascular differentiation in plants: The physiological basis of cambium ontogeny and xylem evolution. In: Savidge R, Barnett J, Napier R, editors. Cell and Molecular Biology of Wood Formation. Oxford, UK: BIOS Scientific; 2000. pp. 223-236

[46] del Rıo JC, Gutie'rrez A, Martınez AT. Identifying acetylated lignin units in non-wood fibers using pyrolysis-gas chromatography/mass spectrometry. Rapid Communications in Mass Spectrometry. 2004;**18**:1181-1185

[47] Boyce CK, Cody GD, Fogel ML, Hazen RM, Alexander CMOD, Knoll AH. Chemical evidence for cell wall lignification and the evolution of tracheids in early Devonian plants. International Journal of Plant Sciences. 2003;**164**:691-702

[48] Boudet AM, Lapierre C, Grima-Pettenati J. Biochemistry and molecular biology of the lignification. New Phytologist. 1995;**129**:203-236

[49] Delwiche CF, Graham LE, Thomson N. Lignin-like compounds and sporopollenin in Coleochaete, an algal model for land plant ancestry. Science. 1989;**245**:399-401

[50] Noeiaghaei T, Mukherjee A, Dhami N, Chae SR. Biogenic deterioration of concrete and its mitigation technologies. Construction and Building Materials. 2017;**149**:575-586

[51] Huang JH, Tian L, Ilgen G. Biogenic arsenic volatilisation from an acidic fen. Science of the Total Environment. 2017;**615**:1470-1477

[52] O'Callaghan K. Technologies for the utilisation of biogenic waste in the bioeconomy. Food Chemistry. 2016;**198**:2-11

[53] Brosowski A, Thran D, Mantau U, Mahro B, Erdmann G, Adler P, Stinner W, Reinhold G, Hering T, Blanke C. A review of biomass potential and current utilisation—A Status quo for 93 biogenic wastes and residues in Germany. Biomass and Bioenergy. 2016;**95**:257-272

[54] Suberu MY, Bashir N, Mustafa MW. Biogenic waste methane emissions and methane optimization for bioelectricity in Nigeria. Renewable and Sustainable Energy Reviews. 2013;**25**:643-654

[55] Mohammed YS, Mokhtar AS, Bashir N, Saidur R. An overview of agricultural biomass for decentralized rural energy in Ghana. Renewable and Sustainable Energy Reviews. 2013;**20**:15-22

[56] Batidzirai B, Smeets EMW, Faaij APC. Harmonising bioenergy resource potentialsd: Methodological lessons from review of state of the art bioenergy potential assessments. Renewable and Sustainable Energy Reviews. 2012;**16**:6598-6630

[57] Authur R, Baidoo MF. Harnessing methane generated from livestock manure in Ghana, Nigeria, Mali and Burkina Faso. Biomass and Bioenergy. 2011;**35**:4647-4656

[58] Huang JH, Matzner E. Dynamics of organic and inorganic arsenic in the solution phase of an acidic fen in Germany. Geochimica et Cosmochimica Acta. 2006;**70**:2023-2033

[59] Groß M, Lima MT, Uhlig M, Ebraheme A, Roeber O, Olschewski B, von Klitzing R, Schomäcker R, Schwarze M. Biopolymers for dye removal via foam separation. Separation and Purification Technology. 2017;**188**:451-457

[60] Kim J. Multifunctional Smart Biopolymer Composites as Actuators. Department of mechanical Engineering: Inha Univeristy, Incheon, South Korea; 2017

[61] Mohiuddin M, Kumar B, Haque S. Biopolymer Composites in Photovoltaics and Photodetectors. School of Engineering, Rmit University, Melbourne, Vic, Australia, Department of Mechanical & Industrial Engineering; University of Illinois at Chicago, Chicago, IL, USA, Department of Electrical & Computer Engineering; Sungkyunkwan University, Seoul, South Korea; 2017

[62] Bhuiyan D, Jablonsky MJ, Kolesov I, Middleton J, Wick TM, Tannenbaum R. Novel synthesis and characterization of a collagen-based biopolymer initiated by hydroxyapatite nanoparticles. Acta Biomaterialia. 2015;**15**:181-190

[63] Rani MSA, Rudhziah S, Ahmad A, Mohamed NS. Biopolymer electrolyte based on derivatives of cellulose from kenaf bast fiber. Polymer. 2014;**6**(9):2371-2385

[64] Nataraj SK, Wang CH, Huang HC, Du HY, Wang SF, Chen YC, Chen LC, Chen KH. Highly proton-selective biopolymer layer-coated ion-exchange membrane for direct methanol fuel cells. ChemSusChem. 2012;**5**(2):392-395

[65] Badawy MEI, Rabea EI. A biopolymer chitosan and its derivatives as promising antimicrobial agents against plant pathogens and their applications in crop protection. International Journal of Carbohydrate Chemistry. 2011;**2011** 29 pages

[66] Kumar P, Sandeep KP, Alavi S, Truong VD, Gorga RE. Preparation and characterization of bio-nanocomposite films based on soy protein isolate and montmorillonite using melt extrusion. Journal of Food Engineering. 2010;**100**(3):480-489

[67] Mahadeva SK, Kim J. Effect of polyethylene oxide-polyethylene glycol content and humidity on performance of electro-active paper actuators based on cellulose/polyethylene oxide-polyethylene glycol microcomposite. Polymer Engineering and Science. 2010;**50**(6):1199-1204

[68] Kim J, Yun S, Ounaies Z. Discovery of cellulose as a smart material. Macromolecules. 2006;**39**:4202-4206

[69] Spillman WB Jr, Sirkis JS, Gardiner PT. Smart materials and structures: What are they? Smart Materials and Structures. 1996;**5**:247-254

[70] Xua J, Nia T. Integrated technological paradigm-based soft paths towards sustainable development of small hydropower. Renewable and Sustainable Energy Reviews. 2017;**70**:623-634

[71] Trillo C, Luigi P. A novel paradigm to achieve sustainable regeneration in Historical Centres with Cultural Heritage. 2nd International Symposium "NEW METROPOLITAN PERSPECTIVES"—Strategic planning, spatial planning, economic programs and decision support tools, through the implementation of Horizon/Europe2020. ISTH2020, Reggio Calabria (Italy), 18–20 May 2016. Procedia—Social and Behavioural Sciences. 2016;**223**:693-697

[72] Bichard E. Developing an Approach to Sustainable Return of Investment in the UK, Brazil and the USA. RICS Research Trust; 2015

[73] Putz M, Stoldt J, Fanghänel C, Bierera A, Schlegel A. Making sustainability paradigms a part of PPC. The 22nd CIRP conference on Life Cycle Engineering. Procedia CIRP. 2015;**29**:209-214

[74] Brousseau E, Barton R, Dimov S, Bigot S. A methodology for evaluating the technological maturity of micro and nano fabrication processes. IFIP. Advances in Information and Communication Technology. 2010;**315**(1):329-336

[75] Fatemi AM, Fooladi IJ. Sustainable finance: A new paradigm. Global Finance Journal. 2013;**24**:101-113

[76] Hwang WS, Shin J. ICT-specific technological change and economic growth in Korea. Telecommunications Policy. 2017;**41**:282-294

[77] Kima DH, Lee H, Kwak J. Standards as a driving force that influences emerging technological trajectories in the converging world of the Internet and things: An investigation of the M2M/IoT patent network. Research Policy. 2017;**46**:1234-1254

[78] Elsevier Editorial. Advancing understanding of the complex nature of urban systems. Ecological Indicators. 2016;**70**:566-573

[79] Ho JC, Lee CS. A typology of technological change: Technological paradigm theory with validation and generalization from case studies. Technological Forecasting and Social Change. 2015;**97**:128-139

[80] Kumar RV, Kidwai SB. Is the fast-paced technological advancement in radiation treatment equipment good for Indian Scenario? No? Journal of Cancer Policy. 2015;**4**:26-30

[81] International Monetary Fund (IMF). World Economic and Financial Surveys. World Economic Outlook Database. IMF; 2013, 2013. Available from: http://www.imf.org/external/pubs/ft/weo/2013/01/weodata/index.aspx [Accessed October 2017]

[82] Araújo E. Investment-specific shocks and real business cycles in emerging economies: Evidence from Brazil. Economic Modelling. 2012;**29**:671-678

[83] Sharif MN. Technological innovation governance for winning the future. Technological Forecasting and Social Change. 2012;**79**:595-604

[84] The Hindu Opinion. Maharashtra: The Dull Days of White Gold. The Hindu. Available from: http://www.hindu.com/2009/04/03/stories/2009040355480800.html; 2009 Accessed: October 2017

[85] Hayashi AM. Technology trajectories and the birth of new industries. Sloan Management Review. 2004;**45**:7-8

[86] Souitaris V. Technological trajectories as moderators of firm-level determinants of innovation. Research Policy. 2002;**31**:877-898

[87] Audretsch DB. Technological regimes, industrial demography and the evolution of industrial structures. Industrial and Corporate Change. 1997;**6**:49-82

[88] Ghisetti C, Quatraro F. Green technologies and environmental productivity: A cross-sectoral analysis of direct and indirect effects in Italian regions. Ecological Economics. 2017;**132**:1-13

[89] Bai C, Kusi-Sarpong S, Sarkis J. An implementation path for green information technology systems in the Ghanaian mining industry. Journal of Cleaner Production. 2017; **164**:1105-1123

[90] Zong B, Sun B, Cheng S, Mu X, Yang K, Zhao J, Zhang X, Wu W. Green production technology of the monomer of Nylon-6: Caprolactam green chemical engineering—Review. Engineering. 2017;**3**:379-384

[91] Korola E, Shushunova N. Benefits of a modular green roof technology world multidisciplinary civil engineering-architecture-urban planning symposium. WMCAUS 2016. Procedia Engineering. 2016;**161**:1820-1826

[92] Allacker K, Mathieux F, Manfredi S, Pelletier N, De Camillis C, Ardente F, Pant R. Allocation solutions for secondary material production and end of life recovery: Proposals for product policy initiatives. Resources, Conservation and Recycling. 2014;**88**:1-12

[93] Rostami R, Khoshnava SM, Lamit H, Streimikiene D, Mardani A. An overview of Afghanistan's trends toward renewable and sustainable energies. Renewable and Sustainable Energy Reviews. 2017;**76**:1440-1464

[94] Bernardi M, Tangorra RR, Palmisano P, Bogliano A. Chemicals From Renewable Sources. Biochemtex – Mossi & Ghisolfi Group. Tortona, Italy: Elsevier Inc; 2016

[95] Arkadiusz Piwowar A, Dzikuć M, Adamczyk J. Agricultural biogas plants in Poland—Selected technological, market and environmental aspects. Renewable and Sustainable Energy Reviews. 2016;**58**:69-74

[96] Jos GJO, Greet JM, Marilena M, Jeroen AHWP. Trends in Global CO_2 Emission: Netherlands Environmental Assessment Agency; 2014 Report. Available from: http://edgar.jrc.ec.europa.eu/news_docs/jrc-2014-trends-in-global-co2-emissions-2014-report-93171.pdf [Accessed 04 October 2017]

[97] Sorensen B. Renewable Energy Conservation, Transmission and Storage. Burlington: Academic Press; 2007

[98] Silverman M, Worthman S. The future of renewable energy industries. The Electricity Journal. 1995:12-31

[99] Eudes A, Liang Y, Mitra P, Loque D. Lignin bioengineering. Current Opinion in Biotechnology. 2014;**26**:189-198

[100] Bonawitz ND, Chapple C. Can genetic engineering of lignin deposition be accomplished without an unacceptable yield penalty? Current Opinion in Biotechnology. 2013;**24**:336-343

[101] Blanch HW, Simmons BA, Klein-Marcuschamer D. Biomass deconstruction to sugars. Biotechnology Journal. 2011;**6**:1086-1102

[102] Li X, Weng JK, Chapple C. Improvement of biomass through lignin modification. The Plant Journal. 2008;**54**:569-581

[103] Raud M, Mitt M, Oja T, Olt J, Orupõld K, Kikas T. The utilisation potential of urban greening waste: Tartu case study. Urban Forestry & Urban Greening. 2017;**21**:96-101

[104] Dehne L, Babarro CV, Saake B, Schwarz KU. Influence of lignin source and esterification on properties oflignin-polyethylene blend. Industrial Crops and Products. 2016;**86**:320-328

[105] Lim JS, Manan ZA, Alwi SRW, Hashim H. A review on utilisation of biomass from rice industry as a source of renewable energy. Renewable and Sustainable Energy Reviews. 2012;**16**:3084-3094

[106] Reston VA. American Society of Civil Engineers (ASCE) Press. Gosselink, R. J. A. (2011). Lignin as a renewable aromatic resource for the chemical industry, PhD thesis. Wageningen, NL: Wageningen University; 2011

[107] Ashleya J, Shahbazia MA, Kanta K, Chidambara VA, Wolff A, Bang DD, Sun Y. Molecularly imprinted polymers for sample preparation and biosensing in food analysis: Progress and perspectives. Biosensors and Bioelectronics. 2017;**91**:606-615

[108] Uzun L, Turner APF. Molecularly-imprinted polymer sensors: Realising their potential. Biosensors and Bioelectronics. 2016;**76**:131-144

[109] Anene A, Hosni K, Chevalier Y, Kalfat R, Hbaieb S. Molecularly imprinted polymer for extraction of patulin in apple juice Samples. Food Control. 2016;**70**:90-95

[110] Wackerlig J, Lieberzeit PA. Molecularly imprinted polymer nanoparticles in chemical sensing—Synthesis, characterisation and application—A review. Sensors and Actuators B. 2015;**207**:144-157

[111] Garcia D, Gomez-Caballero A, Guerreiro A, Goicolea MA, Barrio RJ. Molecularly imprinted polymers as a tool for the study of the4-ethylphenol metabolic pathway in red wines. Journal of Chromatography A. 2015;**1410**:164-172

[112] Manzoor S, Buffon R, Rossi AV. Molecularly imprinted solid phase extraction of fluconazole from pharmaceutical formulations. Talanta. 2015;**134**:1-7

[113] Gao L, Li X, Zhang Q, Dai J, Wei X, Song Z, Yan Y, Li C. Molecularly imprinted polymer microspheres for optical measurement of ultra-trace nonfluorescent cyhalothrin in honey. Food Chemistry. 2014;**156**:1-6

[114] Ruela ALM, Figueiredo EC, Pereira GR. Molecularly imprinted polymers as nicotine transdermal delivery systems. Chemical Engineering Journal. 2014;**248**:1-8

[115] Freudenberg K. Molecular Biology, Biochemistry, and Biophysics. 1968;**2**:47

[116] Martone PT, Estevez JM, Lu F, Ruel K, Denny MW, Somerville C, Ralph J. Current Biology. 2009;**19**:169

[117] Chen F, Dixon RA. Nature Biotechnology. 2007;**25**:759

[118] Lin SY, Dence CW. Methods in Lignin Chemistry. Berlin-Heidelberg: Springer Science & Business Media; 2012

[119] Rencoret J, Ralph G, Marques A, Gutierrez A, Martinez JC d R. Journal of Agricultural and Food Chemistry. 2013;**61**:2434-2445

[120] Barton F. Animal Feed Science and Technology. 1988;**21**(1988):279-286

[121] Sannigrahi P, Ragauskas AJ, Miller SJ. Lignin structural modifications resulting from ethanol organosolv treatment of loblolly pine. Energy & Fuels. 2010;**24**(1):683-689

[122] Somerville C, Bauer S, Brininstool G, Facette M, Hamann T, Milne J, Osborne E, Paredez A, Persson S, Raab T, Vorwerk S, Youngs H: Science (Washington, DC, U.S.) 2004, 306:2206. In: Simmons et al., 2010

[123] De Wild P, Reith H, Heeres E. Biomass pyrolysis for chemicals. Biofuels. 2011;**2**(2):185-208

[124] Huang F. What is biomass. In: Ragauskas AJ, editor. Materials for Biofuels. Singapore: World Scientific Publishing Co.; 2014. pp. 1-26

[125] Bilal M, Asgher M. Degradation of agro-wastes by lignocellulolytic activity of an Oyster Mushroom, Pleurotus Sapidus. Journal of the National Science Foundation of Sri Lanka. 2016;**44**:399-407

[126] Rodriguez S, Toca JL. Industrial and biotechnological applications oflaccases: A review. Biotechnology Advances. 2006;**24**(2006):500-513

[127] Boeriu CG, Bravo D, Gosselink RJA, Jan EG, van Dam JEG. Characterisation of structure-dependent functional properties of lignin with infrared spectroscopy. Industrial Crops and Products. 2004;**20**:205-218

[128] Gallezot P. Catalytic conversion of biomass: Challenges and issues. ChemSusChem. 2008;**1**(8–9):734-737

[129] Gallo JMR, Bueno JMC, Schuchardt U. Catalytic transformations of ethanol for biorefineries. Journal of the Brazilian Chemical Society. 2014;**25**:2229-2243

[130] Suarez PAZ, Meneghetti SMP. Aniversário do Biodiesel em 2007: Evolução Historica e Situaçao Atual no Brasil. Quimica Nova. 2007;**30**:2068-2071

[131] Slater RA, Frederickson J. Composting Municipal Waste in the UK: Some Lessons from Europe. Resources, Conservation and Recycling. 2001;**32**:359-374

[132] Brar SK, Soccol CR, Dhillon GS, editors. Biotransformation of waste biomass into high value biochemicals. New York: Springer; 2014. Cavagnaro, T. R. (2015). Biologically regulated nutrient supply

[133] Ollikainen M. Forestry in bioeconomy e smart green growth for the humankind. Scandinavian Journal of Forest Research. 2014:37-41

[134] Pfau SF, Hagens JE, Dankbaar B, Smits AJM. Visions of sustainability in bioeconomy research. Sustain. 2014;**6**:1222-1249

[135] Johnson TG, Altman I. Rural development opportunities in the bioeconomy. Biomass and Bioenergy. 2014;**63**:341-344

[136] OECD. The Bioeconomy to 2030-Designing a Policy Agenda. Paris, France; 2009

[137] De Besi M, McCormick K. Towards a bioeconomy in Europe: National, regional and industrial strategies. Sustain. 2015;**7**:10461-10478

[138] Giurca A, Metz T. A social network analysis of Germany's wood-based bioeconomy: Social capital and shared beliefs. Environmental Innovation and Societal Transitions. 2017. https://www.researchgate.net/publication/319665464_A_social_network_analysis_of_Germany%27s_wood-based_bioeconomy_Social_capital_and_shared_beliefs

[139] Zwier J, Blok V, Lemmens P, Geerts RJ. The ideal of a zero-waste humanity: Philosophical reflections on the demand for a bio-based economy. Journal of Agricultural and Environmental Ethics. 2015;**28**:353-374

[140] Fritsche U, Iriarte L. Sustainability criteria and indicators for the bio-based economy in Europe: State of discussion and way forward. Energies. 2014;**7**

[141] Keegan D, Kretschmer B, Elbersen B, Panoutsou C. Cascading use: A systematic approach to biomass beyond the energy sector. Biofuels, Bioproducts and Biorefining. 2014;**7**: 193-206

[142] BMBF. National Policy Strategy on Bioeconomy. Berlin: BMBF; 2014

[143] Hellsmark H, Mossberg J, Söderholm P, Frishammar J. Innovation system strengths and weaknesses in progressing sustainable technology: The case of Swedish biorefinery development. Journal of Cleaner Production. 2016;**131**:702-715

[144] Purkus A, Hagemann N, Bedtke N, Gawel E. Towards a sustainable innovation system for the German wood-based bioeconomy: Implications for policy design. Journal of Cleaner Production. 2017;**172**:3955-3968

[145] Bergek A, Hekkert M, Jacobsson S. Functions in innovation systems: a framework for analysing energy system dynamics and identifying goals for systembuilding activities by entrepreneurs and policymakers. In: Foxon T, Köhler J, Oughton C, editors. Innovation for a Low Carbon Economy: Economic. Edward Elgar, Cheltenham: Institutional and Management; 2008. p. 79

[146] Hekkert MP, Suurs RAA, Negro SO, Kuhlmann S, Smits REHM. Functions of innovation systems: A new approach for analysing technological change. Technological Forecasting and Social Change. 2007;**74**:413-432

[147] Ralph J, Lundquist K, Brunow G, Lu F, Kim H, Schatz PF, Marita JM, Hatfield RD, Ralph SA, Christensen JH, Boerjan W. Phytochemistry Reviews. 2004;**3**:29

[148] Hisano H, Nandakumar R, Wang Z-Y. In Vitro Cellular & Developmental Biology—Plant. Vol. 452009. p. 306

[149] Lange H, Silvia Decin S, Crestini C. Oxidative upgrade of lignin—Recent routes reviewed. European Polymer Journal. 2013;**49**:1151-1173

[150] Rippert P, Puyaubert J, Grisollet D, Derrier L, Matringe M. Tyrosine and phenylalanine are synthesized within the plastids in Arabidopsis. Plant Physiology. 2009;**149**:1251-1260

[151] Chapple C. Molecular-genetic analysis of plant cytochrome P450-dependent monooxygenases. Annual Review of Plant Physiology and Plant Molecular Biology. 1998;**49**:311-343

[152] Ro DK, Mah N, Ellis BE, Douglas CJ. Functional characterization and subcellular localization of poplar (Populus trichocarpa 3 Populus deltoides) cinnamate 4-hydroxylase. Plant Physiology. 2001;**126**:317-329

[153] Rasmussen S, Dixon RA. Transgene-mediated and elicitor-induced perturbation of metabolic channelling at the entry point into the phenylpropanoid pathway. Plant Cell. 1999;**11**:1537-1552

[154] Achnine L, Blancaflor EB, Rasmussen S, Dixon RA. Colocalization of L-phenylalanine ammonia-lyase and cinnamate 4-hydroxylase for metabolic channelling in phenylpropanoid biosynthesis. Plant Cell. 2004;**16**:3098-3109

[155] Ehlting J, Mattheus N, Aeschliman DS, Li E, Hamberger B, Cullis IF, Zhuang J, Kaneda M, Mansfield SD, Samuels L, Ritland K, Ellis BE, Bohlmann J, Douglas CJ. Plant. 2005; **42**:618

[156] Vlaminck L, Lingier S, Hufendiek A, Du Prez FE. Lignin inspired phenolic polyethers synthesized via ADMET: Systematic structure-property investigation. European Polymer Journal. 2017

[157] Laurichesse S, Avérous L. Chemical modification of lignins: Towards biobased polymers. Progress in Polymer Science. 2014;**39**(7):1266-1290

[158] Vlaminck L, Lingier S, Hufendiek A, Du Prez FE. Lignin inspired phenolic polyethers synthesized via ADMET: Systematic structure-property investigation. European Polymer Journal. 2017;**95**:503-503. DOI: 10.1016/j.eurpolymj.2017.08.042

[159] Tronchet M, Balague C, Kroj T, Jouanin L, Roby D. Cinnamyl alcohol dehydrogenases C and D, key enzymes in lignin biosynthesis, play an essential role in disease resistance in Arabidopsis. Molecular Plant Pathology. 2010;**11**:83-92

[160] Cano-Delgado A, Penfield S, Smith C, Catley M, Bevan M. Reduced cellulose synthesis invokes lignification and defense responses in *Arabidopsis thaliana*. The Plant Journal. 2003;**34**:351-362

[161] Sticklen MB. Plant genetic engineering for biofuel production towards affordable cellulosic ethanol. Nature Reviews Genetics. 2008;**9**:433-443

[162] Weng JK, Akiyama T, Bonawitz ND, Li X, Ralph J, Chapple C. Convergent evolution of syringyl lignin biosynthesis via distinct pathways in the lycophyte Selaginella and flowering plants. Plant Cell. 2010;**22**:1033-1045

[163] Mansfield SD. Solutions for dissolution—Engineering cell walls for deconstruction. Current Opinion in Biotechnology. 2009;**20**:286-294

[164] Zhou J, Lee C, Zhong R, Ye Z-H. Plant Cell. 2009;**2**:248

[165] Aresta M, Dibenedetto A, Dumeignil F, editors. Biorefinery: From Biomass to Chemicals and Fuels. Berlin/Boston: de Gruyter; 2012

[166] Gosselink R. Co-ordination network for lignin-standardisation, production and applications adapted to market requirements (EUROLIGNIN). Industrial Crops and Products. 2004;**20**(2):121-129

[167] Macfarlane AL, Mai M, Kadla JF. Bio-based Chemicals from Bio-refining: Lignin Conversion and Utilisation. Canada: University of British Columbia; 2014. DOI: 10.1533/9780857097385.2.659

[168] Dos Santos DA, Rudnitskaya A, Evtuguin DV. Modified kraft lignin for bioremediation applications. Journal of Environmental Science and Health Part A. 2012;**47**(2):298-307

[169] Zakzeski J, Bruijnincx PCA, Jongerius AL, Weckhuysen BM. The catalytic valorization of lignin for the production of renewable chemicals. Chemical Reviews. 2010;**110**(6):3552-3599

[170] Kubo S, Kadla JF. Effect of poly(ethylene oxide) molecular mass on miscibility and hydrogen bonding with lignin. Holzforschung. 2006;**60**(3):245-252

[171] Li Z, Ge Y. Extraction of lignin from sugar cane bagasse and its modification into a high performance dispersant for pesticide formulations. Journal of the Brazilian Chemical Society. 2011;**22**(10):1866-1871

[172] Sannigrahi P, Pu Y. Ragauskas A. Cellulosic biorefineries—Unleashing lignin opportunities. Current Opinion in Environment Sustainability. 2010;**2**(5–6):383-393

[173] Werpy TG, Peterson G. Top Value Added Chemicals from Biomass: Results of Screening for Potential Candidate from Sugars and Synthesis Gas. Vol. 1. Pacific Northwest National Laboratory: US Department of Energy; 2004

[174] Smith BR, Rice RW, Ince PJ. Pulp capacity in the United States. General Technical Report FPL-GTR-139. USDA Forest Service; 2000

[175] Mai C, Milsterin O, Huttermann A. Chemoenzymatical grafting of acrylamide onto lignin. Journal of Biotechnology. 2000;**79**:173-183

[176] Lu FJ, Chu LH, Gau RJ. Free radical-scavenging properties of lignin. Nutrition and Cancer. 1998;**30**:31-38

[177] Barclay LRC, Xi F, Norris JQ. Antioxidant properties of phenolic lignin model compounds. Journal of Wood Chemistry and Technology. 1997;**17**:73-90

[178] Stewart D. Lignin as a base material for materials applications: chemistry, application and economics. Industrial Crops and Products. 2008;**27**(2):202-207

[179] Nunn TR, Howard JB, Longwell JP, Peters WA. Product compositions and kinetics in the rapid pyrolysis of milled wood lignin. Industrial & Engineering Chemistry Process Design and Development. 1983;**24**(3):844-852

[180] Ndibewu PP, Mgangira MB, Cingo N, McCrindle RI. Metal and anion composition of two biopolymeric chemical stabilizers and toxicity risk implication for the environment. Journal of Toxicology and Environmental Health. Part A. 2010;**73**(4):261-271. http://dx.doi.org/10.1080/15287390903249164

[181] Nadif A, Hunkeler D, Käuper P. Sulfur-free lignins from alkaline pulping tested in mortar for use as mortar additives. Bioresource Technology. 2002;**84**(1):49-55

[182] Holtman H, Chang M, Jameel H, Kadla JF. Elucidation of lignin structure through degradative methods: Comparison of modified DFRC and thioacidolysis. Journal of Agricultural and Food Chemistry. 2003;**51**(12):3535-3540

[183] Marton J. Lignin Structure and Reactions, Advances in Chemistry. Vol. 59. Washington, DC: American Chemical Society; 1966

[184] Crestini C, Melone F, Sette M, Saladino R. Milled wood lignin: A linear oligomer. Biomacromolecules. 2011;**12**(11):3928-3935

[185] Adler E. Structural elements of lignin: Early review. Industrial & Engineering Chemistry. 1957;**49**(9):1377-1383

[186] Lebo SE Jr, Gargulak JD, TJ MN, editors. Lignin: Kirk-Othmer Encyclopedia of Chemical Technology. 4th ed. New York: John Wiley & Sons, Inc; 2001

[187] Schneiderman, Hillmyer MA. 50th anniversary perspective: There is a great future in sustainable polymers. Macromolecules. 2017;**50**(10):3733-3749

[188] Pouteau C, Dole P, Cathala B, Averous L, Boquillon N. Antioxidant properties of lignin polypropylene. Polymer Degradation and Stability. 2013;**81**:9-18

[189] Global PET Supply to Exceed 24.39 Mln Tonnes in 2015, 2014. www.mcgroup.co.uk [Accessed October 26, 2017]

[190] Llevot A, Grau E, Carlotti S, Grelier S, Cramail H. From lignin-derived aromatic compounds to novel biobased polymers. Macromolecular Rapid Communications. 2016;**37**(1): 9-28

[191] Isikgor FH, Becer CR. Lignocellulosic biomass: A sustainable platform for the production of bio-based chemicals and polymers. Polymer Chemistry. 2015;**6**(25):4497-4559

[192] FDCA and PEF From Plant Based materials to FDCA and PEF. 2016. www.avantium.com. [Accessed September 2017]

[193] Mythili C, Retna AM, Gopalakrishnan S. Synthesis, mechanical, thermal and chemical properties of polyurethanes based on cardanol. Bulletin of Materials Science. 2004;**27**(3):235-241

[194] Yannopoulos JC. Gold ore. In: The Extractive Metallurgy of Gold. Springer Science and Business Media; 2012. pp. 1-2, 11-15, 209

[195] Pyrzynska K. Sorbent materials for separation and preconcentration of gold in environmental geological samples—A review. Analytica Chimica Acta. 2012;**741**:9-14

[196] Antero A, Miraja K, Tinaii S, Monica N. Gold and gold mining. In: Noeberg GF, Fowler BA, Noeberg M, editors. Handbook on the toxicology of metals. 4th ed. Vol. 1. Academic Press; 2014 pp. 818, 820, 821, 827

[197] Marsden J. Ore deposits and process minerology. In: The Chemistry of Gold Extraction. 2nd ed. Society for Mining, Metallurgy and Exploration, Inc.; 2006 pp. 25, 111-113

[198] Micheal WG. Gold. In: US Geological Survey, Mineral Commodity, Summaries. 2015. pp. 66-67

[199] Umeda H, Sasaki A, Takahashi K, Haga K, Shibayama A. Recovery and concentration of precious metals from strong acidic wastewaters. Materials Transactions. 2011;**52**(7):1462-1470

[200] Baralam V. Recent advances in the determination of PGEs in exploration studies—A review. Journal of Geological Society of India. 2008;**72**(11):661-677

[201] Ahamed MEH, Mbianda XY, Mulaba-Bafubianda AF. Development of molecularly imprinted polymers for selective separation and recovery of gold and silver from industrial and mining waste waters [PhD thesis]. Aukland Park, Johannesburg: University of Johannesburg; 2017. pp. 11 (draft)

[202] Nicol MJ, Fleming CA, Paul RL. The Chemistry of the Extraction of Gold. 2001

[203] Balcerzak M. Sample digestion methods for the determination of traces of precious metals by spectrometric techniques. Analytical sciences. Japan Society of Analytical Chemistry. 2002;**18**:737-750

[204] Schmidbaur H, Cronje S, Djorjevic B, Schuster O. Understanding gold chemistry through relativity. Chemical Physics. 2005;**311**:151-161

[205] Bardi U. Extracted: How the Quest for Mineral Wealth Is Plundering the Planet. The Past, Present, and Future of the Global Mineral Depletion e a Report to the Club of Rome. USA: Chelsea Green Publishing; 2014. ISBN: 978-1-60358-541-5 (pbk)

[206] Habib K, Hamelin L, Henrik Wenzel H. A dynamic perspective of the geopolitical supply risk of metals. Journal of Cleaner Production. 2016;**133**(2016):850-858

Chapter 11

Application of Lignins in Formulation and Manufacturing Bio-Based Polyurethanes by ^{31}P NMR Spectroscopy

Behzad Ahvazi and Tri-Dung Ngo

Additional information is available at the end of the chapter

http://dx.doi.org/10.5772/intechopen.73287

Abstract

The application of several industrial and technical biomass lignins from forestry and agricultural residues extracted by different chemical pulping processes in view of bio-based polyurethanes were studied. Several qualitative and quantitative methods were utilized to characterize and elucidate lignin profiles by evaluating their physical and chemical properties as a potential substitute for petroleum-based polyols. The morphology, molecular weight distributions, elemental compositions, glass transition temperature and several important functional groups containing hydroxyl units were classified and their contents were determined quantitatively. In this chapter, lignin substitution in polyurethane by (i) direct substitution, (ii) with combination of polyols or (iii) chemically modified was reviewed. The emerging information aims at addressing a number of pressing issues relevant to the scientific development for value-added applications from lignins during industrial production of biofoams.

Keywords: lignins, characterization, polyols, chemical modification, polyurethane

1. Introduction

Polymers play an essential and ubiquitous role in our daily lives due to their extraordinary wide variety of both physical and chemical properties [1]. However, with ever increasing global concern over rapid depletion of non-renewable fossil fuel-based resources, uncertainties of petroleum prices and the high energy intensity in production of chemicals and synthetic polymers, there is a strong interest worldwide in developing suitable technologies that can derive chemicals and materials from renewable biomass [2]. In recent years, the conversion of renewable lignocellulosic resources into chemicals, liquids fuels and feed supplement has gained considerable attention. Modification and characterization of bio-based polymers for preparation of higher valued green chemicals and bio-based products favors the future use of lignin biomass components with substantial environmental and economical benefits [3–6].

Lignin is the second most naturally abundant biopolymer substance in plant cell walls, exceeded only by cellulose [7, 8]. Its main function is to impart rigidity to the cell walls and, acting as an intercellular binder, impart to woods resistance to compression, impact and bending.

Lignin is an amorphous, highly branched polyphenolic macromolecule of complex structure with high molecular weight. The chemical structure of lignin is highly irregular and extremely challenging. Lignin polymer consists primarily of phenyl propanoid units mainly *p*-coumaryl alcohol (H), coniferyl alcohol (G) and/or sinapyl alcohol (S). These components are cross-linked together in three dimensions via a radical coupling process during its biosynthesis [9–15]. Its physical and chemical properties are highly dependent on the wood species, botanic region and the isolation processes [16, 17] (**Figure 1**).

Lignin is available in large quantities as a byproduct of the pulp and paper and hydrolytic industries. The most abundant industrial lignins are from kraft and sulfite pulping processes. Lignins obtained under these processes undergo significant structural changes and are no longer identical to their original native structures [10, 18–20]. Due to its complex nature and undefined chemical structure, the industrial applications of lignin are rather limited. Lignin is utilized almost exclusively as fuel to power the evaporators of the chemical recovery processes and liquor concentration system of pulp mills [21]. However, based on its interesting functionalities and properties, lignin offers perspective for higher added value applications in renewable products [22].

The overall reactivity of different lignins depends on their chemical structures [23]. The lignin polymer contains a large number of several important functional groups [24] including

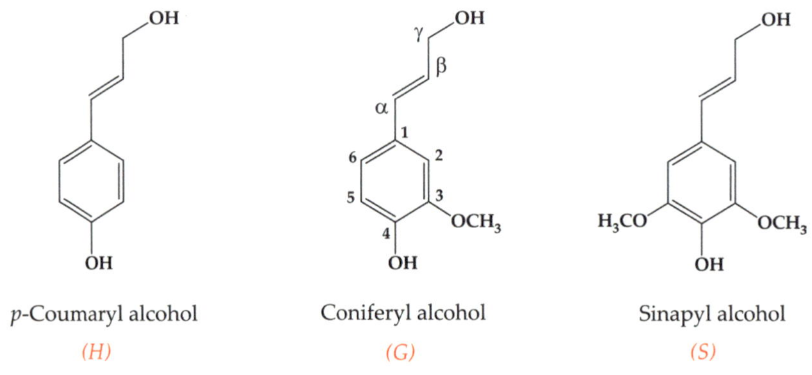

Figure 1. Chemical structure of lignin monomeric phenylpropanoid units.

phenolic hydroxyl, aliphatic hydroxyl and carboxylic acid groups. Different lignin samples contain different types and quantities of these functional groups that could be utilized in making and/or replacing a majority of petroleum-based products in the manufacturing of industrial biomaterials and bio-composites [25–27] such as polyurethanes [28–30]. Traditional polyurethane foam is made from petroleum-based polyols that biodegrades slowly if at all. The main advantage of lignin polymer group lies, on one hand, in the sustainability of renewable raw materials and on the other hand in the 'zero waste' principle resulting from their bio-degradability.

Polyurethanes (PU) are considered as one of the most useful multi-purpose polymers with formulations covering a wide range of products and applications. PU can be utilized as sheets, foams or paints in various products such as flexible foam seating, rigid foam insulating panels, high performance adhesives and sealants [31].

PU polymers are formed through step-growth polymerization by reacting a monomer which has two or more isocyanate functional groups with another monomer containing two or more hydroxyl groups in the presence of a catalyst [32].

The first monomeric unit, isocyanates are classified as aromatic or aliphatic. Aromatic diisocyanates such as toluene diisocyanate (TDI) and methylene diphenyl diisocyanates (MDI) are mostly utilized in formation of polyurethanes since the aromatically linked isocyanate group is much more reactive than the aliphatic one. In addition, the aromatic isocyanate is more economical. However, the aliphatic isocyanates such as hexamethylene diisocyanates (HDI) and isophorone diisocyanates (IPDI) are used only if special properties are required for the final product [33–36].

Polyols, the other monomeric component, are classified as polyether or polyester polyols. Polyols can be further classified according to their end use as flexible or rigid, depending on the functionality of the initiator and their molecular weight. Flexible polyols have molecular weights ranging from 2000 to 10,000 containing 18–56 hydroxyl units while rigid polyols have molecular weights ranging from 250 to 700 with 300–700 hydroxyl units. Polyols with 700–2000 molecular weights and 60–280 hydroxyl units are used to improve stiffness or flexibility to base system [33–36].

Lignin contains considerable amount of aliphatic hydroxyl groups on the propionic side chain of its several different monomeric units. These units contain both primary and secondary hydroxyl groups with properties similar to polyols. However, unlike petroleum polyols, these propionic hydroxyl groups are not the major factor for order of lignin reactivity. Other lignin features attribute and influence lignin performance in PU formulation.

Many polymer properties such as Tg, modulus, tensile strength, etc. are directly dependent upon their molecular weights. Molecular weight determinations on a variety of isolated lignins would provide essential information on the molecular size or size distribution of different polymeric fragments in lignin and their physical and chemical properties during copolymerization [37]. The lignin properties are related not only to the nature of the hydrophilic groups, which depends upon the pulping process, but also to the size of the polymeric pieces, which in turn depends on the recovery techniques used to process the black liquors. The molecular weight for a number of lignin will help to identify the most suitable lignin(s) for a PU application.

Lignins extracted under different processes contain different chemical and physical properties. Therefore, in order to increase the reactivity of lignin and improve their compatibilities with other polymers, certain functional groups may be selectively modified. For example, free phenolic-OH groups are found to be more reactive [38, 39] than benzylic-OH except toward diisocyanates [40, 41]. In this case, the reaction of lignin with alkene oxide would result in the formation of lignin-polyol derivatives, which in turn improve the solubility, and uniformity of lignin [39, 42, 43]. Modification of specific functional group(s) may be performed in accordance to the results obtained from polyurethane polymerization using lignin. Characterization, selection and modification of lignin polymer as raw materials would then pave the way to prepare a bio-based polymer with similar characteristics to polyols that could be utilized in substituting polyols for industrial applications during the polyurethane production.

In this chapter, physical and chemical properties such as functional groups, molecular weights and chemical modification of differently isolated lignins from agriculture residues and forestry were investigated for their potential as a substitute for polyols during polyurethane productions.

A number of industrial, technical and commercially available lignin samples were characterized by employing different qualitative and quantitative techniques [44, 45]. In these studies, different lignin samples were examined to elucidate their physical and chemical profiles in view of biomaterials applications. The L1 and L2 samples were hardwood lignins extracted by kraft and organosolv pulping processes, respectively. The subsequent four samples were all recovered from softwoods. The L3 and L4 were extracted by sulfite while L5 and L6 were isolated by kraft process. However, the L6 lignin sample was further modified with low sulfonated sodium salt. The L7 and L8 samples were soda lignin samples extracted from wheat straw. Detailed information regarding their suppliers, as well as their moisture and ash contents corresponding to each lignin sample, is reported in **Table 1**.

The experimental data show the L2 sample has the lowest, while the L4 and L6 have the highest moisture and ash content. This is not surprising since the L2 was extracted by organosolv process. The process is known to yield lignin which is hydrophobic and has the higher purity with lowest ash content among other lignin samples. Meanwhile, the L4 and L6 were isolated by sulfite process where the resulting lignosulfonates are known to be hydrophilic lignin and prone to absorb moisture more readily. More specifically, in the case of L6, the percent moisture content was found to increase by more than twofolds from $3.70 \pm 0.09\%$ to $8.20 \pm 0.08\%$ during two and a half years of storage time. However, the most striking feature was its high concentration of inorganic contaminants. Repeated measurements showed that almost two thirds of L6 contains ash and non-organic materials. In general, the lignosulfonates and sulfonated kraft lignin were found to have the highest ash content.

Sample	Lignin source	Feedstock	Isolation process	Moisture (%)	Ash (%)
L1	Canadian Mill	Hardwoods	Kraft	6.05 ± 0.09	0.45 ± 0.04
L2	Aldrich, Catalog No. 371017	Hardwoods	Organosolv	2.40 ± 0.00	0.11 ± 0.01
L3	Tembec, Arbo SO1	Softwoods	Sulfite	4.66 ± 0.03	24.15 ± 0.27
L4	Aldrich, Catalog No. 471038[a]	Softwoods	Sulfite	6.81 ± 0.04	20.02 ± 0.08
L5	Indulin AT, MeadWestvaco	Softwoods	Kraft	4.66 ± 0.12	3.06 ± 0.30
L6	Aldrich, Catalog No. 471003[b]	Softwoods	Kraft, low sulfonated	3.70 ± 0.09	66.19 ± 0.18
L7	Protobind 2400, Green Value	Wheat straw	Soda	2.48 ± 0.07	1.12 ± 0.01
L8	Protobind 3000, Green Value	Wheat straw	Soda	6.45 ± 0.02	2.12 ± 0.03

[a]Modified sodium lignosulfonate.
[b]Sodium salt of low sulfonated kraft lignin.

Table 1. Specification of various types of lignin and their moisture and ash content [44, 45].

The elemental compositions for various types of lignins are exclusively different; the percent elemental composition for each lignin sample in terms of carbon, hydrogen, nitrogen, oxygen, sulfur and methoxy content and the empirical formulae and the molecular weight of each lignin per C9 unit was reported in **Table 2** [44, 45]. The lignosulfonate samples have the lowest carbon, hydrogen and methoxy content followed by the sulfonated kraft lignin. The oxygen, sulfur and the molecular weights for these samples were found to be at the highest. This is somewhat to be expected since during the sulfite pulping process, the substituted sulfonate groups mostly at Cα and Cγ of lignin propanoic side-chains bear three oxygen atoms, leading to higher molecular weight. The organosolv lignin sample was found to contain the highest percentage of carbon and methoxy contents even when compared to its hardwood counterpart kraft lignin sample. Considering the monomeric units of both L1 and L2 samples that bear two methoxy groups, the L2 was found to have the lowest oxygen content with the lowest molecular weight among other lignins. Apart from lignosulfonates and sulfonated kraft lignin (L6), the sulfur content for kraft lignins (L1) and Indulin AT (L5) were found to be 1.93 and 1.32%, respectively. Determination of nitrogen content in lignin is also important since its quantity varies for different types of industrial lignins. The percent elemental nitrogen in order of descending was found for the L3, L7, L5 and L8 samples to be 1.09, 0.79, 0.59 and 0.31%, respectively. However, the nitrogen content was calculated in the order of L8, L3, L7 and L5 to contain 0.43, 0.22, 0.11 and 0.08 nitrogen element per C9 unit, respectively.

Lignin sample	Elemental analysis							
	%C	%H	%N	%O	%S	$\%OCH_3$	C9 structural unit	MW (g/mol)
L1	58.12	6.18	<0.50	33.05	1.93	17.09	$C_9H_{9.40}O_{3.18}S_{0.13}(OCH_3)_{1.156}$	208.37
L2	65.56	5.81	<0.50	27.65	<0.50	17.37	$C_9H_{7.51}O_{2.15}(OCH_3)_{1.028}$	181.92
L3	40.99	4.65	1.09	37.26	6.63	6.74	$C_9H_{11.16}O_{5.95}N_{0.22}S_{0.58}(OCH_3)_{0.612}$	255.23
L4	41.98	4.75	<0.50	39.71	6.59	7.96	$C_9H_{10.96}O_{6.18}S_{0.57}(OCH_3)_{0.713}$	258.53
L5	58.86	6.24	0.59	30.82	1.32	11.72	$C_9H_{10.06}O_{3.08}N_{0.08}S_{0.08}(OCH_3)_{0.751}$	194.67
L6	44.15	5.20	<0.50	38.08	4.16	8.77	$C_9H_{11.44}O_{5.56}S_{0.34}(OCH_3)_{0.750}$	242.93
L7	61.55	6.32	0.79	29.44	1.19	17.21	$C_9H_{9.07}O_{2.53}N_{0.11}S_{0.07}(OCH_3)_{1.092}$	195.54
L8	59.75	6.16	0.31	28.6	<0.50	11.2	$C_9H_{9.81}O_{2.78}N_{0.43}(OCH_3)_{0.704}$	190.41

Table 2. Elemental analysis, OCH_3 content, empirical formulae and molecular weight per C_9 lignin unit for different types of lignin [44, 45].

2. Lignin substitution in PU formulations

Lignin incorporation in polyurethane production has been one of the most intensively investigated applications [26–30, 46–55]. Lignin utilization modifies the polyurethane cure rate by contributing aromatic groups and increasing the degree of cross-linking. Lignin acts as a reinforcing agent which adds rigidity to the polymeric matrix [29]. Several studies have suggested that i) the solubility and ii) the uniformity of the lignin are the foremost important key parameters affecting its reactivity for polyols substitution in polyurethane production [52, 46–58]. This in turn is directly dependent upon different classes of hydroxyl groups that may be present in lignin causing both electronic and steric factors affecting the lignin reactivity [59]. Lignin substitution in polyurethane could be achieved by either i) direct substitution, ii) with combination of polyols [37, 46, 47, 49] or iii) chemically modified [28, 60–62].

2.1. Direct lignin substitution

There are three major types of hydroxyl groups in lignin. They are phenolic hydroxyl, α-hydroxyl and γ-carbinol group. Generally, phenolic —OH units are the most acidic and reactive groups. They play an important role in alkali-catalyzed reactions. Under acidic conditions, α-hydroxyls are transformed into active benzylic carbon cation leading to a variety of addition or transformation reactions while phenolic γ-carbinol is released as formaldehyde under both acidic and alkaline conditions [63] (**Figure 2**).

Different lignin samples contain different types and content of phenolic hydroxyl units depending on the pulping and isolation processes [64, 65]. Lignin extracted under soda pulping is reported to contain less phenolic hydroxyl groups and higher condensed structures than kraft lignin [66]. On the other hand, kraft lignin, as compared to MWL, is suggested by

http://dx.doi.org/10.5772/intechopen.73287

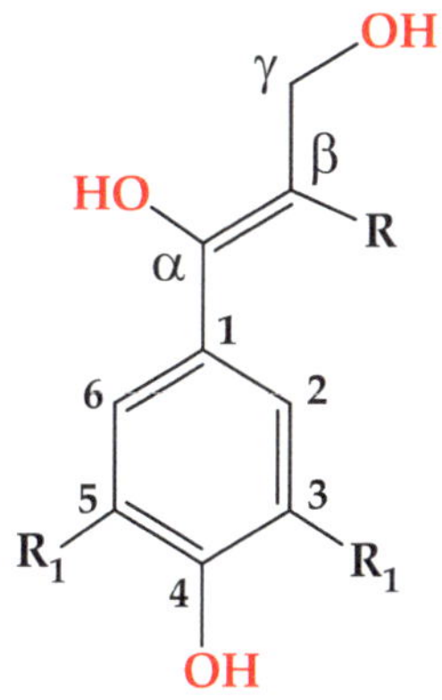

Figure 2. Three major types of hydroxyl groups in a monomeric structure in lignin.

Lai [63] to have higher content of phenolic hydroxyl group, α-carbonyl and stilbene structure with lower aliphatic hydroxyl group content. Steam explosion lignin and acid hydrolysis lignin have been suggested to be the most and the least compatible in preparation of polyurethane network films. Lignin isolated from other processes such as organosolv, kraft and MWL are reported to decrease in the order of their compatibility during polyurethane copolymerization [63]. Despite the extensive research on different types of lignins, there is limited information on utilization of soda lignin from agriculture residues in polyurethane application.

An important factor for incorporation of lignin in preparation of PU is the absolute determination of aliphatic —OH groups for calculating the NCO/OH during the formulation.

Several important hydroxyl-containing functional groups in lignins such as phenolic and aliphatic hydroxyl groups as well carboxylic acids are classified and their contents are determined quantitatively by employing P-31 NMR spectroscopy. The content and classification of these functional groups play an important role in determining their compatibilities during the process of chemical polymerization.

The technique has been successfully implemented for the absolute determination of various phenolic and non-phenolic structures in several native and technical lignins [67–72]. The availability of a phosphitylating reagent [71] that permits excellent resolution and the quantitative spectroscopic detection of several phenolic moieties with varying substitution patterns provide an excellent opportunity for systematically exploring the complex reactions that occurs in lignin during several different pulping processes. The phosphitylating reagent (2-chloro-1,3,2-dioxaphospholane) or reagent I provides essential information about primary hydroxyl groups along with both *erythro* and *threo* diastereomers of the arylglycerol-β-aryl ether structures. The phosphitylating reagent (2-chloro-4,4,5,5-tetramethyl-1,3,2-dioxaphospholane) or TMDP is a powerful analytical tool for distinguishing between different phenolic condensed units such as diphenylmethanes (DPM), diaryl ethers (4-O-5′) and biphenolics (5–5′) as well as benzylic and terminal carboxylic acids in lignin (**Figure 3**).

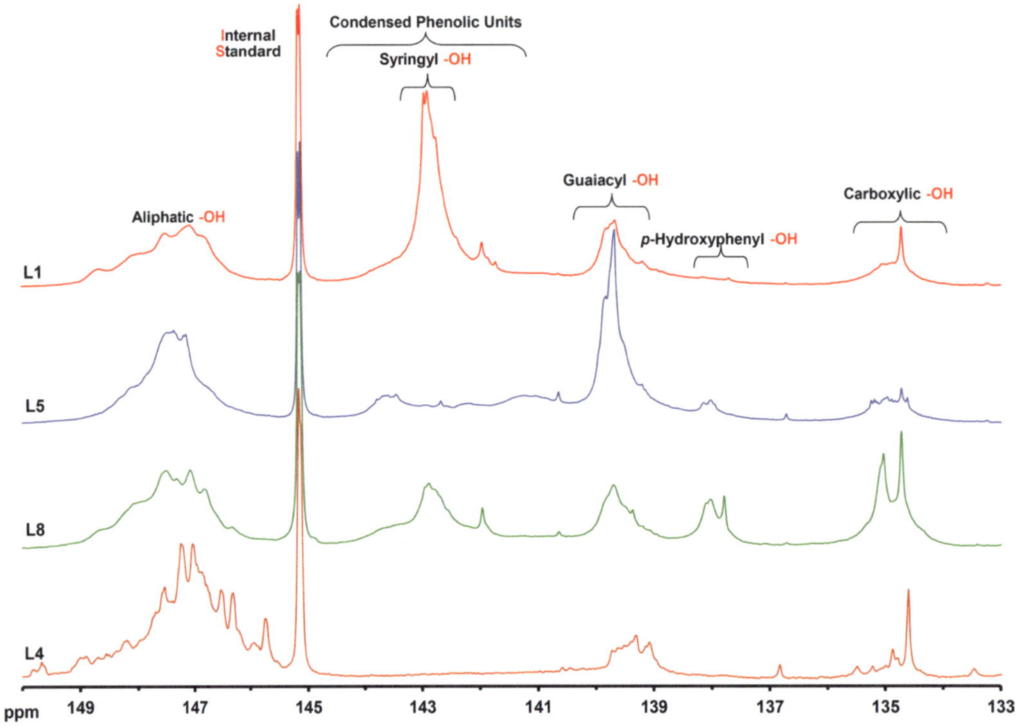

Figure 3. ^{31}P NMR spectra and signal assignments of hardwood kraft (L1), softwood Indulin AT (L5), Aldrich (L4) sodium lignosulfonates and wheat straw soda (L8) lignins phosphitylated with TMDP.

The ^{31}P NMR spectral analysis of all four lignin samples features distinct differences in terms of the nature and the content of various lignin moieties with hydroxyl groups. The presence of both syringyl (S) and guaiacyl (G) units in L1 sample clearly confirms the lignin was recovered from hardwoods. Since softwood lignins contain almost exclusively guaiacyl, the absence of syringyl signal in L5 and L4 suggest both lignin samples were extracted from softwoods. However, L5 was extracted under alkaline kraft while L4 was extracted under acid sulfite pulping processes. The ^{31}P NMR signal for aliphatic hydroxyl groups isolated from kraft lignin is often recorded as a smooth broad band. The same NMR signal for lignosulfonates is significantly different. The signal is always branched and lopsided. The L8 lignin was recovered from agro-residues. Its ^{31}P NMR spectrum exhibited high concentration of carboxylic acids, typical of soda lignin [73, 74] with a similar ratio of the benzylic to the terminal type. In addition, the lignin contains several distinct signals corresponding to all three different (G:S:H) monomeric components of lignin.

In addition to quantitative determination of several different classes of hydroxyl groups in several lignin samples, their quantities were also normalized. This is a recommended approach for determining the contribution of lignin hydroxyl units toward the reaction with other polymeric precursors in consideration of biomaterial fabrications. **Table 3** represents the percent distribution of four major lignin functional groups by quantitative ^{31}P NMR spectroscopy.

³¹P NMR analysis mmol/g & (%)	L1	L2	L3	L4	L5	L6	L7	L8
Carboxylic –OH								
	0.34(6.1%)	***0.24***(4.8%)	***0.51***(6.8%)	***0.10***(4.9%)	***0.35***(6.1%)	***0.40***(10.4%)	***0.75*** (13.6%)	***0.91***(17%)
Non-condensed phenolic –OH								
G[a]	0.85(15.4%)	0.93(18.7%)	0.87(11.6%)	0.21(10.3%)	1.97(33.9%)	0.61(15.8%)	0.96 (17.5%)	0.74(13.8%)
S[b]	2.22(40.1%)	1.72(34.5%)	0.00	0.00	0.00	0.60(15.6%)	0.87 (15.8%)	0.77(14.4%)
H[c]	0.06(1.1%)	0.15(3.0%)	0.03(0.4%)	0.00	0.26(4.5%)	0.05(1.3%)	0.33 (6.0%)	0.41(7.7%)
Sum	***3.13***(56.6%)	***2.80***(56.2%)	***0.90***(12.0%)	***0.21***(10.3%)	***2.23***(38.4%)	***1.26***(32.7%)	***2.16*** (39.3%)	***1.92***(35.9%)
Condensed phenolic –OH								
DPM[d]	N/A	0.28(5.6%)	0.02(0.3%)	0.00	0.34(5.8%)	0.14(3.6%)	0.15 (2.7%)	0.23(4.3%)
4-O-5′[e]	N/A	N/A	0.15(2.0%)	0.00	0.45(7.7%)	N/A	N/A	N/A
5–5′[f]	0.36(6.5%)	0.37(7.4%)	0.12(1.6%)	0.00	0.41(7.0%)	0.49(12.7%)	0.30 (5.5%)	0.26(4.9%)
Sum	***0.36***(6.5%)	***0.65***(13.0%)	***0.29***(3.9%)	***0.00***	***1.20***(20.5%)	***0.63***(16.3%)	***0.45*** (8.2%)	***0.49***(9.2%)
Aliphatic–OH	1.70(30.7%)	1.29(25.9%)	5.83(77.4%)	1.73(84.8%)	2.04(35.1%)	1.56(40.5%)	2.14 (38.9%)	2.03(37.9%)
Total	**5.53**	**4.98**	**7.53**	**2.04**	**5.82**	**3.85**	**5.50**	**5.35**

[a]Guaiacyl –OH.
[b]Syringyl –OH including diphenylmethane (DPM) and/or diaryl ether (4-O-5′) for L1, L2, L6, L7.
[c]*p*-Hydroxyphenyl –OH.
[d]Diphenylmethane.
[e]Diaryl ether.
[f]Biphenolic 5–5′.

Table 3. Quantitative ³¹P NMR determination of carboxylic, phenolic, condensed structures and aliphatic hydroxyl groups present in the lignins [44, 45].

The percent aliphatic hydroxyl units were found to be the highest for both the L4 and L3 sample of lignosulfonates, respectively, even though the L3 contained 4.1 mmole/g more aliphatic hydroxyl units than L4 sample. Both hardwood lignin samples (L1 and L2) contained the most non-condensed hydroxyl units due to the presence of both guaiacyl and syringyl units while the most condensed units were present in the Indulin AT. Soda lignins from the agricultural residues were found to have most carboxylic acids.

It is imperative to point out that despite the higher reactivity of phenolic hydroxyl groups, the benzylic hydroxyl groups were shown to be more reactive than phenolic hydroxyl groups toward diisocyanates [40, 41]. Studies with several lignin-like model compounds and aliphatic isocyanate, lignin polymer and aliphatic isocyanate, and lignin models and polyisocyanates have all lend a hand to similar observation [75, 76]. These experimentations indicate the most suitable lignin for partial polyols substitution should contain high benzylic hydroxyl group (α-hydroxyl and γ-carbinol) content and have little or no phenolic hydroxyl groups.

Direct lignin substitution in formulation of biofoams was achieved successfully [77]. The physical properties of the obtained PU foams are reported in **Table 4**. Since the density of the PU biofoams with lignin was comparable with the PU foam reference thus it was possible to compare their opening cell content and thermal conductivity. The opening cell content of the PU biofoams were reported to be quite similar with that of the reference indicating that lignin did not modify the cellular structure and did not cause cellular wall collapse as shown in **Figure 4**. It was interesting to observe that the thermal conductivity of the lignin-PU foams was much closed to that of the PU reference.

Direct substitution of lignin in PU biofoams formulation is not very effective. The extent of its substitution is limited by the active aliphatic hydroxyl groups, their abundancy and order of their reactivity which influence lignin solubility. Other factors such as the presence of phenolic groups, carbohydrate impurities, ash content, particle size, molecular weight, rheology, dispersity and the type of lignin recovery play important role in its performance as a precursor during manufacturing of biofoams.

The application of lignosulfonate as a partial substitute for petroleum-based polyol during preparation of PU biofoam was evaluated. Three different PU biofoams were prepared with direct substitution of lignosulfonates at 10–12, 25 and 50%wt in lieu of conventional polyols. The PU biofoams corresponding to each concentration were dispersed in aqueous solution at room temperature over 10 days. The supernatant were freeze-dried and the weight of the leached lignosulfonates was determined. The weight was compared to the dried PU biofoams and a mass balance was established. The image of **Figure 5** and the data in **Table 5** clearly show that in spite of formation of PU biofoams, lignosulfonate did not completely undergo reaction as polyols with aromatic diisocyanates but rather participated as fillers. This was concluded due to low pKa of sulfonates (pKa $\leq$ 2), and could only be evaluated in this fashion since lignosulfonates are water soluble lignins.

Sample	Density (kg/m^3)	Opening cell (%)	Thermal conductivity (W/m.K)
Without lignin	33.06	10.7	0.0250
10% lignin	33.42	10.8	0.0255
20% lignin	34.04	10.9	0.0252

Table 4. Physical properties of the rigid PU foams [77].

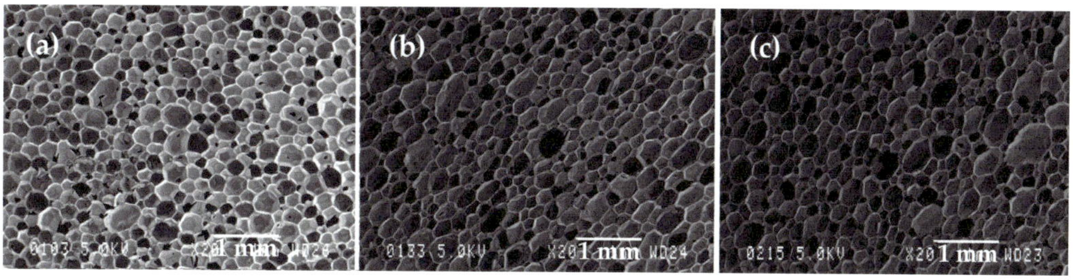

Figure 4. SEM image of PU (a) without lignin, (b) 10 wt% and (c) 20 wt% lignin [77].

Figure 5. Images of lignosulfonates leaching out of PU biofoam in water.

Lignin concentration (wt%)	Lignin loss (w/w%) (extracted lignin)	Lignin loss (w/w%) (PU weight)	Lignin loss (w/w%) (mass balance)
10–12	2.5 ± 0.4	3.0 ± 0.2	2.5–3.0
25	8.7 ± 1.8	9.5 ± 1.5	8.7–9.5
50	23.1 ± 0.7	28.4 ± 0.5	23.1–28.4

Table 5. Determination of leached lignosulfonates in water from the rigid PU biofoams.

2.2. Lignin and polyols

Addition of lignin affects both thermal and mechanical properties of polyurethane. The extent of changes in polyurethane networks properties is suggested to be related to the preparation method and the lignin content [78, 79].

The performance of lignin and polyurethane products may be improved by the addition of other polyols [80, 81] such as polyethylene glycols (PEG), polypropylene glycols (PPG, and other chain length of the polyether polyols during the copolymerization process [82].

The addition of PEG and PPG affects the flexibility and rigidity of the polyurethanes. In particular, the molecular weight of PEG influences tensile strength of polyurethanes. Both thermal and mechanical properties of the polyurethanes can be controlled by changing the amounts of lignin and polyol [28, 83]. In this method, lignin is often dissolved in polyols before reacting with diisocyanates. The isocyanate functional group would act as cross-linking agent, connecting the lignin and polyols. To obtain greater cross-link density and better mechanical properties, greater NCO/OH ratio is required and this parameter becomes more important as the lignin concentration increases [84, 85]. The key parameter to this reaction is highly dependent on the lignin hydroxyl groups and their interactions with polyols.

The reactivity of polyol with lignin was examined by addition of polyester polyol to lignin. A sample of commercially available polyester polyol, which is currently utilized in PU production, was examined for its chemical compositions and interaction with several lignin samples. The ^{31}P NMR spectroscopy technique for quantitative determination and classification of the polyol hydroxyl content was performed. In this study, two different phosphitylating reagents were utilized to elucidate hydroxyl contribution of polyol toward diisocyanates during PU polymerization. The total —OH content were quantitatively determined to be 7.47 and 7.50 mmol/g. The ^{31}P NMR spectra of polyol with TMDP reagent as illustrated in **Figure 6** shows a major signal (triplet) in the aliphatic region of ^{31}P NMR spectra corresponding to the presence of aliphatic hydroxyl groups. A closer examination of the ^{31}P NMR spectra also revealed the presence of additional aliphatic hydroxyl units in the same region as well as trace

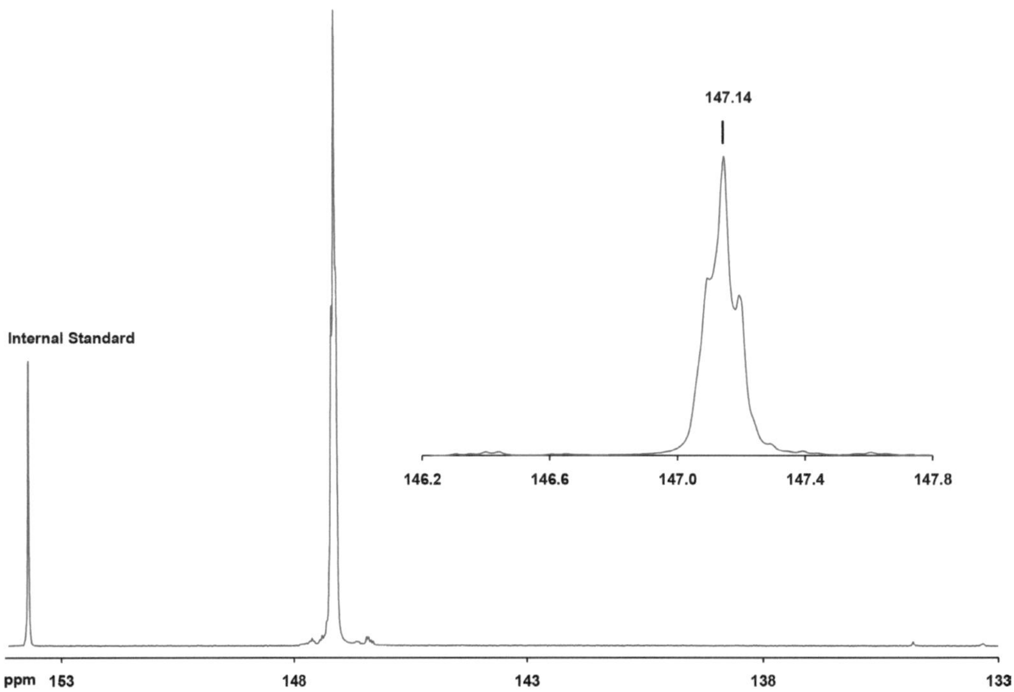

Figure 6. ^{31}P NMR spectra of polyester polyol with TMDP.

amount of carboxylic acids at 134.7 ppm. This information coupled with FTIR [86] confirmed that the commercially available polyol is a polyester-based polyol.

In a typical experiment, lignin samples were mixed and were allowed to react with the commercially available polyester polyol. The isolated products were then washed several times, purified and examined by employing ^{31}P NMR spectroscopy for quantitative determination of different classes of hydroxyl groups. The ^{31}P NMR spectra of these samples as shown in **Figure 6** exhibit a sharp and distinct signal in the aliphatic region. The ^{31}P NMR chemical shift of this new and sharp signal in all lignin samples at about 147.1 ppm strongly suggest the formation of lignin-polyols.

It is noteworthy to mention the ^{31}P NMR spectrum of the original Protobind 2400 lignin sample presented a significant feature. The spectrum exhibited presence of unusual new and sharp peak at 147.04 ppm in the aliphatic region. The nature and the shape of such a signal in that region strongly suggest that aliphatic group in the original lignin has been chemically modified by the supplier.

The quantitative determination of different classes of —OH groups by ^{31}P NMR spectroscopy for all treated lignins with polyol are reported in **Table 6**. The comparison of hydroxyl contents

Lignin samples	Protobind 2400		Indulin AT		Arbo™ SO1		Protobind 3000	
	L7	PL7	L5	PL5	L3	PL3	L8	PL8
^{31}P NMR analysis	mmol/g & (%)							
COOH	0.75(13.6%)	0.52(9.6%)	0.39(6.1%)	0.23(4.2%)	0.51(6.8%)	0.09(1.9%)	0.91(16.9%)	0.70(13.3%)
Non-condensed phenolic —OH								
G[a]	0.96(17.5%)	0.64(11.8%)	2.05(32.1%)	1.26(22.8%)	0.87(11.6%)	0.54(11.7%)	0.74(13.8%)	0.57(10.8%)
S[b]	0.87(15.8%)	0.62(11.4%)	0.00	0.00	0.00	0.00	0.77(14.3%)	0.63(12.0%)
H[c]	0.33(6.0%)	0.19(3.5%)	0.20(3.1%)	0.11(2.0%)	0.03(0.4%)	0.02(0.4%)	0.41(7.6%)	0.26(4.9%)
Condensed phenolic —OH	0.45(8.2%)	0.34(6.3%)	1.32(20.7%)	0.89(16.1%)	0.28(3.7%)	0.27(5.8%)	0.50(9.3%)	0.42(8.0%)
Aliphatic —OH	2.14(38.9%)	3.13(57.5%)	2.42(37.9%)	3.04(55.0%)	5.83(77.5%)	3.71(80.1%)	2.04(38.0%)	2.69(51.0%)
Total phenolic —OH	2.61(47.5%)	1.79(32.9%)	3.57(56.0%)	2.26(14.9%)	1.18(15.7%)	0.83(17.9%)	2.42(45.1%)	1.88(35.7%)
Total —OH	5.50	5.44	6.38	5.53	7.52	4.63	5.37	5.27

[a]Guaiacyl —OH.
[b]Syringyl —OH.
[c]*p*-Hydroxyphenyl —OH.

Table 6. Quantitative determination of several hydroxyl groups in lignin before and after treatment with polyol by ^{31}P NMR spectroscopy [45].

of all the lignin samples before (L) and after reaction with the commercially available polyester polyol (PL) shows significant difference in several –OH contents in lignins. The quantities of the carboxylic acids, phenolic hydroxyl and condensed phenolic units were all found to have decreased, while the primary hydroxyl content increased considerably after blending with the polyol. This information, coupled with the quantitative determination of the different classes of hydroxyl contents (**Table 6**), suggests that either lignin hydroxyl units cross-linked with the polyol during the premixing stage or, in spite of thorough washing, trace amounts of polyol were detected in all of the lignin samples, which affected the concentration of hydroxyls in the lignin [45].

Ahvazi *et al.* investigated physical and chemical characteristics of several lignin-polyol blends by several qualitative and quantitative methods in view of bio-based polyurethane applications [45]. During this study, differently isolated biomass lignins from forestry and agricultural residues were blended with polyester polyol and one with polyethylene glycol. The isolated products were examined thoroughly to elucidate subsequent lignin and polyol interactions during the premixing stage of bio-based polyurethanes formulation. They reported that polyol was detected in lignin even after vigorous washings with several organic solvents and soxhlet extraction as shown in **Figure 7**. The experimental data coupled with two-dimensional

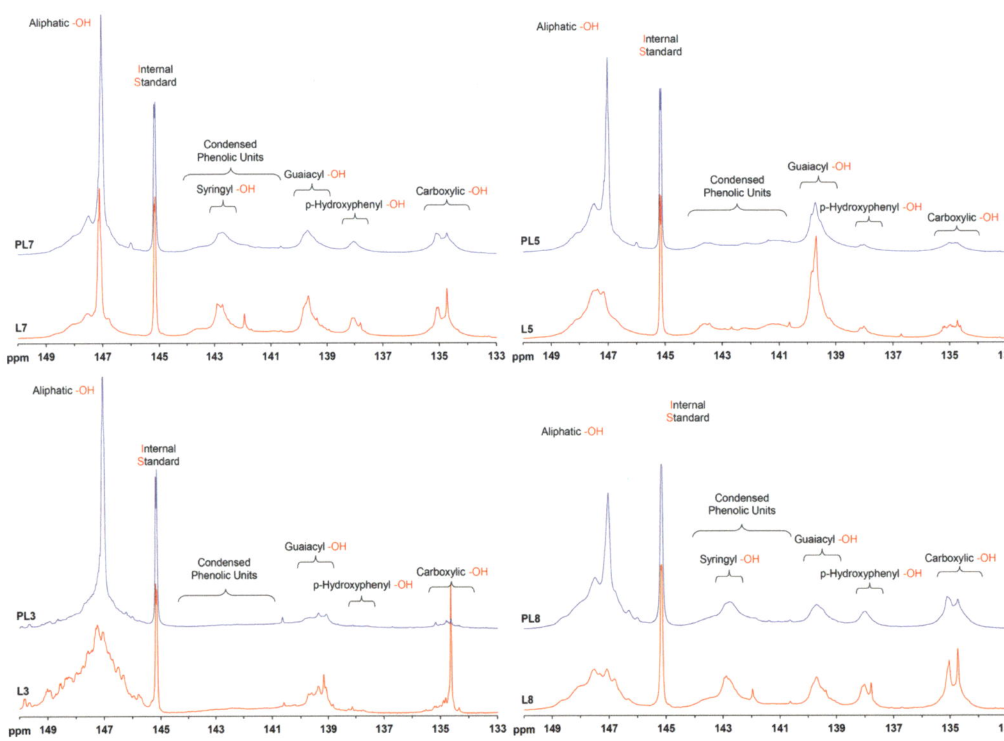

Figure 7. ^{31}P NMR spectra and signal assignments of the Protobind 2400, Indulin AT, Arbo SO1, Protobind 3000 before, (L) and after (PL) treatment with polyester polyol phosphitylated with TMDP [45].

http://dx.doi.org/10.5772/intechopen.73287

heteronuclear multiple quantum coherence (HMQC) NMR spectroscopy confirmed the formation of lignin-polyol via strong intermolecular attractions as well as some linkages between several lignin hydroxyl and polyol functional groups.

Lignin-polyol was prepared and reported in the literature [87]. **Figure 8a** and **b** shows the optical images different solubility and compatibility of lignin in polyol. SEM images of the foams are shown in **Figure 8c** and **d** illustrating the cellular morphology at different magnifications, respectively. The cellular structure of the foam is identical in both cases, with and without lignin, which consists of regular round and closed cells. There is no significant change in the cellular size and shape as well as cellular collapse with the presence of lignin. This indicates the presence of lignin in the appropriate formulation does not alter the formation of the foam cells. In addition, there is no lignin particles could be found in the biofoams. One could speculate that lignin has been participated into the PU network structure [87].

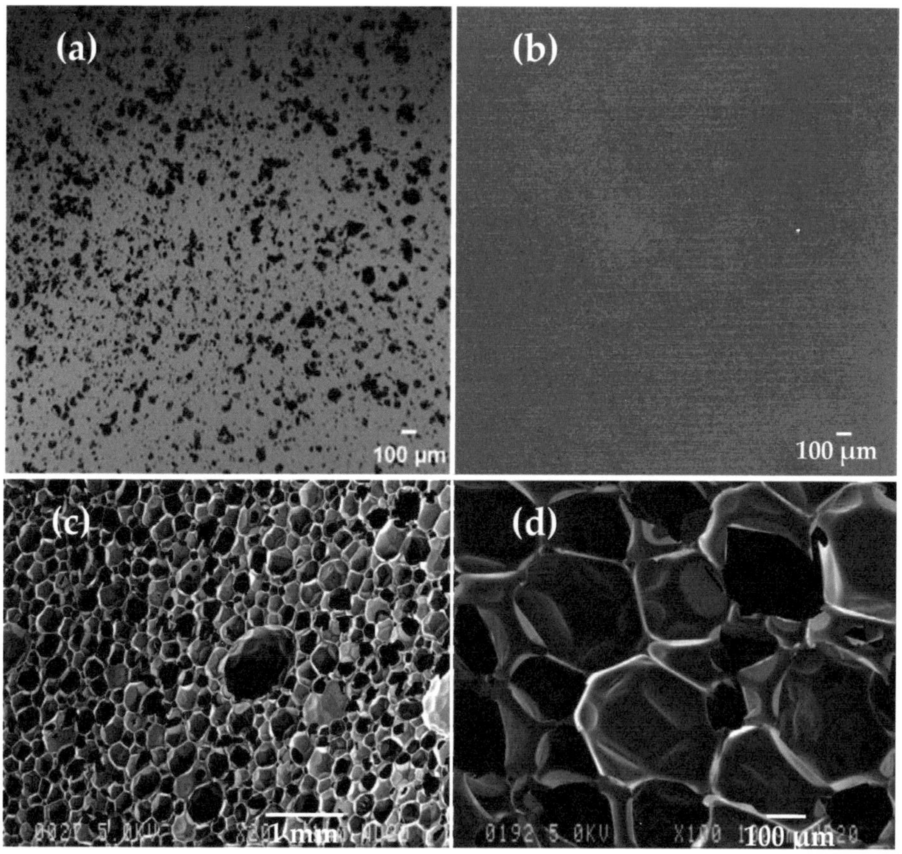

Figure 8. OM image of lingo-polyol (a) insoluble lignin, (b) soluble lignin in polyol, and SEM image for the PU foam with soluble lignin at low and high magnifications (c), (d) [87].

2.3. Chemically modified lignin

The objective of lignin modification is to increase reactivity of the lignin by modifying specific reactive groups and enhance the polycondensation process during the polyurethane production [30, 61, 62]. Chemically modified lignins provide several advantages in replacing polyols. During lignin modification, the majority of phenolic hydroxyl groups are converted to aliphatic hydroxyl units and more reactive hydroxyl groups become readily available. In addition, the steric and/or electronic constraints are also eliminated. The major drawback may be associated with its cost effectiveness. The procedure involves the conversion of phenolic hydroxyl groups to the aliphatic and more reactive hydroxyl groups. This can be achieved by either (i) direct oxyalkylation [39, 42, 88–92] or by (ii) two-steps reaction of lignin with maleic anhydride [93, 94] followed by oxyalkylation [95, 96]. The oxyalkylation are more reactive with alcohols than phenol and least with carboxylic acids in accordance to the order of nucleophilicity. During oxyalkylation, the reaction of lignin with alkene oxide would lead to chain-extended hydroxypropyl lignin and the formation of lignin-polyol derivatives [39, 42, 43]. The resulting polyols were expected to be more fluid, to possess secondary hydroxyl groups predominantly and therefore to react at lower, more convenient rates with an isocyanate-terminated prepolymer to form urethane foams with modified properties. The oxyalkylation of lignin with propylene oxide improve its solubility better than ethylene oxide.

The oxypropylated lignins were reported by Nadji et al. [62]. They noticed the oxypropylation of lignins is a straightforward process and the ensuing polyols were viscous products with functional characteristics very similar to those of conventional compounds used in the manufacture of polyurethanes that constitute the most relevant results of their investigation. They also reported that the rigid foams obtained from the polyols showed good thermal properties and dimensional stability, even after aging. These results are very encouraging, since they concern a rational valorization of an abundant renewable industrial byproduct.

Berrima et al. reported that soda lignin (L) was converted into liquid polyols by oxypropylation with propylene oxide (PO) and potassium hydroxide as a catalyst [97]. The obtained polyols were viscous liquids and possessed relatively high hydroxyl index. These two parameters were close to those of commercial polyols used in a similar context and opened the way of testing them in this field of application. The study also noticed that rigid lignin-based polyurethane foams were elaborated and displayed very promising properties as shown in **Table 7**. These materials can be used for thermal and acoustic insulation.

Ahvazi et al. reported that chemical modification of the lignin with propylene oxide to form lignin-polyol derivatives [98]. In their study, the chemical modification was performed by a two-step reaction of lignin with maleic anhydride followed by propylene oxide and by direct oxyalkylation under acidic and alkaline conditions. The physical and chemical properties of lignin-polyols from each method and the subsequent chain-extended hydroxyl groups were evaluated. They concluded that direct oxyalkylation of lignin under alkaline conditions was more efficient than acidic conditions and more effective than the two-step process for

Parameters	L	L + MA	L + MA + PO
^{31}P NMR analysis	mmol/g & (%)		
COOH	0.97(17.2%)	2.87(76.5%)	0.18(6.9%)
Non-condensed phenolic –OH			
G[a]	0.79(14.0%)	0.28(7.5%)	0.13(5.0%)
S[b]	0.82(14.6%)	0.23(6.1%)	0.16(6.1%)
H[c]	0.45(8.0%)	0.16(4.3%)	0.06(2.3%)
Condensed phenolic –OH	0.48(8.5%)	0.13(3.5%)	0.18(6.9%)
Aliphatic –OH			
1°	1.61(28.6%)	0.07(1.9%)	0.72(27.6%)
2° *(erythro & threo)*	0.51(9.1%)	0.01(0.3%)	1.18(45.2%)
Total phenolic –OH	2.54(45.1%)	0.80(21.3%)	0.53(20.3%)
Total –OH	**5.63**	**3.75**	**2.61**
Size exclusion chromatography (GPC)	**(Da × 10^3)**		
Mw	229.8	237.7	309.4
Mn	136.9	90.51	260.1
Polydispersivity			
Mw/Mn	1.68	2.36	1.19
dn/dc	0.151	0.150	0.102
Differential scanning calorimetry			
ΔCp (J/g·°C)	1.258	2.322	0.986
Tg (°C)	104.39	89.22	59.00

[a]Guaiacyl –OH.
[b]Syringyl –OH.
[c]p-Hydroxyphenyl –OH.

Table 7. Characterization of modified wheat straw soda lignin with the two-step oxyalkylation method [98].

preparing lignin-polyol with higher aliphatic hydroxyl contents. The overall lignin modification would convert phenolic hydroxyl groups to a longer branched polyether with a hydroxyl group at the end as illustrated in **Figure 9**.

The comparison of ^{31}P NMR spectra of two-step modification (L + MA + PO) to the starting material and the direct maleated lignin (L + MA) demonstrates two significant phenomena [98]. The L + MA + PO spectrum coupled with the quantitation data of **Table 7** indicates that carboxylic acids did not completely undergo transformation to their corresponding aliphatic groups during hydroxypropylation. This is expected because oxyalkylation is less reactive

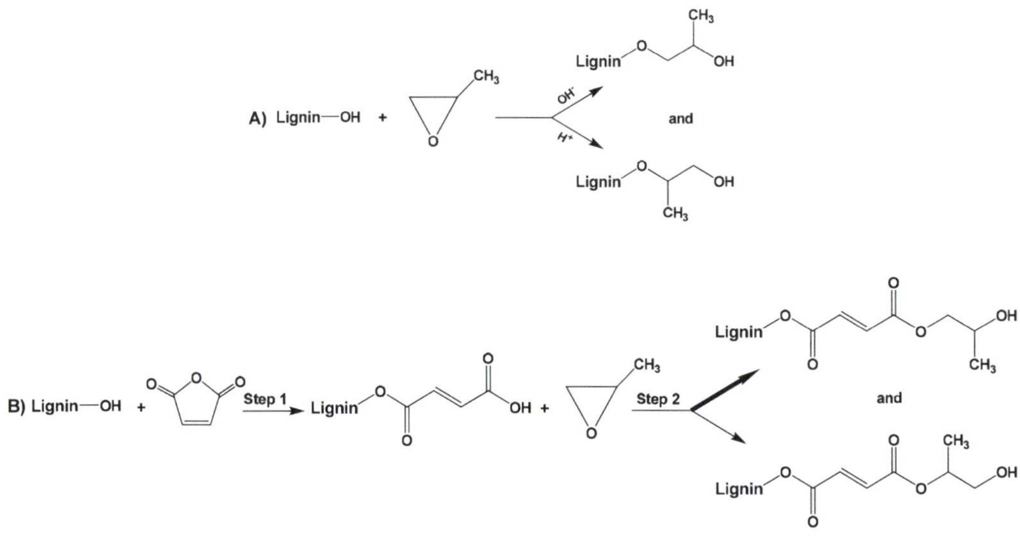

Figure 9. Schematic diagrams of direct and two-step oxyalkylation methods of hydroxyl groups in lignin [98].

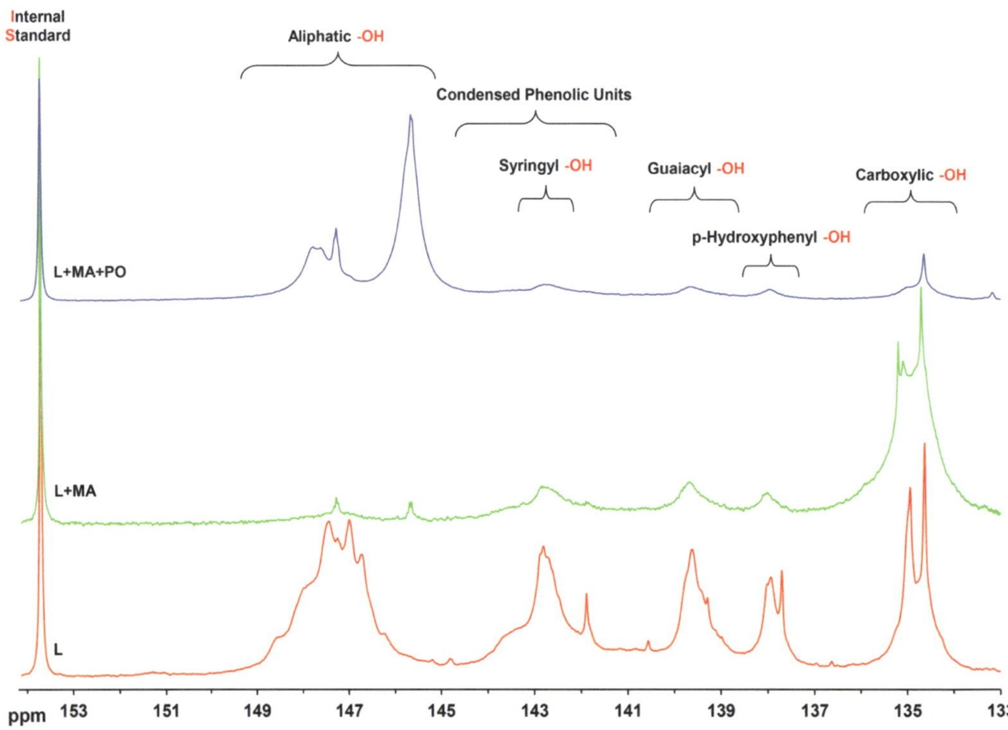

Figure 10. ^{31}P NMR spectra and signal assignments of wheat straw lignin (L) derivatized with maleic anhydride (L + MA) followed by propylene oxide (L + MA + PO) with TMDP [98].

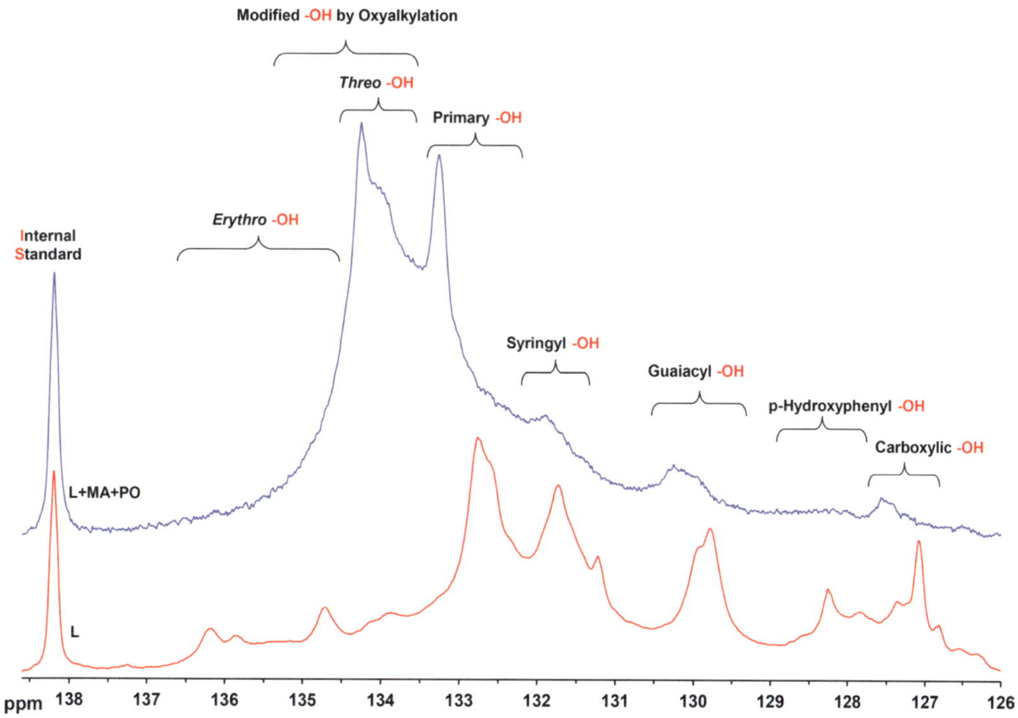

Figure 11. ^{31}P NMR spectra and signal assignments of wheat straw lignin before (L) and after modification with maleic anhydride and propylene oxide (L + MA + PO) with reagent I [98].

with carboxylic acids than the other functional groups, as 0.18 mmol/g or 6.9% of carboxylic acids remained unreacted. Indeed, the signal assignment of remaining terminal carboxylic acids recorded at 134.6 ppm in the ^{31}P NMR spectrum (**Figure 10**) shows that among the two types of carboxylic acid groups, the benzylic type was more reactive than the terminal type toward subsequent chemical modification with propylene oxide. The other phenomenon was the appearance of two different and distinct broad signals in the aliphatic region of ^{31}P NMR spectra recorded between 149.5 and 146.7 ppm and between 146.5 and 144.1 ppm. The signal recorded between 149.5 and 146.7 ppm appears in the same region as the aliphatic hydroxyl groups of the starting material. However, the signal recorded between 146.5 and 144.1 ppm was shifted upfield from the aliphatic region, and it is seemingly attributed to the hydroxypropylation reaction. To further help identify and classify these two signals, L + MA + PO was examined with reagent I and compared to the spectrum of the starting material as presented in **Figure 6**. The ^{31}P NMR spectral analysis of L + MA + PO with reagent I clearly reveal the presence of two distinct signals in the aliphatic region. The signal recorded between 133.6 and 132.1 ppm (primary —OH) appears in the same region as the aliphatic hydroxyl groups of the starting material. However, the other signal recorded between 136.6 and 133.6 ppm indicates the formation of new hydroxyl groups recorded in the area that is designated only to the secondary alcohols and assigned to both *erythro* and *threo* diastereomers of the arylglycerol-β-aryl ether structures (Cα-aryl) in lignin [29] (**Figure 11**).

3. Conclusion

In this study, two ^{31}P NMR reagents were utilized effectively to quantify several different —OH functional groups in different lignin samples from forestry and agriculture residues. The experimental data suggest that the application of lignin as a substitute for petroleum-based polyols in formulation of PU is limited not only by its aliphatic hydroxyl content but also with other features. Factors such as lignin extraction method, impurities, molecular weight, ash content and other features attribute to its solubility and compatibility with polyols. To obtain superior PU network film, lignin could be mixed with polyols to improve its solubility and homogeneity prior to reaction with diisocyanates. Chemical modification with two separate methods has shown that direct PO modification of lignin is more efficient and economically feasible. This method is highly to generate secondary —OH groups in lignin.

Author details

Behzad Ahvazi[1,2*] and Tri-Dung Ngo[1,3]

*Address all correspondence to: behzad.ahvazi@innotechalberta.ca

1 InnoTech Alberta Inc., Edmonton, Alberta, Canada

2 Department of Mechanical Engineering, Faculty of Engineering, 10-203 Donadeo Innovation Centre for Engineering, University of Alberta, Edmonton, Alberta, Canada

3 Department of Civil and Environmental Engineering, Faculty of Engineering, 7-203 Donadeo Innovation Centre for Engineering, University of Alberta, Edmonton, Alberta, Canada

References

[1] Stevens MP. Polymer Chemistry: An Introduction. 2nd ed. New York: Oxford University Press; 1990. DOI: 10.1002/pi.4990250415

[2] Kleinert M, Barth T. Towards a lignincellulosic biorefinery: Direct one-step conversion of lignin to hydrogen-enriched biofuel. Energy & Fuels. 2008;**22**:1371-1379. DOI: 10.1021/ef700631w

[3] Glasser WG, Barnett CA, Muller PC, Sarkanen KV. The chemistry of several novel bioconversion lignins. Journal of Agricultural and Food Chemistry. 1983;**31**(5):921-930. DOI: 10.1021/jf00119a001

[4] Borges da Silva EA, Zabkova M, Araújo JD, Cateto CA, Barreiro MF, Belgacem MN, Rodrigues AE. An integrated process to produce vanillin and lignin-based polyurethanes from kraft lignin. Chemical Engineering Research and Design. 2009;**87**:1276-1292. DOI: 10.1016/j.cherd.2009.05.008

[5] Wise DL. Organic Chemicals from Biomass. Biotechnology Series. 4. California: The Benjamin/Cummings Publishing Company, Inc.; 1983. ISBN-10: 0805390405

[6] Argyropoulos DS. Materials, Chemicals and Energy from Forest Biomass. Washington: Oxford University Press; 2008. DOI: 10.1021/bk-2007-0954

[7] Smook GA. Handbook for Pulp & Paper Technologists. 2nd ed. Vancouver: Angus Wilde Publications; 1992. Chapter 1. p. 5. ISBN-10: 0969462816

[8] Hergert HL. Infrared spectra. In: Sarkanen KV, Ludwig CH, editors. Lignins. Occurrence, Formation, Structure and Reactions. New York: Interscience publisher, John Wiley & Sons; 1971. pp. 267-297. ISBN 0471754226

[9] Freudenberg K. The constitution and biosynthesis of lignin. In: Kleinszeller A, Springer GF, Wittmann HG, editors. Molecular Biology Biochemistry and Biophysics. Vol. 2. Berlin-Heidelberg: Springer-Verlag; 1968. pp. 47-122. DOI: 10.1126/science.165.3895.784

[10] Adler E. Lignin chemistry- past, present and future. Journal of Wood Chemistry and Technology. 1977;**11**:169-218. DOI: 10.1007/BF00365615

[11] Sakakibara A. Chemistry of lignin. In: Hon DNS, Shiraishi N, editors. Wood and Cellulose Chemistry. New York: Marcel Dekker Inc.; 1991. pp. 113-175. ISBN 0-8247-0024-4

[12] Lai YZ, Sarkanen KV. Isolation and structural studies. In: Sarkanen KV, Ludwig CH, editors. Lignins. Occurrence, Formation, Structure and Reactions. New York: Interscience Publisher, John Wiley & Sons; 1971. pp. 165-240. DOI: 10.1002/pol.1972.110100315

[13] Chen CL. Lignin: Occurrence in wood tissues, isolation, reactions and structure. In: Lewis M, Goldstein IS, editors. Wood Structure and Composition. New York: Marcel Dekker Inc.; 1991. pp. 183-261

[14] Neish AC. Monomeric intermediates in the biosynthesis of lignin. In: Freudenberg K, Neish AC, editors. Constitution and Biosynthesis of Lignin. Berlin: Springer-Verlag; 1968. pp. 5-43. DOI: 10.1126/science.165.3895.784

[15] Wardrop AB. Occurrence and formation in plants. In: Sarkanen KV, Ludwig CH, editors. Lignins. Occurrence, Formation, Structure and Reactions. New York: Interscience Publisher, John Wiley & Sons; 1971. pp. 19-41. DOI: 10.1002/pol.1972.110100315

[16] Brauns FE, Brauns DA. The Chemistry of Lignin, Supplement Volume. New York: Academic Press; 1960

[17] Gross GG. Biosynthesis of lignin and related monomers. In: Loewus FA, Runeckles VC, editors. The Structure, Biosynthesis and Degradation of Wood. Recent Adv. In Phytochem. 11. New York: Plenum Press; 1977. pp. 141-184

[18] Chiang VL, Puumala RJ, Takeuchi H, Eckert RC. Comparison of sotwood and hardwood kraft pulping. Tappi. 1988;**71**(9):173-176

[19] Sarkanen KV. Precursors and their polymerization. In: Sarkanen KV, Ludwig CH, editors. Lignins. Occurrence, Formation, Structure and Reactions. New York: Interscience Publisher, John Wiley & Sons; 1971. pp. 95-163

[20] Gellerstedt G, Lindfors EL, Lapierre C, Monties B. Structural changes in lignin during kraft cooking. Part 2. Characterization by acidolysis. Svensk. Papperstidn. 1984;**87**(9):R61-R67

[21] Smook GA. Handbook for Pulp & Paper Technologists. 2nd ed. Vancouver: Angus Wilde publications, Chapter 10; 1992. pp. 123-152. ISBN-10: 0969462816

[22] Thielemans W, Wool RP. Lignin esters for use in unsaturated thermosets: Lignin modification and solubility modeling. Biomacromolecules. 2005;**6**(4):1895-1905. DOI: 10.1021/bm0500345

[23] Gierer J. Chemistry of delignification. Part 1. General concept and reactions during pulping. Journal of Wood Chemistry and Technology. 1985;**19**:289-312. DOI: 10.1007/BF00350807

[24] Adler E. Structural elements of lignin. Industrial and Engineering Chemistry. 1957;**49**(9): 1377-1383. DOI: 10.1021/ie50573a031

[25] Feldman D. Lignin and its polyblends – A review. In: Hu TQ, editor. Chemical Modification, Properties, and Usage of Lignin. New York: Kluwer Academic/Plenum Publishers; 2002. pp. 81-99. ISBN 978-1-4615-0643-0. DOI: 10.1007/978-1-4615-0643-0

[26] Feldman D, Lacasse M, Beznaczuk LM. Lignin-polymer systems and some applications. Progress in Polymer Science. 1986;**12**:271-299. DOI: 10.1016/0079-6700(86)90002-X

[27] Wang J, St J, Manley R, Feldman D. Synthetic polymer-lignin copolymers and blends. Progress in Polymer Science. 1992;**17**:611-646. DOI: 10.1016/0079-6700(92)90003-H

[28] Hatakeyama H. Polyurethanes containing lignin. In: Hu TQ, editor. Chemical Modification, Properties, and Usage of Lignin. New York: Kluwer Academic/Plenum Publishers; 2002. pp. 41-56. ISBN 978-1-4615-0643-0. DOI: 10.1007/978-1-4615-0643-0

[29] Feldman D, Lacasse M, St. J Manley R. Polyurethane-based sealant modified by blending with kraft lignin. Journal of Applied Polymer Science. 1988;**35**:247-257. DOI: 10.1002/app.1988.070350118

[30] Cateto C, Barreiro F, Rodrigues A, Belgacem N. Rigid polyurethane foam from lignin based-polyols. IVth international conference on times of polymers (TOP) and composites. AIP Conference Proceedings. 2008;**1042**:243-245

[31] Stevens MP. Polymer Chemistry: An Introduction. 2nd ed. New York: Oxford University Press; 1990. pp. 440-446. ISBN 0-19-506646-2

[32] Ashida K. Polyurethane and Related Foams: Chemistry and Technology. Boca Raton: CRC Press, Taylor & Francis Group; 2006. ISBN 9781587161599

[33] Saunders JH, Frisch KC. Polyurethanes: Chemistry and Technology. New York: Wiley-Interscience; Part1, 1962; Part 2; 1964

[34] Phillips LN, Parker DB. Polyurethanes-Chemistry, Technology, and Properties. New York: Gordon and Breach; 1964

[35] Edwards KN. Urethane Chemistry and Applications. ACS Symp. Ser.172. Washington, D. C: American Chemical Society; 1981

[36] Doyle EN. The Development and Use of Polyurethane Products. New York: McGraw-Hill; 1969

[37] Yoshida H, Mörck R, Kringstad KP, Hatakeyama H. Kraft lignin in polyurethanes II. Effects of the molecular weight of the kraft lignin on the properties of polyurethanes from kraft lignin-polyether triol-polymeric MDI system. Journal of Applied Polymer Science. 1990;**40**:1819-1832. DOI: 10.1002/app.1990.070401102

[38] Mozheiko LN, Gromova MF, Bakalo LA, Sergeeva VN. Polyurethanes prepared from oxy-propylated lignin. Polymer. Science. U.S.S.R. 1981;**23**(1):141-149. DOI: 10.1016/0032-3950(81)90307-5

[39] Wu LCF, Glasser WG. Engineering plastics from lignin. I. Synthesis of hydroxypropyl lignin. Journal of Applied Polymer Science. 1984;**29**:1111-1123. DOI: 10.1002/app.1984.070290408

[40] Pizzi A, Walton T. Non-emulsifiable, water-based, mixed diisocyanate adhesive systems for exterior plywood - part I. Novel reaction mechanisms and their chemical evidence. Holzforschung. 1992;**46**:541-547. DOI: 10.1515/hfsg.1992.46.6.541

[41] Zhuang JM, Steiner PR. Thermal reactions of diisocyanate (MDI) with phenols and benzylalcohols: DSC study and synthesis of MDI adducts. Holzforschung. 1993;**47**:425-434. DOI: 10.1515/hfsg.1993.47.5.425

[42] Hornof V, Hombek R. Surface-active agents based on propoxylated lignosulfonate. Journal of Applied Polymer Science. 1990;**41**:2391-2398. DOI: 10.1002/app.1990.070410939

[43] Jain RK, Glasser WG. Lignin derivatives. II. Functional ethers. Holzforschung. 1993;**47**:325-332. DOI: 10.1515/hfsg.1993.47.4.325

[44] Ahvazi B, Cloutier E, Wojciechowicz O, Ngo TD. Lignin profiling: A guide for selecting appropriate lignin as precursors in biomaterials development. ACS Sustainable Chemistry & Engineering. 2016;**4**:5090-5105. DOI: 10.1021/acssuschemeng.6b00873

[45] Ahvazi B, Wojciechowicz O, Xu P, Ngo TD, Hawari J. Formation of ligno-polyols: Fact or fiction. BioResourcces. 2017;**12**(3):6629-6655. DOI: 10.15376/biores.12.3.6629-6655

[46] Evtuguin DV, Andreolety JP, Gandini A. Polyurethanes based on oxygen-organosolv lignin. European Polymer Journal. 1998;**34**(8):1163-1169. DOI: 10.1016/S0014- 3057(97)00245-0

[47] Vanderlaan MN, Thring RW. Polyurethanes from Alcell lignin fractions obtained by sequential solvent extraction. Biomass and Bioenergy. 1998;**14**(5/6):525-531. DOI: 10.1016/S0961-9534(97)10058-7

[48] Yoshida H, Mörck R, Kringstad KP. Kraft lignin in polyurethanes I. Mechanical properties of polyurethanes from kraft lignin-polyether triol-polymeric MDI system. Journal of Applied Polymer Science. 1987;**34**:1187-1198. DOI: 10.1002/app.1987.070340326

[49] Cateto CA, Barreiro MF, Rodrigues AE. Monitoring of lignin-based polyurethane synthesis by FTIR-ATR. Industrial Crops and Products. 2008;**27**(2):168-174. DOI: 10.1016/j.indcrop.2007.07.018

[50] Natansohn A, Lacasse M, Banu D, Feldman D. CP-MAS NMR spectra of polyurethane-lignin blends. Journal of Applied Polymer Science. 1990;**40**:899-904. DOI: 10.1002/app.1990.070400523

[51] Saraf VP, Glasser WG. Engineering plastics from lignin. III. Structure property relationship in solution cast polyurethane films. Journal of Applied Polymer Science. 1984;**29**: 1831-1841. DOI: 10.1002/app.1984.070290534

[52] Saraf VP, Glasser WG, Wilkes GL, McGrath JE. Engineering plastics from lignin. VI. Structure property relationship of PEG-containing polyurethane networks. Journal of Applied Polymer Science. 1985;**30**:2207-2224. DOI: 10.1002/app.1985.070300533

[53] Feldman D, Lacasse M. Polymer-filler interaction in polyurethane kraft lignin polyblends. Journal of Applied Polymer Science. 1994;**51**:701-709. DOI: 10.1002/app.1994.070510416

[54] Feldman D, Lacasse M. Swelling and adhesion characteristics of lignin-filled polyurethane sealant. Journal of Adhesion Science and Technology. 1994;**8**(5):473-484. DOI: 10.1163/156856194X00177

[55] Feldman D, Lacasse M. Mechanical characteristics of sealants based on polyurethane-lignin polyblends. Journal of Adhesion Science and Technology. 1994;**8**(9):957-969. DOI: 10.1163/156856194X00843

[56] Hsu OHH, Glasser WG. Polyurethane adhesives and coatings from modified lignin. Wood Science. 1976;**9**(2):97-103

[57] Rials TG, Glasser WG. Engineering plastics from lignin - XIII. Effect of lignin structure on polyurethane network formation. Holzforschung. 1986;**40**:353-360. DOI: 10.1515/hfsg.1986.40.6.353

[58] Mörck R, Reimann A, Kringstad KP. Fractionation of kraft lignin by successive extraction with organic solvents. III. Fractionation of kraft lignin from birch. Holzforschung. 1988;**42**:111-116. DOI: 10.1515/hfsg.1998.42.2.111

[59] Kelley SS, Glasser WG, Ward TC. Multiphase materials with lignin. IX. Effect of lignin content on interpenetrating polymer network properties. Polymer. 1989;**30**(12):2265-2268. DOI: 10.1016/0032-3861(89)90259-0

[60] Glasser WG. Novel structure materials from lignin. In: Kennedy JF, Phillips GO, Williams PA, editors. Wood Processing and Utilization. Chichester, U.K.: Ellis Hardwood Limited; 1989. pp. 163-173. ISBN 0470216123, 9780470216125

[61] Gandini A, Belgacem MN, Guo ZX, Montanari S. Lignin as macromonomers for polyesters and polyurethanes. In: Hu TQ, editor. Chemical Modification, Properties, and Usage of Lignin. New York: Kluwer Academic/Plenum Publishers; 2002. pp. 57-80. ISBN 978-1-4615-0643-0. DOI: 10.1007/978-1-4615-0643-0

[62] Nadji H, Bruzzèse C, Belgacem MN, Benaboura A, Gandini A. Oxypropylation of lignins and preparation of rigid polyurethane foams from the ensuing polyols. Macromolecular Materials and Engineering. 2005;**290**(10):1009-1016. DOI: 10.1002/mame.200500200

[63] Lai YZ. Reactivity and accessibility of cellulose, hemicelluloses, and lignins. In: Hon N-SD, editor. Chemical Modification of Lignocellulosic Materials. New York: Marcel Dekker, Inc.; 1996. pp. 67-81. ISBN 9780824794729

[64] Evstigneyev E, Maiyorova H, Platonov A. Lignin functionalization and the alkaline delignification rate. Tappi. 1992:177-182

[65] Marton J. Reaction in alkaline pulping. In: Sarkanen KV, Ludwig CH, editors. Lignins. Occurrence, Formation, Structure and Reactions. New York: Interscience Publisher, John Wiley & Sons; 1971. pp. 639-694. DOI: 10.1002/pol.1972.110100315

[66] Ahvazi BC, Pageau G, Argyropoulos DS. On the formation of diphenylmethane structures in lignin under kraft, emcc, and soda pulping conditions. Canadian Journal of Chemistry. 1998;**76**:506-512. DOI: 10.1139/v98-060

[67] Argyropoulos DS, Hortling B, Poppius-Levlin K, Sun Y, Mazur M. MILOX pulping, lignin characterization by ^{31}P NMR spectroscopy and oxidative degradation. Nordic Pulp & Paper Research Journal. 1995;**10**(1):68-73. DOI: 10.3183/NPPRJ-1995-10-01-p068-073

[68] Sun Y, Argyropoulos DS. Fundamentals of high pressure oxygen and low pressure oxygen-peroxide (EOP) delignification of softwood and hardwood kraft pulps; a comparison. Journal of Pulp and Paper Science. 1995;**21**(5):185-190

[69] Froass PM, Ragauskas AJ, Jiang J. Chemical structure of residual lignin from kraft pulp. Journal of Wood Chemistry and Technology. 1996;**16**:347-365. DOI: 10.1080/02773819608545820

[70] Akim L, Argyropoulos D, Jouanin L, Leple′ JL, Pilate G, Pollet B, Lapierre C. Quantitative ^{31}P NMR spectroscopy of lignins from transgenic poplars. Holzforshung. 2001;**55**:386-390. DOI: 10.1515/HF.2001.064

[71] Granata A, Argyropoulos D. 2-Chloro-4,4,5,5-tetramethyl- 1,3,2-dioxaphospholate, a reagent for the accurate determination of the uncondensed and condensed phenolic moieties in lignins. Journal of Agricultural and Food Chemistry. 1995;**43**(6):1538-1544. DOI: 10.1021/jf00054a023

[72] Argyropoulos D. Quantitative phosphorus-31 NMR analysis of lignin: A new tool for the lignin chemist. Journal of Wood Chemistry and Technology. 1994;**14**(1):45-63. DOI: 10.1080/02773819408003085

[73] Gosselink RJA, Abächerli A, Semke H, Malherbe R, Käuper P, Nadif A, Van Dam JEG. Analytical protocols for characterization of sulfur-free lignin. Industrial Crops and Products. 2004;**19**(3):271-281. DOI: 10.1016/j.indcrop.2003.10.008

[74] Mousavioun P, Doherty WOS. Chemical and thermal properties of fractionated bagasse soda lignin. Industrial Crops and Products. 2010;**31**:52-58. DOI: 10.1016/j.indcrop.2009.09.001

[75] Forbes CP, van der Klashorst GH, Psotta K. Lignosulphonate crosslinking reactions. 6. The reactions of lignosulphonate and lignosulphonate model compounds with cyanuric chloride. Holzfoschung. 1984;**38**:43-46

[76] van der Klashorst GH, Forbes CP, Psotta K. Lignosulphonate crosslinking reactions - 5. The reactions of lignosulphonate and lignosulphonate model compounds with acid chlorides. Holzfoschung. 1983;**37**:279-286

[77] Ton-That MT, Ngo TD, Bélanger C, Langlois A, Drouin M, Ahvazi B, Hawari J. Rigid polyisocyanurate (PIR) biofoams from non-food grade and renewable biopolyols. Polyurethanes 2011-Technical conference, Gaylord Opryland, Nashville, Tennessee, September 26-28; 2011

[78] Ciobanu C, Ungureanu M, Ignat L, Ungureanu D, Popa VI. Properties of lignin-polyurethane films prepared by casting method. Industrial Crops and Products. 2004;**20**(2):231-241. DOI: 10.1016/j.indcrop.2004.04.024

[79] Hatakeyama H, Kosugi R, Hatakeyama T. Thermal properties of lignin-and molasses-based polyurethane foams. Journal of Thermal Analysis and Calorimetry. 2008;**92**(2): 419-424. DOI: 10.1002/s109-007-8963-1

[80] Saraf VP, Glasser WG, Wilkes GL, McGrath JE. Engineering plastics from lignin. VI. Structure-properties relationship of peg-containing polyurethane network. Journal of Applied Polymer Science. 1985;**30**:2207-2224. DOI: 10.1002/app.1985.070300533

[81] Mörck R, Reimann A, Kringstad KP. Fractionation of kraft lignin by successive extraction with organic solvents. III. Fractionation of kraft lignin from birch. Holzforschung. 1988;**42**:111-116. DOI: 10.1515/hfsg.1988.42.2.111

[82] Hatakeyama T, Asano Y, Hatakeyama H. Mechanical and thermal properties of rigid polyurethane foams derived from sodium lignosulfonate mixed with diethylene-, triethylene- and polyethylene glycols. Macromolecular Symposia. 2003;**197**:171-180. DOI: 10.1002/masy.200350716

[83] Chahar S, Dastidar MG, Choudhary V, Sharma DK. Synthesis and characterization of polyurethanes derived from waste black liquor lignin. Journal of Adhesion Science and Technology. 2004;**18**(2):169-179. DOI: 10.1163/156856104772759386

[84] Thring RW, Vanderlaan MN, Griffin SL. Polyurethanes from Alcell lignin. Biomass and Bioenergy. 1997;**13**(3):125-132. DOI: 10.1016/S0961-9534(97)00030-5

[85] Rials T, Glasser WG. Engineering plastics from lignin. IV: Effects of cross-link density on polyurethane film properties-variations in NCO:OH ratio. Holzforschung. 1984; **38**(4):191-199. DOI: 10.1515/hfsg.1984.38.4.191

[86] Cateto CA, Barreiro MF, Rodrigues AE, Belgacem MN. Optimization study of lignin oxypropylation in view of the preparation of polyurethane rigid foam. Industrial and Engineering Chemistry Research. 2009;**48**:2583-2589. DOI: 10.1021/ie801251r

[87] Ton-That MT, Ngo TD, New biopolyols from non-food grade and renewable resources. Utech Polyurethane Technical Conference 2012, Maastricht, Netherlands, April 17-19; 2012

[88] Glasser WG, Barnett C, Rials T, Saraf VP. Engineering plastics from lignin. II: Characterization of hydroxyalkyl lignin derivatives. Journal of Applied Polymer Science. 1984; **29**:1815-1830. DOI: 10.1002/app.1984.070290533

[89] Pavier C, Gandini A. Oxypropylation of sugar beet pulp. 2. Separation of the grafted pulp from the propylene oxide homopolymer. Carbohydrate Polymers. 2000;**42**:13-17. DOI: 10.1016/S0144-8617(99)00124-1

[90] Mozheiko LN, Gromova MF, Bakalo LA, Sergeyeva VN. Polyurethanes prepared from oxypropylated lignin. Polymer Science U.S.S.R. 1981;**23**(1):141-149. DOI: 10.1016/0032-3950(81)90307-5

[91] Kurple KR, Lignin Based Polyols, United States Patent Number 6,025,452; 2000

[92] Cateto C, Barreiro MF, Rodrigues AE, Belgacem MN. Oxypropylation of lignins and characterization of the ensuing polyols. Proceedings of 8th ILI Forum; 2007. pp. 115-119

[93] Feldman D, Luchian C, Banu D, Lacasse M. Polyurethane-maleic anhydride grafted lignin polyblends. Cellulose Chemistry and Technology. 1991;**25**:163-180. ISSN 2457-9459

[94] Thielemans W, Wool RP. Lignin esters for use in unsaturated thermosets: Lignin modification and solubility modeling. Biomacromolecules. 2005;**6**:1895-1905. DOI: 10.1021/bm0500345

[95] Glasser WG, Barnett C, Rials T, Kelley S, Synthesis and characterization of several different hydroxyalkyl lignin derivatives. Proc. Int. Symp. Wood Pulping Chem.; Tsukuba. Japan; 1983. pp. 89-94

[96] Glasser WG, Hsu OHH, Polyurethane intermediates and products and methods of producing same from lignin, US Patent Number 4017474, 1977

[97] Berrima B, Mortha G, Boufi S, Aloui EE, Belgacem MN. Oxypropylation of soda lignin: Characterization and application in polyurethane foams production. Cellulose Chemistry and Technology. 2016;**50**(9-10):941-950 ISSN 2457-9459

[98] Ahvazi B, Wojciechowicz O, Ton-That MT, Hawari J. Preparation of lignopolyols from wheat straw soda lignin. Journal of Agricultural and Food Chemistry. 2011;**59**:10505-10516. DOI: 10.1021/jf202452m